MST210

Mathematical methods, models and modelling

Book C

Cover image: This shows arrows representing a vector field and a corresponding contour map. The vector field shows the direction of change for a model of two competing populations of animals. You will meet this model in Unit 12.

This publication forms part of an Open University module. Details of this and other Open University modules can be obtained from the Student Registration and Enquiry Service, The Open University, PO Box 197, Milton Keynes MK7 6BJ, United Kingdom (tel. +44 (0)845 300 6090; email general-enquiries@open.ac.uk).

Alternatively, you may visit the Open University website at www.open.ac.uk where you can learn more about the wide range of modules and packs offered at all levels by The Open University.

To purchase a selection of Open University materials visit www.ouw.co.uk, or contact Open University Worldwide, Walton Hall, Milton Keynes MK7 6AA, United Kingdom for a brochure (tel. +44 (0)1908 858779; fax +44 (0)1908 858787; email ouw-customer-services@open.ac.uk).

The Open University, Walton Hall, Milton Keynes, MK7 6AA.

First published 2014.

Edited, designed and typeset by The Open University, using the Open University TEX System.

Printed in the United Kingdom by Hobbs the Printers Ltd, Totton, Hampshire.

The Open University has had Woodland Carbon Code Pending Issuance Units assigned from Doddington North forest creation project (IHS ID103/26819) that will, as the trees grow, compensate for the greenhouse gas emissions from the manufacture of the paper in MST210 Block C. More information can be found at https://www.woodlandcarboncode.org.uk/

ISBN 978 1 7800 7869 4

1.1

Contents

Unit 8

Mathematical modelling

Introduction

Applied mathematics is concerned not only with the development of *mathematical methods* but also with the application of these methods in *mathematical models*, which are idealised representations of aspects of the real world. This text aims to develop your appreciation of the role of mathematics in understanding and predicting the behaviour of the real world (as opposed to the world of mathematical theories). This process is called *mathematical modelling*.

For example, some mathematical methods are dealt with in Unit 1, whereas established mathematical models are the subject of Units 2 and 3.

There was a time when the application of mathematics was restricted more or less to physics and engineering. Those days are long gone, and now the application of mathematics is widespread. This unit introduces skills that will enable you to develop your own mathematical models for simple real-world situations. The mathematical modelling process starts with a problem in the real world. This problem is translated into a mathematical model, whose solution may provide insights into the original real-world problem. The mathematical model may also help to predict what will happen in the real world if changes are made. The key stages in the mathematical modelling process are as follows.

We may indicate these stages in abbreviated form in the margin, as we have done for procedures elsewhere.

1. **Specify the purpose of the model:**
   define the problem;
   decide which aspects of the problem to investigate.
2. **Create the model:**
   state assumptions;
   choose variables and parameters;
   formulate mathematical relationships.
3. **Do the mathematics:**
   solve equations;
   draw graphs;
   derive results.
4. **Interpret the results:**
   collect relevant data;
   describe the mathematical solution in words;
   decide what results to compare with reality.
5. **Evaluate the model:**
   test the model by comparing its predictions with reality;
   criticise the model.

The essential difference between a variable and a parameter is that a variable is a quantity whose values change during the situation described by the model, while a parameter is a constant of the model (for a given situation).

The mathematical modelling process is sometimes referred to as the *mathematical modelling cycle*.

The diagram in Figure 1 may help you to remember the five key stages in the mathematical modelling process.

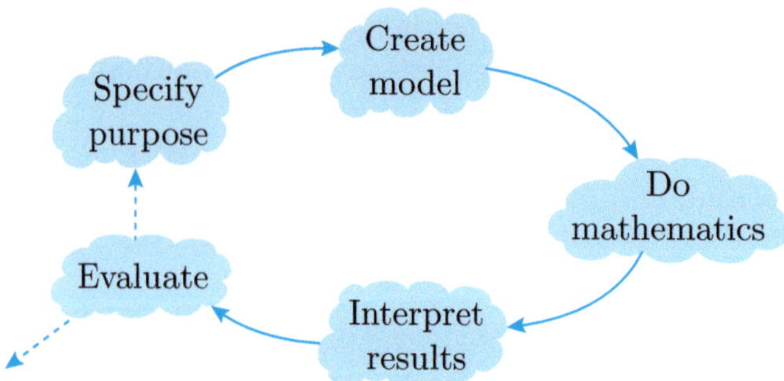

Figure 1 Mathematical modelling process

In developing a mathematical model, you may need to go around the loop in Figure 1 several times, improving your model each time. When you start to create a mathematical model for a real problem, begin with a simple model, in order to obtain a feel for the problem. Often a simple model that gives a reasonable approximation to the real world is more useful than a complicated model. A more complex model may give a better fit to the available data, but the key processes that are being modelled may be more difficult to identify, and the mathematical problem may be harder to solve. If, in evaluating a model, you find that it is not satisfactory for its purpose, then identify why it is deficient and try to include additional features that address those deficiencies in your next pass through the modelling loop.

As you become more experienced as a mathematical modeller, you will probably find that the above schema for mathematical modelling is insufficiently flexible for the mathematical model that you are developing. The purpose of specifying the modelling process clearly is to try to bring some structure to the subject, but the schema should not be seen as a straitjacket. However, it is recommended that you use this schema when tackling the modelling problems that you will encounter in this module.

Mathematical modelling is unlike other branches of mathematics. The mathematical models that you develop depend critically on the original problem and on the simplifying assumptions that you make about the system being modelled. Consequently, there are many ways of tackling any modelling problem and none can be expected to yield an exact solution. When you attempt a modelling exercise in this text, your mathematical model may differ from the one suggested in our solution, but provided that your model is justified by reasoned arguments based on reasonable assumptions, it may be just as valid as the one developed here.

1 Pollution in the Great Lakes

This section explores a real-world system where mathematical modelling has been used to aid understanding of what is happening and to predict what will happen if changes are made. The system concerned is extremely complex, but by keeping things as simple as possible, we will extract sufficient information to obtain a mathematical model of the system. Refinements to this simple model will be made in Unit 18.

The Great Lakes of North America are shown in Figure 2. Lake Superior and Lake Michigan drain into Lake Huron, which drains via Lake Erie into Lake Ontario, and this lake drains into the St Lawrence River and thence into the Atlantic Ocean.

Figure 2 The Great Lakes of North America

The Great Lakes provide drinking water for tens of millions of people who live in the surrounding area. They also provide a source of food, transport and recreation. In the first half of the twentieth century they were used for dumping sewage and other pollutants. Sources of pollution include industrial waste, agricultural chemicals, acid rain, and oil and chemical spills. Our case study concerns the construction of a mathematical model of how the pollution level in a lake varies over time.

The information for this case study comes from two main sources: R.H. Rainey (1967) 'Natural displacement of pollution from the Great Lakes', *Science*, **155**, 1242–3; R.V. Thomann and J.A. Mueller (1987) *Principles of Surface Water Quality Modeling and Control*, Harper and Row.

The mathematical modelling process can be applied to the problem of predicting future pollution levels in the Great Lakes; the details are given below.

◀ Specify purpose ▶

1 Specify the purpose of the model

Define the problem

The problem is to predict how long it will take for the level of pollution in a lake to reduce to a target level if all sources of pollution are eliminated. It is intended to use a mathematical model to investigate pollution levels in any one of the Great Lakes, although the model could also be used for any other polluted lake.

Decide which aspects of the problem to investigate

We want to investigate how the pollution level varies with time as clean water flows into the lake and polluted water flows out.

◀ Create model ▶

2 Create the model

State assumptions

The model makes the following assumptions.

(a) All sources of pollution have been removed.

(b) A pollutant does not biodegrade in the lake or decay through any other biological, chemical or physical process.

(c) A pollutant is evenly dispersed within the lake at all times.

(d) Water flows into and out of the lake at the same constant rate (so all seasonal effects can be ignored).

(e) All other water gains and losses (e.g. rainfall, evaporation, extraction and seepage) can be ignored.

(f) The volume of water in the lake is constant.

The statement that the volume is constant is a consequence of the previous two assumptions, but it is included as an assumption in its own right because of its importance in the modelling process.

(g) If the mathematical model is to be used for the downstream lakes (Huron, Erie and Ontario), then negligible pollution is flowing into them from the upstream lakes.

Choose variables and parameters

The variables in the model are:

t the time, in seconds, since all sources of pollution were removed;

$m(t)$ the mass, in kilograms, of pollutant in the lake at time t;

$c(t)$ the concentration, in kilograms per cubic metre, of pollutant in the lake at time t.

The parameters in the model are:

c_{target}	the target concentration level, in kilograms per cubic metre, of pollutant in the lake;
T	the time taken, in seconds, to reduce the concentration of pollutant to the target level c_{target} (i.e. $c(T) = c_{\text{target}}$);
V	the volume, in cubic metres, of water in the lake;
r	the water flow rate, in cubic metres per second, into and out of the lake;
$k = r/V$	the **proportionate flow rate**, in seconds^{-1}.

Formulate mathematical relationships

By Assumption (c), the pollution concentration is uniform, so the relationship between concentration and mass is given by

$$c(t) = \frac{m(t)}{V}.$$

The **input–output principle**,

$$\boxed{\text{accumulation}} = \boxed{\text{input}} - \boxed{\text{output}},$$

is applied to the mass of pollutant during the time interval $[t, t+\delta t]$. This principle is often used in mathematical modelling.

The input–output principle was introduced in Unit 1.

By Assumptions (a) and (g), the mass of pollutant entering the lake is zero, so the *input* is zero. By Assumption (b), the pollutant leaves the lake only through the outflow of water. In the time interval $[t, t+\delta t]$, the volume of water leaving the lake is $r\,\delta t$, where r, the flow rate, is constant because of Assumptions (d), (e) and (f). Multiplying this by the concentration of pollutant, which is uniform by Assumption (c), gives the mass of pollutant that leaves the lake in that time interval as $r\,c(t)\,\delta t$ or $(r/V)\,m(t)\,\delta t$. This is the *output*.

Note how the assumptions are identified while the mathematical models is developed.

The *accumulation* of the mass of pollutant within the lake, over the time interval $[t, t+\delta t]$, is the difference between $m(t+\delta t)$ and $m(t)$, that is, $m(t+\delta t) - m(t)$. Applying the input–output principle gives

$$m(t+\delta t) - m(t) \simeq 0 - \frac{r}{V}\,m(t)\,\delta t.$$

It is sometimes useful to replace a group of parameters by a single parameter, particularly when the same group of parameters is often repeated; this is called **reparametrisation**. Replacing r/V by k, dividing by δt and then letting $\delta t \to 0$, leads to the differential equation

$$\frac{dm}{dt} = -k\,m(t), \quad \text{where } k = \frac{r}{V}. \tag{1}$$

◀ Do mathematics ▶

3 Do the mathematics

Solve equations

The solution of the differential equation for m is

$$m(t) = m(0)\, e^{-kt}, \quad \text{where } m(0) \text{ is the initial mass of pollutant.}$$

Since $c(t) = m(t)/V$, the pollutant concentration c is given by

$$c(t) = c(0)\, e^{-kt}, \quad \text{where } c(0) \text{ is the initial concentration of pollutant.}$$

The time T taken for the pollutant concentration to reduce to the target level $c_{\text{target}} = c(T) = c(0)\, e^{-kT}$ is given by

$$T = -\frac{1}{k} \ln\left(\frac{c_{\text{target}}}{c(0)}\right).$$

Draw graphs

Typical graphs of concentration against time are shown below.

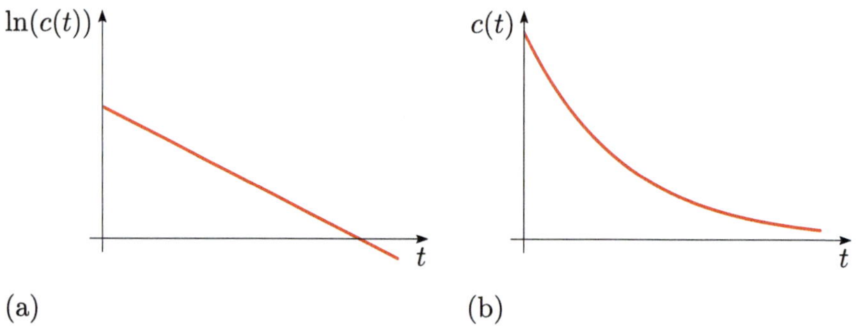

Figure 3 (a) Log concentration versus time; (b) concentration versus time

Derive results

Suppose that a political decision has been taken to reduce the level of pollution to a tenth of its initial level, so that $c_{\text{target}}/c(0) = \frac{1}{10}$. The corresponding length of time required is

$$T = -\frac{1}{k} \ln\left(\tfrac{1}{10}\right) = \frac{\ln 10}{k} \simeq \frac{2.30}{k},$$

so T is inversely proportional to the proportionate flow rate k.

◀ Interpret results ▶

4 Interpret the results

Collect relevant data

The only data required are the values of k. These can be determined for each lake from its volume V and water flow rate r, since $k = r/V$. The data are given in Table 1.

Table 1 Data for the Great Lakes

Lake	Volume V ($10^{12}\,\mathrm{m}^3$)	Water flow rate r ($10^3\,\mathrm{m}^3\,\mathrm{s}^{-1}$)	$k = r/V$ ($10^{-9}\,\mathrm{s}^{-1}$)
Superior	12.10	2.01	0.166
Michigan	4.92	1.57	0.319
Huron	3.54	5.10	1.441
Erie	0.48	5.90	12.292
Ontario	1.64	6.78	4.134

The units given mean, for example, that the volume of Lake Superior is $12.10 \times 10^{12}\,\mathrm{m}^3$, its water flow rate is $2.01 \times 10^3\,\mathrm{m}^3\,\mathrm{s}^{-1}$, and its proportionate flow rate is $0.166 \times 10^{-9}\,\mathrm{s}^{-1}$.

Describe the mathematical solution in words

For Lake Superior, for example, the model predicts that a reduction in the level of pollution to a tenth of its initial level will take a time (in seconds)

$$T = \frac{\ln 10}{k} = 13.9 \times 10^9.$$

Dividing T by $60 \times 60 \times 24 \times 365.24$ to convert it into years, the model predicts that it will take 440 years for the pollution level in Lake Superior to drop by a factor of ten. A similar calculation for Lake Michigan predicts that it will take 7.22×10^9 seconds or 229 years for the pollution level to drop by a factor of ten.

According to Assumption (g), only clean water enters the downstream lakes Huron, Erie and Ontario from the upstream lakes, so the model can be used for these lakes as well. Similar calculations then predict that the times required to reduce the pollution level to a tenth of its initial level are 51 years for Lake Huron, 6 years for Lake Erie and 18 years for Lake Ontario.

However, it may be necessary to revisit this assumption later.

The reason for the long time periods that are predicted for purging the two upper lakes is their rather low proportionate flow rates. The quantity $V/r = 1/k$ is the *average retention time* for water in the lake, that is, the length of time that it takes for an amount of water equal to the volume of water in the lake to flow out of the lake, and hence the time that any molecule of water would expect to remain in the lake. For Lake Superior, the average retention time is 191 years, and for Lake Michigan it is 99 years. For the downstream lakes the average retention times are much shorter.

Decide what results to compare with reality

It is perhaps fortunate that the two most polluted lakes in the 1960s were Lake Erie and Lake Ontario, which have relatively small volumes of water and fairly high water flow rates. It was thus possible, as predicted by the model, to clean up these lakes fairly quickly.

You will see a model of the cumulative effects of pollution levels on the whole system in Unit 18.

The model predicts that if significant amounts of pollution are allowed to enter Lake Superior or Lake Michigan, then it will take hundreds of years before the lakes can recover. It could be argued that the mathematical model is quite conservative in its estimation, since it assumes that all sources of pollution have stopped and that the pollutant is completely dispersed in the water, rather than in the flora and fauna or in the sediment at the bottom of the lake. However, since most pollutants biodegrade, it could also be argued that the estimate is rather pessimistic. Even so, a major contamination of either of these lakes, particularly by a pollutant that does not biodegrade, would be a catastrophe for which there would, apparently, be no short-term solution.

It is salutary to note that in recent years there have been fish consumption advisory warnings in place for many areas, and in particular for Lake Michigan. People are advised not to eat some fish because of the high levels of toxic chemicals in them. Although these chemicals are present in small quantities in the lakes, they tend to accumulate in the fish and then on through the ecosystem. The battle to control pollution in these lakes has not yet been won.

Summary

The development of the pollution model for the Great Lakes has highlighted a number of important aspects of mathematical modelling. These include the following.

This concludes the summary of the first four modelling stages for this problem. The fifth stage, 'Evaluate the model', will be addressed in Unit 18.

- By representing quantities using symbols, it is possible to develop models that can be applied in a variety of different situations.
- The simplifying assumptions must be relevant to the model, and should be linked to the mathematical relationships between the variables and parameters.
- Data for the model should be collected after it has been established what data are required.

Exercise 1

In your own words, and without using any equations or symbols, give an outline, or description, of the formulation of the model in this section. (As a guide the outline should be between 50 and 100 words.)

Exercise 2

(a) Suppose that pollution continues to enter a lake at a constant rate q (in $\text{kg}\,\text{s}^{-1}$), and assume that this pollutant instantaneously diffuses throughout the lake. How would the mathematical model given by equation (1) change?

(b) Show that the mass of pollution in the lake is now modelled by

$$m(t) = \frac{q}{k} + \left(m(0) - \frac{q}{k}\right) e^{-kt}.$$

What is the long-term effect on the mass of pollution in the lake?

(c) What happens if initially there is no pollution in the lake?

(d) What is the corresponding mathematical model for pollutant concentration?

(e) Sketch the graph of $m(t)$ against t if (i) $m(0) > q/k$, (ii) $m(0) < q/k$.

2 Analysing a model: skid marks

The Introduction described the stages of the mathematical modelling process, and in Section 1 you saw how some of those stages can be applied in the context of modelling pollution in the Great Lakes. In this section you are asked to relate the stages of the mathematical modelling process to a previously formulated mathematical model. This model was not developed using the structure for modelling described in the Introduction, and consequently some of the modelling ideas have been obscured. This example is typical of accounts of modelling that you may see in books, or produced in the workplace. The aim of this section is to help you to draw out and clarify mathematical modelling ideas by considering the example.

The solution in the following example includes all of the elements of mathematical modelling, but you may have to search a little to find them. The basic mathematical idea used is an equation for motion in one dimension with constant acceleration.

First read the whole description of the mathematical model. Then turn to the exercises, which consist of a number of questions about the model, drawing out the main modelling points. You will need to refer back to the modelling description in the example as you work through the exercises.

As you read through the description of the model, bear in mind the points made in the previous section. Look out for the purpose of the model, the system that is being modelled, the simplifying assumptions, definitions of the variables and the derivation of relations between them, and the conclusions that are drawn from the model.

Example 1

When the police are investigating a road accident, skid marks made by a car can be very informative. From the length of the skid marks, the police can estimate the speed at which the car was travelling before the wheels locked and the car went into the skid.

To assist in this estimation, the police sometimes drive a similar 'test' car with similar tyres and under similar road conditions, and cause it to skid at the same place, but at a lower speed. They then compare the skid marks produced by the test car with the original ones.

Solution

The police assume that the frictional force between the wheels and the road is dependent not on the speed of the car, but only on the mass of the car, the condition of the road and the type of surface. The frictional force is assumed to be proportional to the mass of the car, as are any accelerating or decelerating forces due to gravity if the car is going up or down a hill. The decelerations of the original car and the test car are the same according to these assumptions.

So it is necessary to use the test car's results to calculate the deceleration and then to use the calculated deceleration to estimate the speed of the original car.

For the test car, let us call the initial speed u_{test} and the distance that the car travels when skidding x_{test}. The final speed v is 0, so

$$u = u_{\text{test}}, \quad v = 0, \quad x = x_{\text{test}}, \quad x_0 = 0,$$

a_{test} is to be found.

The equation linking u, v, x, x_0 and constant acceleration a is

$$v^2 = u^2 + 2ax, \tag{2}$$

This equation is derived in Unit 3 as

$$v^2 = v_0^2 + 2a_0(x - x_0),$$

where x_0 is the initial position, v_0 is the initial velocity and a_0 is the constant acceleration.

so

$$0 = u_{\text{test}}^2 + 2a_{\text{test}}x_{\text{test}},$$

that is,

$$a_{\text{test}} = -\frac{u_{\text{test}}^2}{2x_{\text{test}}}.$$

For the original car, if it skidded to rest over a distance x_{car}, assuming that its deceleration a_{car} is the same as that for the test car a_{test}, then

$$a = a_{\text{car}} = a_{\text{test}} = -\frac{u_{\text{test}}^2}{2x_{\text{test}}}, \quad v = 0, \quad x = x_{\text{car}},$$

$u = u_{\text{car}}$ is to be found.

Using the constant acceleration equation again gives

$$0 = u_{\text{car}}^2 - \frac{2u_{\text{test}}^2 x_{\text{car}}}{2x_{\text{test}}},$$

that is,

$$u_{\text{car}} = u_{\text{test}}\sqrt{\frac{x_{\text{car}}}{x_{\text{test}}}}. \tag{3}$$

So the speed of the car before the accident can be estimated. Bear in mind that the speed of the original car would be greater than u_{car} if it had crashed, rather than stopped, at the end of the skid.

Exercise 3

(a) State in your own words the purpose of this model, and say when it may be useful.

◀ Specify purpose ▶

(b) What is the role of the test car? Why could the police not just be issued with tables giving the speed in terms of the length of the skid marks?

(c) The basic concern of mathematical modelling is with finding relationships between variables that specify the system under consideration; in other words, to find formulas that enable you to calculate something if you know the value of something else, or that tell you how something varies with something else.

What are the appropriate 'something' and 'something else' in this model (in words)? (The second sentence of the description of the model may be of help here.) On the basis of what you know about skids, say what you can, in the simplest terms, about the nature of the variation.

Exercise 4

(a) There are several symbols appearing in the description of the model, but according to part (c) of Exercise 3 we are really concerned with the relationship between just two key variables. The symbols conveniently form four groups: symbols for the two key variables; symbols for the data; symbols introduced to make the calculations easier; and symbols used in a general formula employed in the model. Classify the symbols according to this scheme.

◀ Create model ▶

(b) No units of measurement are given anywhere in the definitions of the symbols. This is not the practice adopted in this module, and you are usually advised not to follow it. But does it matter in this instance?

Exercise 5

(a) What is the basic model that underlies the whole discussion?

(b) The basic model identified in your answer to part (a) assumes particle motion with constant acceleration in a straight line. Are these assumptions mentioned in the example? What other assumptions are mentioned or implicit in the example?

(c) Using Newton's second law and the properties of sliding friction, justify the assumption of constant acceleration (i.e. constant deceleration for a skidding car) for the case where the road is flat.

You may find it helpful to draw a force diagram. Sliding friction is discussed in Unit 3.

Exercise 6

(a) Justify the 'assumption' that the decelerations of the two cars are the same.

(b) The possibility that the accident took place on a slope is alluded to in the description of the model, but it is not stated explicitly whether the model applies to such a situation. Does it?

(c) Decide whether the model would apply to the following situations.
  (i) A crashed Rolls-Royce
  (ii) A crash into a strong headwind
  (iii) A crash in a shower of rain, if the road dries out before the police arrive on the scene

Exercise 7

(a) The model, so far as the original car is concerned, might be summarised as follows.

> We want to determine the initial speed u_{car} in terms of the length of the skid x_{car} (which is the distance that the car travels from the initial point of the skid mark before coming to rest). Under the assumption of constant deceleration in a straight line, this is given by equation (2) as

The car's acceleration a_{car} is negative because the car is decelerating, so $-2a_{\text{car}}x_{\text{car}} > 0$.

$$u_{\text{car}}^2 + 2a_{\text{car}}x_{\text{car}} = 0, \quad \text{or equivalently,} \quad u_{\text{car}} = \sqrt{-2a_{\text{car}}x_{\text{car}}}.$$

Confirm that this result agrees with intuitive expectations, as given in the answer to part (c) of Exercise 3. Explain why this is not the end of the story, and how the test car is involved.

◀ Do mathematics ▶

(b) By combining the formulas for the original car and the test car, the final formula (equation (3))

$$u_{\text{car}} = u_{\text{test}}\sqrt{\frac{x_{\text{car}}}{x_{\text{test}}}}$$

is obtained. From this formula, the example says, 'the speed of the car before the accident can be estimated'. If the actual skid marks are four times as long as those of the test car, and the test car was going at 40 mph when it skidded, was the original car breaking the 70 mph speed limit?

Exercise 8

◀ Interpret results ▶

(a) Suppose that you are a police instructor. Explain to a police officer, who is about to go out on an accident investigation for the first time, how to estimate the speed of the car before the accident.

(b) Explain the final statement of the example: 'the speed of the original car would be greater than u_{car} if it had crashed, rather than stopped, at the end of the skid'.

Reading and interpreting descriptions of mathematical models can be hard work: there is usually a lot of information to absorb, and it can be difficult to focus on what is really important. For example, in the Great Lakes model, the names and volumes of the lakes do not help our understanding of the mathematics (though they are of crucial importance to a geographer). The process of coming to terms with a mathematical model is one of digging away until you find what lies at the root of it all. In the Great Lakes model it is the differential equation $dm/dt = -k\,m(t)$ that lies at the root, whereas in the skid marks model it is the formula relating initial and final velocity to distance, for motion with constant acceleration in a straight line.

See equation (2).

One of the basic skills of mathematical modelling (as you will find when you come to construct models for yourself) is to formulate the fundamental equation or relationship that describes the process in which you are interested. There are several points in the two accounts seen so far that show how this can be done.

- It is important to pick appropriate variables and parameters, and to define them carefully.
- It is necessary to simplify matters in order to make progress. For example, it was recognised in creating the Great Lakes model that the problem of seasonal variations in water level could be postponed, if not ignored.
- It is good practice to record the assumptions that you make in deriving the model. This certainly helps the reader, but also the written record is then clearer and suggests ways of developing the model later.
- It is important to collect relevant data, both to check the predictions of the model and to furnish the values of any parameters that are needed to apply the model to any particular situation. The same data cannot be used for parameter estimation and the validation of the model.

Exercise 9

This exercise is based on the skid marks model discussed above.

(a) Discuss the advantages and disadvantages of using compound symbols for variables, such as u_{test} and u_{car}, instead of single symbols such as u and U.

(b) Why is it useful to classify parameters and variables as distinct categories?

(c) Do you consider the following statement to be a useful assumption? 'The value of the coefficient of friction is not greater than 1.0.'

(d) Data are necessary both to provide the values of parameters and to validate the model. What data do you think will be required to provide values of the parameters, and to validate the model?

(e) Do you consider the skid marks modelling report (Example 1) to be easy to follow?

---

3 The skills of modelling

Mathematical modelling involves many different skills. To be good at mathematical modelling you have to be able to do some mathematics, but you need to be capable of doing other things as well. If you tend to think of mathematics as a set of procedures (such as the procedures for solving differential equations), then you may not regard these additional mathematical modelling skills as being relevant to mathematics. Most practising mathematicians, concerned with pure as well as with applied mathematics, would disagree with that interpretation. One of the most powerful motivations for studying mathematics is the desire to solve *new* problems for which there is no known solution procedure. The skills required in mathematical modelling include many general problem-solving skills. To be able to deal with mathematical modelling problems is more generally useful, and more difficult, than (say) being able to solve first-order differential equations by the integrating factor method. It calls on skills of creativity, analysis and interpretation that apply to all sorts of problems, not just mathematical ones.

Here is a list of skills that may be required in the solution of a modelling problem, placed in the order of the modelling framework introduced earlier. You need to be able to:

- specify the purpose of the model, by defining or interpreting the problem that you are investigating
- create the model by
    - simplifying the problem (by means of appropriate assumptions)
    - choosing appropriate variables and parameters
    - formulating relationships between the variables
- use mathematics to find a solution from the relationships
- interpret the results by describing them in words (or otherwise) so that they can be understood by a possible user
- evaluate the model by
    - checking that the mathematical relationships and the solution make sense
    - comparing the results with reality
    - checking their sensitivity to changes in the data (this will be discussed in Unit 18).

For the problem to be considered shortly, the 'Do mathematics' stage of modelling is very brief and is therefore included in the subsection for 'Create the model'.

When tackling a modelling problem in earnest, you have to call on these skills repeatedly, in a complicated and interactive way. It seems wise, therefore, to practise the skills individually at first, and this is the aim of the current section. Each subsection deals with a subgroup of skills from the list above, and is illustrated by reference to the two models considered in Sections 1 and 2. You are also asked to try out the individual skills by applying them to the modelling problem in the following example.

Example 2

What shape should a tin can be? Most cans are cylindrical, so suppose that the best tin can is cylindrical. What shape of cylinder is best: short and fat, long and thin, or somewhere in between? Think of the standard kind of can in which soup, beans and other foodstuffs are preserved. Many of the cans in supermarkets contain about 400 grams of food and are rather taller than they are wide.

Although cans are made from a variety of materials, they are usually called *tin cans* because they were first made from tinned steel.

There is a degree of uniformity in the general shape of such cans, which is quite surprising when there are so many different brands and they contain so many different things; imported cans seem to have a shape similar to those that are manufactured in the UK. Is this traditional, or is it because they all conform to some ideal shape, or because the same machine makes most of them? And if the shape is not ideal, could there be advantages in changing it?

In this section you are asked to investigate the best shape for a cylindrical tin can. This is a modelling problem for you to do, but you are not left entirely on your own to tackle it: the problem has been broken down into steps, corresponding to the modelling skills listed above.

There is no unique correct answer in modelling; if you reach a different answer to the one given, that does not mean that your answer is wrong. However, for the purposes of moving the story forward at each stage, the model will be developed in the text from the solution given to the preceding exercise.

3.1 Specify the purpose of the model

◂ Specify purpose ▸

In mathematical modelling, problems are rarely posed in a way that can be translated directly into mathematical form. For example, from the description of the treatment of pollution in the account of the Great Lakes model, you might have thought that in order to construct a mathematical model, you would need to know quite a lot about pollution. However, it was possible to produce a straightforward non-technical statement of the underlying problem that could serve as the basis for a model, namely:

> to investigate how the pollution level varies with time as clean water flows into the lake and polluted water flows out.

It is important to establish at the outset a clear statement of the purpose of a model. For example, the purpose of the Great Lakes model is to discover the time that it will take for the pollution level in a lake to reduce to a given proportion of its initial pollution level. Such a statement is typical of the approach that is required in order to start a modelling problem.

Although a clear statement of the problem is necessary, it may change as the model develops, and the final statement may be different from that conceived at the outset. However, it is important to have a target at which to aim, even if this target changes during the process.

It is worth bearing in mind that some models, created for a specific purpose, may be applicable in other situations. It may also be possible to save time by modifying a model that has been used for one situation so that it can be applied to another. For example, the mathematical model developed to predict how long it would take for pollution in the Great Lakes to reduce to a target level could be adapted for use in drug therapy, for the cleaning of milk churns by running water through them, or for the distillation of whisky if a stream feeding the distillery is becoming polluted.

Exercise 10

Consider the problem of finding the best shape for a cylindrical tin can that is to contain a specified quantity of baked beans.

(a) The idea of 'best' occurs frequently in mathematical modelling, and its meaning needs to be made precise. What should the word 'best' mean in the phrase 'the best shape for a cylindrical tin can'?

(b) Try to formulate a clear statement of a suitable modelling problem, based on your answer to part (a).

3.2 Create the model

◀ Create model ▶

There are a number of skills that are needed when building a sensible model that approximates a real situation. In creating a model, these skills may be required at a variety of stages and not necessarily in the order in which they are presented here.

Simplify the problem

The skid marks model depended on the results that the deceleration of a skidding car is constant and that, for given conditions, different cars have the same deceleration while skidding. These follow from assumptions that underpin a well-established theory of sliding friction, but hold only if (for example) air resistance is ignored. To ignore air resistance is justified on two counts: first, its effects are probably small compared with those of sliding friction; second, the resulting model is relatively easy to analyse, and may provide some insight into the problem. In modelling you should always look for as simple a model as possible, consistent with the principal features of the problem. (To have ignored the effects of friction would obviously have been counter-productive.)

In Unit 18, you will see how we can check whether simplifying assumptions are justifiable.

It is important to be clear about the simplifying assumptions that have been made in order to arrive at the model. Recording an explicit list of the assumptions makes it easier for the reader to follow the development of the model, and should you need to improve your model, you then have an obvious place to start: review the assumptions, and ask which should be modified or relaxed.

Exercise 11

Continuing with the tin can problem as specified in Exercise 10(b), what simplifying assumptions, if any, need to be made?

Choose appropriate variables and parameters

Identification of the key variables is of paramount importance. If you can summarise the problem in terms of describing roughly how one quantity varies with another, or several others, then you should have no difficulty in identifying the key variables. Once these key variables have been identified, it should be possible to obtain relationships between them, which may throw up other variables and/or parameters. It is good practice to keep a list of all the variables and parameters, adding to it as necessary, to ensure that all of them have been consistently defined and used. It pays to be careful in defining variables and parameters, to avoid confusion later. For example, 'time since all pollution ceased' is clearer, and less likely to be misinterpreted, than just 'time'.

It is possible that not all the variables and parameters are identified before the relationships are formed, and you may need to add more to your list as the model progresses. You may also find it convenient to replace a group of parameters with a symbol in the mathematical formulation of the model; then the symbol must be added to the list.

In the case of the Great Lakes model, there are three key variables: the time (in seconds) since all sources of pollution were removed, the mass (in kilograms) of pollutant in the lake, and the concentration (in kilograms per cubic metre) of pollutant in the lake. The creation of the mathematical model involves identifying relationships between these key variables. In writing down the relationships, five parameters were identified: the target concentration level (in kilograms per cubic metre) of pollutant in the lake; the time taken (in seconds) to reduce the concentration of pollutant to the target level; the volume (in cubic metres) of water in the lake; the water flow rate (in cubic metres per second); and the proportionate flow rate (in seconds^{-1}).

In the skid marks model, there are two key variables: the length of the skid for the original car, and its initial speed, each in appropriate units. In writing down the relationship between these two key variables, two further variables and a parameter were identified: the length of the skid for the test car, its initial speed, and the common deceleration of the two cars. The units for these quantities need to match the units chosen for the key variables. (The final speeds of the two cars could be regarded as additional parameters.)

Exercise 12

Continuing with the tin can problem, define the variables and parameters that you think will be needed, giving appropriate units.

Formulate relationships

The use of the input–output principle to formulate relationships in creating the Great Lakes model is a good illustration of a quite common modelling technique. Another common technique for formulating relationships, in the case of mechanics problems, is to make use of Newton's laws of motion, although their use may be hidden, as in the skid marks model.

Often it is helpful to draw a diagram. Not only does a diagram help in the definition of the variables and parameters, but it tends to help in gathering together some key factors.

A picture is said to be worth a thousand words.

Exercise 13

Draw a diagram to help with the tin can problem.

Exercise 14

(a) Write down formulas that relate the variables in the tin can problem. You should explain on which assumptions any formula is based.

(b) Derive a formula that relates the area A of the can to its radius r, where the volume V is a parameter.

(c) Are there any assumptions that have not been used in the formulation? Are they needed?

(d) Are there any variables or parameters that have not been used in the derivation?

Find a solution

◀ Do mathematics ▶

Typical mathematical techniques used in simple models are solving algebraic equations, solving differential equations and finding an optimum value.

Once the formulas that relate variables have been derived, some mathematics will probably be needed to find a solution to the model. In the skid marks model, the solution for u_{car} of the equation

$$0 = u_{\text{car}}^2 - \frac{2u_{\text{test}}^2 x_{\text{car}}}{2x_{\text{test}}}$$

was required. In the Great Lakes model a differential equation was solved, an initial condition was used, and an algebraic equation had to be solved to find the target time.

For the tin can problem, Exercise 14 shows that we need to find the value(s) of r for which

$$A = 2\left(\frac{V}{r} + \pi r^2\right) \qquad (4)$$

is a minimum (as specified in the solution to Exercise 10(b)).

Exercise 15

Use formula (4) to find 'the best shape for a cylindrical tin can'.

3.3 Interpret the results, evaluate the model

Obtaining a mathematical solution to a modelling problem is not the end of the modelling process. The solution needs to be interpreted in terms of the original problem posed, and a number of checks should be made. This subsection outlines some of the techniques used to interpret the solution and to check its reliability.

Check that the model and solution make sense

◀ Interpret results ▶

It is sometimes possible to take some particular values for the variables, and so make a quick check on the correctness of the model. In the skid marks example, it makes sense that the longer the skid mark of the car, the faster the speed that it was travelling beforehand, and this is borne out by the solution. Also, if the length of the skid mark is zero, then the model and intuition both give the same value, zero, for the speed of the car. In the Great Lakes model, the time taken to reach the target pollution level is increased if the target level is decreased, and this makes sense. It is always worth investigating the solution with checks such as these in mind.

Exercise 16

(a) How would you expect the surface area of the tin can to change as the radius becomes very small? Is this what the model predicts?

(b) How would you expect the surface area of the tin can to change as the radius becomes very large? Is this what the model predicts?

(c) Is there any other test that can be applied to check whether the solution is reasonable?

Compare the results with reality

◀ Evaluate model ▶

A check that the model predicts the kind of results that one would expect from common sense and from experience, as described above, is one type of comparison with reality. Beyond that, if possible, one should validate the model by comparing its predictions with data from an experiment or other reliable source. It is good practice to try to reformulate the results so that this check turns into something simple such as drawing a straight line. You may also require some data to give an explicit numerical solution to the problem, such as the values of lake volume and water flow rate in the Great Lakes problem. The practicality of the skid marks model lies in the use of the test car to provide these data.

Exercise 17

(a) Summarise the solution obtained for the tin can problem descriptively, giving the optimal shape of the can in terms of the ratio of height to radius.

(b) Measure the radius and volume of some tin cans. Does the shape of can predicted by your solution correspond to the actual shapes of tin cans?

The volume can be an estimate based on measurements of the diameter and height.

Exercise 18

In your own words, and without using any equations or symbols, give an outline of the formulation of the model.

---

4 Dimensions and units

The choice of unit of measurement does not affect the physical quantity being measured. For example, the width of this page is the same irrespective of whether we measure it in millimetres or inches. Similarly, no matter what units we use, speed must always be expressed as length divided by time. To measure speed we could use metres per second, centimetres per hour or yards per day, but it has to be a unit-of-length per unit-of-time.

4.1 Dimensional consistency

In modelling, we attempt to establish relationships between variables by using assumptions to develop equations. In this subsection we investigate a technique used to verify that these equations make sense, in that they are *dimensionally consistent.* Such an analysis can never establish that an equation is correct, but it provides a useful check and can sometimes tell us that an equation is wrong.

In reality, the mass of an egg white varies slightly from egg to egg.

For example, suppose that you are cooking a meringue using egg whites and sugar. In order to determine the mass m of the meringue, you would first need to determine the mass m_1 of an egg white and the mass m_2 of the sugar, all measured in the same units. Then, assuming that no mass is lost during the cooking process, the mass of a meringue using n egg whites is given by $m = nm_1 + m_2$. Each additive term on the right-hand side of this equation (nm_1 and m_2) is a mass, and the equation makes sense only if each of these terms is measured in the same units, giving the overall mass m in those units. This is the essence of dimensional consistency.

In terms of dimensional consistency, the important property of volume is that it is length cubed, and the important property of speed is that it is length divided by time. In these examples, we can think of length and time as being fundamental properties, referred to as *base dimensions.*

> Mass, length and time are fundamental properties and are referred to as **base dimensions**. The base dimension mass is denoted by M, length by L, and time by T.

The dimensions of all physical quantities (in this module) can be expressed in terms of these three base dimensions. For example, the dimensions of speed, namely length divided by time, are $\mathrm{L\,T^{-1}}$. To write this more succinctly, we introduce square brackets to mean 'the dimensions of', and write

$$[\text{speed}] = \mathrm{L\,T^{-1}}.$$

This is read as 'the dimensions of speed are $\mathrm{L\,T^{-1}}$'. Similarly,

$$[\text{volume}] = \mathrm{L}^3.$$

The dimensions of other physical quantities can be expressed in terms of M, L and T by working from their units of measurement.

Example 3

Find the dimensions of the magnitude of acceleration and of the magnitude of force.

Solution

The SI unit for the magnitude of acceleration is $\mathrm{m\,s^{-2}}$, so

$$[\text{acceleration}] = \mathrm{L\,T^{-2}}.$$

The SI unit for the magnitude of force is the newton, where $1\,\mathrm{N} = 1\,\mathrm{kg\,m\,s^{-2}}$, therefore

$$[\text{force}] = \mathrm{M\,L\,T^{-2}}.$$

As Example 3 has illustrated, we can multiply, divide and take powers of dimensions using the usual rules of arithmetic and algebra. However, dimensions cannot be added or subtracted.

Note that some quantities and units are **dimensionless**. A number is dimensionless, and we write $[\text{number}] = 1$.

Exercise 19

(a) What are the dimensions of area?

(b) Given that in SI units density is measured in $\mathrm{kg\,m^{-3}}$, what are the dimensions of density?

(c) The size of an angle θ, measured in radians, can be determined by assuming that the angle lies at the centre of a circle and that the two lines defining the angle are radii of the circle, as shown in Figure 4. If the radii have length r and the arc defined by θ has length l, then $\theta = l/r$. What are the dimensions of angle?

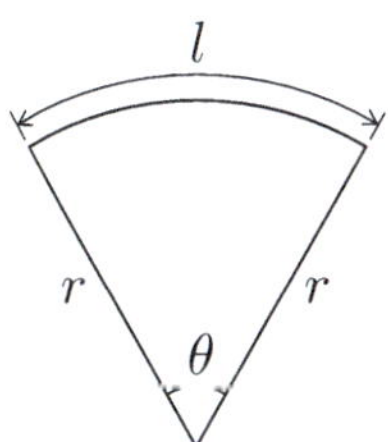

Figure 4 Size of an angle

The dimensions of some of the physical quantities used in this module are given in Table 2. For vector quantities, we give the dimensions of the *magnitude* of the vector.

Table 2 Dimensions

Physical quantity	Dimensions
Number	1
Angle	1
Mass	M
Length	L
Time	T
Area	L^2
Volume	L^3
Speed	$\mathrm{L\,T}^{-1}$
Angular speed	T^{-1}
Acceleration	$\mathrm{L\,T}^{-2}$
Angular acceleration	T^{-2}
Force	$\mathrm{M\,L\,T}^{-2}$
Energy	$\mathrm{M\,L}^2\,\mathrm{T}^{-2}$
Pressure	$\mathrm{M\,L}^{-1}\,\mathrm{T}^{-2}$
Torque	$\mathrm{M\,L}^2\,\mathrm{T}^{-2}$
Density	$\mathrm{M\,L}^{-3}$

Note that energy and torque have the same dimensions although they are different physical quantities. For example, a torque is a vector, and energy is a scalar quantity.

In modelling, any equation must be **dimensionally consistent**: the dimensions must be the same for each of the additive terms on either side of the equation. If they are, then our confidence in the model is increased; if they are not, then we know that we have made a mistake somewhere in deriving the equation.

Example 4

The period τ of small oscillations of a particle of mass m suspended from a fixed point by a light inextensible string of length l is given by

$$\tau = 2\pi\sqrt{\frac{l}{g}}.$$

Show that this equation is dimensionally consistent.

Solution

The period τ has dimensions T. For the right-hand side of the equation we have

$$\left[2\pi\sqrt{l/g}\right] = [2\pi]\,([l]/[g])^{1/2} = 1 \times (\mathrm{L}/(\mathrm{L\,T}^{-2}))^{1/2} = (\mathrm{T}^2)^{1/2} = \mathrm{T}.$$

So both sides of the equation have dimensions T, and the equation is dimensionally consistent.

Example 5

In equation (4) the area of the tin plate used in making a tin can was given as

$$A = 2\left(\frac{V}{r} + \pi r^2\right).$$

Establish that this equation is dimensionally consistent.

Solution

All three terms in the equation, A, $2V/r$ and $2\pi r^2$, must have the same dimensions if the equation is to be dimensionally consistent. We have

$$[A] = \mathrm{L}^2, \quad [2V/r] = [2]\,[V]/[r] = \mathrm{L}^3/\mathrm{L} = \mathrm{L}^2 \quad \text{and} \quad [2\pi r^2] = \mathrm{L}^2,$$

so dimensional consistency is assured.

Example 5 illustrates that when checking for dimensional consistency, each additive term must be checked separately: although we can multiply, divide and take powers of dimensions, we can add or subtract only terms that have the same dimensions.

Exercise 20

Show that the equation

$$v^2 = v_0^2 + 2a_0x,$$

which relates the speed of something moving with constant acceleration a_0 to the distance x travelled, is dimensionally consistent.

Exercise 21

The mathematical model of pollution in a lake in Section 1 led to the differential equation

$$\frac{dm}{dt} = -k\,m(t),$$

where $m(t)$ is the mass of pollutant in the lake at time t, and $k = r/V$, where r is the volume flow rate of water through the lake, and V is the volume of water in the lake.

Show that this equation is dimensionally consistent.

(*Hint*: A derivative is essentially a ratio, or fraction, so its dimensions are those of its numerator divided by those of its denominator.)

Exercise 22

One term in the following equation is in error:

$$\frac{p}{\rho} + \tfrac{1}{2}u - gz = \text{constant},$$

where p is the pressure of a fluid, ρ is its density, u is its speed, g is the magnitude of the acceleration due to gravity, and z is the height above some given level. By checking the dimensions of each term of the equation, find which term is in error, and suggest how it might be made dimensionally correct. (The dimensions of pressure are $\mathrm{M\,L^{-1}\,T^{-2}}$.)

Dimensional consistency can also be used to determine the units of parameters of models.

Exercise 23

See Unit 3, Subsection 4.2.

The magnitude R of the air resistance force on an object can be modelled, in certain circumstances, as

$$R = c_2 D^2 v^2,$$

where c_2 is a constant, D is the effective diameter of the object, and v is its speed.

Use dimensional consistency to determine the dimensions of the constant c_2. What SI units would be used to measure c_2?

We have seen how to check for dimensional consistency in equations where the additive terms involve multiplying, dividing or taking powers. But many of the models that you have seen in the module involve functions such as exp, ln, sin, cos, and so on. How does one check for dimensional consistency in such cases? For example, the solution of the differential equation $dm/dt = -k\,m(t)$ in the lake pollution model is

$$m(t) = m(0)\,e^{-kt}, \tag{5}$$

where $m(0)$ is the initial mass of pollutant. To be able to check for dimensional consistency in this equation, we need to be able to find the dimensions of e^{-kt}. Now we know that exp, ln, sin, cos, and so on, all have real numbers as their domains and image sets. Therefore for dimensional consistency, the argument x of the function must be dimensionless, and we can take e^x, $\ln x$, $\sin x$, $\cos x$, and so on, to be dimensionless.

So the key to checking for dimensional consistency in equations involving such functions is to ensure that their arguments are dimensionless. In the above case, we know from the equation $k = r/V$ given in Exercise 21 that

$$[k] = [r/V] = [r][V]^{-1} = (\mathrm{L^3\,T^{-1}}) \times (\mathrm{L^3})^{-1} = \mathrm{T^{-1}},$$

so $[-kt] = 1$. Hence $[e^{-kt}] = 1$, and equation (5) is dimensionally consistent, since $[m(t)] = [m(0)] = \mathrm{M}$.

Sometimes, however, the arguments of functions are *not* dimensionless. For example, in obtaining equation (5), one might have begun by using the separation of variables method to solve the differential equation $dm/dt = -k\,m(t)$ and obtained

$$\ln(m(t)) = -kt + C, \qquad (6)$$

where C is an arbitrary constant. Now $m(t)$ is not dimensionless (it has dimensions M), so we cannot determine the dimensions of $\ln(m(t))$. However, putting $t = 0$ into equation (6) gives $C = \ln(m(0))$, and this enables us to rewrite equation (6) as

$$\ln\left(\frac{m(t)}{m(0)}\right) = -kt, \qquad (7)$$

where $[m(t)/m(0)] = \mathrm{M}/\mathrm{M} = 1$. So now the argument of ln *is* dimensionless, and it is an easy matter to check for dimensional consistency (both sides of equation (7) have dimensions 1).

Very often, if the argument of a function in an equation is not dimensionless, some simple manipulation of the equation can render it so and hence enable dimensional consistency to be checked.

Exercise 24

The motion of a particle in a certain problem can be modelled as

$$x(t) = A\cos(\omega t + \phi),$$

where $x(t)$ measures the displacement at time t, and ϕ is a constant. What dimensions for A, ϕ and ω will ensure dimensional consistency?

4.2 Change of units

There is now almost universal acceptance of the SI system in professional scientific and engineering circles. This system is based on seven **base units**, of which the following four are used in this module.

Table 3 Base units

Physical quantity	Unit	Abbreviation
Length	metre	m
Mass	kilogram	kg
Time	second	s
Temperature	kelvin	K

The other three base units are the ampere (unit of electric current), the candela (unit of luminous intensity) and the mole (unit of the amount of substance).

In addition, there are various **derived units** in common use, which can be formed by combining the above base units. For example, we have already introduced the *newton* (N) as the unit of force, and the *joule* (J) is the unit of energy. In terms of the base units, $1\,\mathrm{N} = 1\,\mathrm{kg\,m\,s^{-2}}$ and $1\,\mathrm{J} = 1\,\mathrm{kg\,m^2\,s^{-2}}$.

To avoid very large or very small numbers, we also use multiple or fractional units. For example, the distance between London and Edinburgh is more conveniently expressed as 665 km (kilometres) rather than $6.65 \times 10^5\,\mathrm{m}$. Similarly, the thickness of copper used in a central heating pipe is usually stated to be 1 mm (millimetres) rather than $1 \times 10^{-3}\,\mathrm{m}$. The most important prefixes for forming these units are given in Table 4.

Table 4 Prefixes

Multiplication factor	Prefix	Symbol
$10^9 = 1\,000\,000\,000$	giga	G
$10^6 = 1\,000\,000$	mega	M
$10^3 = 1\,000$	kilo	k
$10^{-2} = 0.01$	centi	c
$10^{-3} = 0.001$	milli	m
$10^{-6} = 0.000\,001$	micro	μ
$10^{-9} = 0.000\,000\,001$	nano	n

For example, the pressure of the atmosphere, which is about $10^5\,\mathrm{N\,m^{-2}}$, is sometimes written as $100\,\mathrm{kN\,m^{-2}}$, that is, 100 kilonewtons per square metre. The conversion between such alternative units is quite straightforward but does need some care. For example, to convert 6 kilometres into metres, we know that

$$1\,\mathrm{km} = 10^3\,\mathrm{m},$$

so the conversion factor from kilometres to metres is

$$\frac{10^3\,\mathrm{m}}{1\,\mathrm{km}} \quad (= 1).$$

This example may seem trivial, but the method used here can be used in more complicated problems.

Using this conversion factor, we have

$$6\,\mathrm{km} = 6\,\mathrm{km} \times \left(\frac{10^3\,\mathrm{m}}{1\,\mathrm{km}}\right) = 6 \times 10^3\,\mathrm{m}.$$

In this calculation we have, in essence, treated the units in a similar way to algebraic quantities and 'cancelled' the 'km' between the numerator and denominator.

Example 6

The speed of a car is $50\,\mathrm{km\,h^{-1}}$. Express this speed in the SI units of $\mathrm{m\,s^{-1}}$.

Solution

We have

$$1\,\text{km} = 10^3\,\text{m}$$

and

$$1\,\text{h} = 60\,\text{min} = 60 \times 60\,\text{s}.$$

So

$$\begin{aligned} 50\,\text{km}\,\text{h}^{-1} &= 50\,\text{km}\,\text{h}^{-1} \times \left(\frac{10^3\,\text{m}}{1\,\text{km}}\right) \times \left(\frac{60 \times 60\,\text{s}}{1\,\text{h}}\right)^{-1} \\ &= 50 \times 10^3 \times 60^{-1} \times 60^{-1}\,\text{m}\,\text{s}^{-1} \\ &\simeq 13.89\,\text{m}\,\text{s}^{-1}. \end{aligned}$$

So the speed of the car is $13.9\,\text{m}\,\text{s}^{-1}$ to one decimal place.

Exercise 25

The density of a metal is $13.546\,\text{g}\,\text{cm}^{-3}$. Express this density in the SI units of $\text{kg}\,\text{m}^{-3}$.

Unfortunately, the metric system of units is not the one in everyday use everywhere, particularly in the UK and the USA, where lengths may be measured in inches (in), feet (ft) and miles, masses in ounces (oz), pounds (lb) and tons, and temperatures in degrees Fahrenheit. The method that we have used above can equally be used to convert between such units and SI units. However, when converting between Imperial and metric systems of units, the conversion factors are not 'nice round numbers'. For example,

$$1\,\text{lb} = 0.453\,592\,\text{kg}.$$

Some of the common conversion factors between Imperial and SI units are given in Table 5.

Table 5 Imperial units

Imperial unit	SI unit
1 in	$2.54 \times 10^{-2}\,\text{m}$
1 ft	$0.304\,8\,\text{m}$
1 mile	$1.609\,344 \times 10^3\,\text{m}$
1 pint	$5.682\,613 \times 10^{-4}\,\text{m}^3$ (= 0.568 litres)
1 gallon	$4.546\,09 \times 10^{-3}\,\text{m}^3$ (= 4.546 litres)
1 oz	$2.834\,952 \times 10^{-2}\,\text{kg}$
1 lb	$0.453\,592\,\text{kg}$
1 ton	$1.016\,047 \times 10^3\,\text{kg}$

Example 7

The speed of a car is 30 miles per hour (mph). Express this speed in SI units.

Solution

We have

$$1 \text{ mile} = 1.609\,344 \times 10^3\,\text{m}$$

and

$$1\,\text{h} = 60 \times 60\,\text{s}.$$

So

$$\begin{aligned} 30\,\text{mph} &= 30\,\text{miles h}^{-1} \times \left(\frac{1.609\,344 \times 10^3\,\text{m}}{1 \text{ mile}}\right) \times \left(\frac{60 \times 60\,\text{s}}{1\,\text{h}}\right)^{-1} \\ &= 30 \times 1.609\,344 \times 10^3 \times 60^{-1} \times 60^{-1}\,\text{m s}^{-1} \\ &\simeq 13.41\,\text{m s}^{-1}. \end{aligned}$$

So 30 mph is equivalent to $13.4\,\text{m s}^{-1}$ to one decimal place.

Exercise 26

The pressure on a piston is 61 lb per square inch. Express this pressure (to 4 s.f.) in the SI units of kg m^{-2}.

(In the SI system, pressure is *force* per unit area, not *mass* per unit area. To convert our answer here to the correct units, we would have to multiply it by g, the magnitude of the acceleration due to gravity.)

Exercise 27

Before a recent holiday to South America, I was unable to buy local currency in the UK. So I bought US dollars in the UK, at the exchange rate £1 = \$1.75. In Bolivia, I exchanged my dollars for the local currency, at the rate of exchange \$1 = 7.75 boliviano. How much was 1 boliviano worth in £?

Learning outcomes

After studying this unit, you should be able to:

- create a simple model, given a clear statement of a problem
- write down the simplifying assumptions that underpin a model
- identify the key variables and the parameters of a model
- apply the input–output principle to obtain a mathematical model, where appropriate
- obtain mathematical relationships between variables, based on or linking back to the simplifying assumptions
- interpret the mathematical solution to a modelling problem in terms of the original statement of the problem
- understand the processes involved in evaluating a model by comparing with reality
- appreciate the role of data in testing the model and, if necessary, in providing parameter values for the model
- understand that the purpose of a model is the measure used to judge the suitability of the model
- appreciate that simple models can be as useful as more complex models, if they serve their intended purpose
- check that mathematical relationships between variables and parameters are dimensionally consistent
- convert the values of physical quantities from one system of units to another.

Solutions to exercises

Solution to Exercise 1

In describing the formulation of the model, it may be a help to write down some words or phrases first, and then to combine them into a paragraph.

Possible key words or phrases are: pollution has ceased; the water flowing from the lake is polluted; concentration of the pollutant is uniform; small time interval; mass of pollutant; one lake; no change in volume of the lake; input–output principle.

A possible description might read:

> This model considers a polluted lake to which no further pollutant is added. The lake is of constant volume. The water flowing from the lake is polluted, and this is the only way that the pollutant leaves the lake. The input–output principle is applied to the mass of pollutant within the lake, based on the change over a small time interval.

In a modelling report, the description of the formulation should be placed before the actual formulation, in order to guide the reader. However, it could be written (or revised) *last*, after the formulation has been completed.

Solution to Exercise 2

(a) The input–output principle is applied to the mass of pollutant within the lake. In the time interval $[t, t+\delta t]$, the input (mass entering the lake) is $q\,\delta t$, while the output is $(r/V)\,m(t)\,\delta t$, as shown in the text. The accumulation over this time interval is $m(t+\delta t) - m(t)$. Hence the change in the mass of pollutant in this interval is given by

$$m(t+\delta t) - m(t) \simeq q\,\delta t - \frac{r}{V}\,m(t)\,\delta t.$$

Putting $k = r/V$, this leads to the differential equation

$$\frac{dm}{dt} = q - k\,m(t).$$

(b) The integrating factor method gives the general solution of the above equation as

$$m(t) = \frac{q}{k} + Ce^{-kt},$$

where C is an arbitrary constant. Rearranging the solution, and using the initial condition at $t = 0$ to evaluate C, gives the required result:

$$m(t) = \frac{q}{k} + \left(m(0) - \frac{q}{k}\right) e^{-kt}.$$

In the long term, the mass of pollution tends to q/k.

(c) If there is initially no pollution in the lake, then $m(0) = 0$ and we have

$$m(t) = \frac{q}{k}\left(1 - e^{-kt}\right).$$

The mass of pollution increases from zero up to the steady-state level q/k.

(d) Using the equation found in part (b), with $m(t) = V\,c(t)$, and putting $kV = r$, we have

$$c(t) = \frac{q}{k} + \left(c(0) - \frac{q}{r}\right) e^{-kt}.$$

This is the corresponding mathematical model for pollutant concentration.

(e) Typical graphs are shown below.

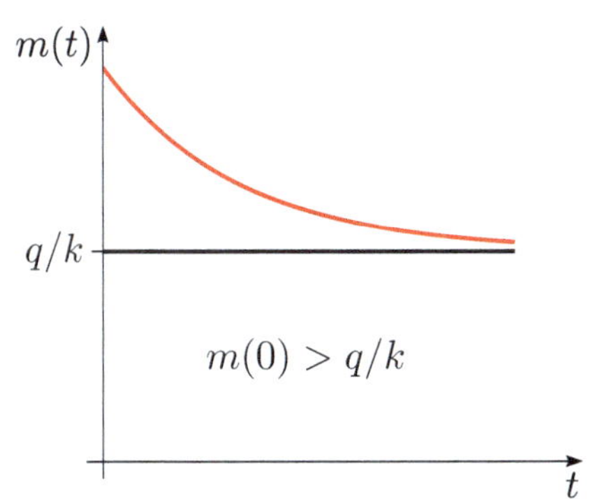

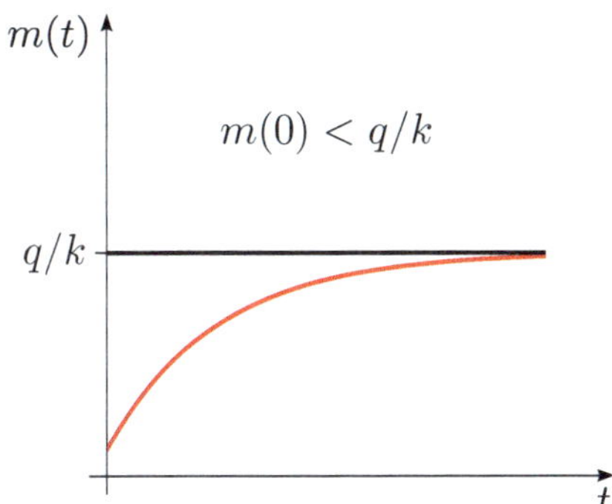

Solution to Exercise 3

(a) When a car skids, its tyres may leave marks on the road. The purpose of the model is to find a method of using the lengths of such skid marks to work out how fast the car was travelling when it began to skid. This information may be useful to the police after a road accident, when they try to find out what happened.

(b) The test car provides data in addition to the length of the skid marks in the accident. These data allow a direct comparison to be made between the length of the skid marks in the accident, for which the speed of the car is unknown, and the length of skid marks made by a car whose speed at the onset of the skid is known. In effect, the data for the test car are used to estimate the deceleration of the car involved in the accident while it was skidding. This is possible because the conditions under which the test takes place are similar to those for the accident, apart from the speed of the car.

The gradient and state of the road surface, the condition of the car's tyres, and other circumstances of the accident have a significant effect on the skid. They may vary widely from accident to accident. It is more reliable to reproduce the conditions in a test than it is to make allowance for them in a table.

(c) The relationship required is one that gives the speed of the vehicle at the onset of the skid in terms of the length of the skid marks. The longer the skid marks, the faster the car was travelling when it began to skid (all other things being equal), so the speed is an increasing function of the length of the skid marks.

Solution to Exercise 4

(a) The key variables are the original car's initial speed u_{car}, and the length of its skid x_{car}. The data, which are the corresponding quantities for the test car, are represented by the symbols u_{test} (initial speed) and x_{test} (length of skid). All of the symbols appearing in the final formula (3) have now been identified, but there are still others. Two that are used to make the calculations easier are a_{car}, the acceleration of the original car, and a_{test}, the acceleration of the test car. (The final speed is zero in each case, so there was no need to introduce symbols for the final speed of either car.) There are also the four symbols v, u, a and x that are used in the general equation (2) for constant acceleration.

(b) No, it does not matter in this instance. One way of seeing why is to rewrite formula (3) as

$$\frac{u_{\text{car}}}{u_{\text{test}}} = \sqrt{\frac{x_{\text{car}}}{x_{\text{test}}}}.$$

The left-hand side of this equation is the *ratio* of two speeds, while the right-hand side is the square root of the *ratio* of two distances. Now the ratio of two speeds is the same value whether the speeds are measured in $\text{m}\,\text{s}^{-1}$, mph or leagues per century (provided that both speeds are measured in the same units). Likewise, the ratio of two distances takes the same value whatever unit of measurement is used (again provided that both distances are measured in the same units).

The omission of units in the description of the variables is therefore not important in this case, although you are usually encouraged to state units of measurement when defining variables.

Solution to Exercise 5

(a) The model that underlies the whole discussion is that of the motion of a particle moving in a straight line with constant acceleration (which you met in Unit 3). The formula required here is that relating the initial and final velocity, acceleration and position, namely

$$v^2 = v_0^2 + 2a_0x,$$

which becomes the same as equation (2) when u is substituted for v_0 and a for a_0.

(b) The assumption of particle motion in a straight line is not mentioned. (It may be reasonable, based on the police's knowledge of skids and the record left by skid marks, but it is not mentioned explicitly as an assumption.)

The assumption of constant acceleration is not mentioned either.

Two other assumptions are mentioned explicitly. The first is that 'the frictional force between the wheels and the road is dependent not on the speed of the car, but only on the mass of the car, the condition of the road and the type of surface'. The second is that the frictional force is 'proportional to the mass of the car, as are any accelerating or decelerating forces due to gravity'. From these two assumptions it is deduced that 'The decelerations of the original car and the test car are the same' (although this is restated as an assumption shortly after equation (2)).

Implicit in what follows, though not obvious from any part of the example, is the assumption that the only motive or resistive forces acting on a car are friction and 'accelerating and decelerating forces due to gravity' (i.e. weight). Thus, for example, it has been assumed that air resistance can be ignored. It could be argued that the lack of mention of forces other than friction and weight amounts to an assumption that no other motive or resistive forces are acting. In sum, then, the only forces assumed to be acting on a car are friction, its weight and, of course, the normal reaction.

Another assumption, implicit in carrying out the test skid, is that the road and tyre conditions are the same for both cars. (It is not clear whether the masses of the cars are assumed to be the same.)

(c) In the light of the answer to part (b), the only forces that are assumed to be acting on a car are its weight $\mathbf{W}$, the friction $\mathbf{F}$, and the normal reaction $\mathbf{N}$. Since the motion is assumed to be in a straight line, we need only two axes. The direction of motion is horizontal, because the road is flat. Modelling the car as a particle of mass m produces the force diagram in the figure in the margin.

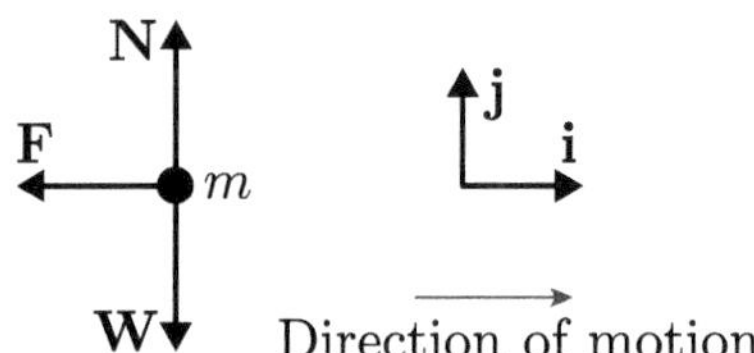

Expressing the forces in components gives

$$\mathbf{F} = |\mathbf{F}|(-\mathbf{i}), \quad \mathbf{N} = |\mathbf{N}|\,\mathbf{j}, \quad \mathbf{W} = mg(-\mathbf{i}).$$

Newton's second law gives

$$\mathbf{F} + \mathbf{W} + \mathbf{N} = m\mathbf{a},$$

where $\mathbf{a} = a\mathbf{i}$ is the acceleration of the car. Resolving in the $\mathbf{i}$-direction gives

$$-|\mathbf{F}| = ma.$$

Resolving in the $\mathbf{j}$-direction gives

$$-|\mathbf{W}| + |\mathbf{N}| = 0.$$

Now $\mathbf{W} = -mg\mathbf{j}$, so $|\mathbf{W}| = mg$. Hence we obtain $|\mathbf{N}| = |\mathbf{W}| = mg$. Since the car is skidding (sliding), we have

$$|\mathbf{F}| = \mu'|\mathbf{N}|,$$

where μ' is the coefficient of sliding friction. Hence $|\mathbf{F}| = \mu' mg$. Using the equation for friction, it follows that $ma = -\mu' mg$, leading to

$$a = -\mu' g.$$

For any given set of road and tyre conditions, μ' is constant. This equation justifies the assumption of constant acceleration (i.e. deceleration).

(Note that if air resistance, or any other force that depends on velocity, were included, then the acceleration a would not be constant. Hence the assumption of constant acceleration implies that all such forces can be ignored.)

Solution to Exercise 6

(a) Since the road and tyre conditions are assumed to be the same for both cars, μ' is the same for both. Hence, by the equation for acceleration $a = -\mu' g$ derived in Exercise 5(c), a is also the same for both, so the 'assumption' that the decelerations of the two cars are the same is justified.

(Note that m does not appear in the equation for determining the acceleration. Hence the conclusion of equal decelerations is independent of the masses of the two cars, which do not therefore need to be taken into account.)

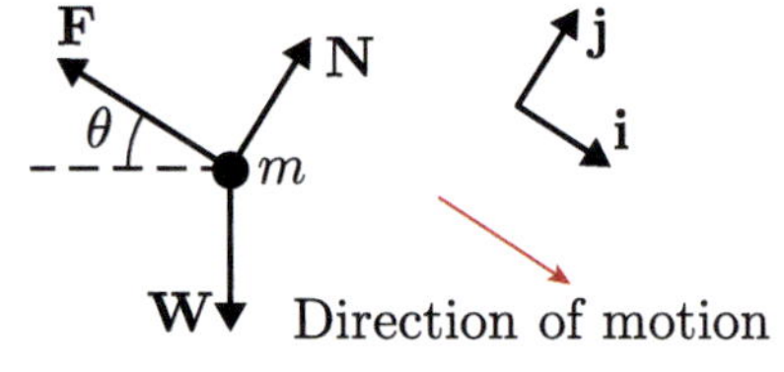

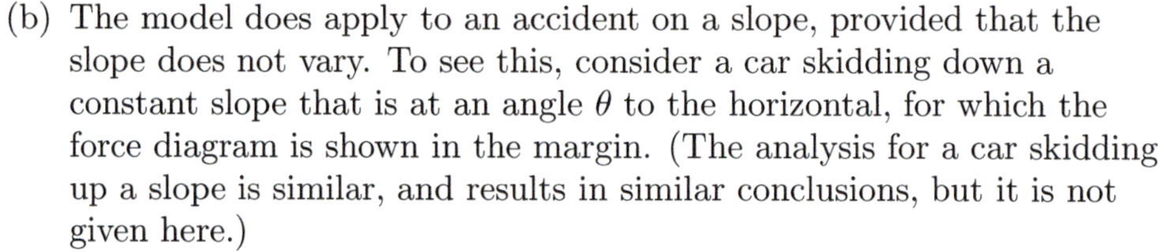

(b) The model does apply to an accident on a slope, provided that the slope does not vary. To see this, consider a car skidding down a constant slope that is at an angle θ to the horizontal, for which the force diagram is shown in the margin. (The analysis for a car skidding up a slope is similar, and results in similar conclusions, but it is not given here.)

Expressing the forces in components gives

$$\mathbf{F} = |\mathbf{F}|(-\mathbf{i}), \quad \mathbf{N} = |\mathbf{N}|\,\mathbf{j},$$
$$\mathbf{W} = mg\,(\sin\theta\,\mathbf{i} + \cos\theta\,(-\mathbf{j})) = mg(\sin\theta\,\mathbf{i} - \cos\theta\,\mathbf{j}).$$

Newton's second law gives

$$\mathbf{F} + \mathbf{W} + \mathbf{N} = m\mathbf{a},$$

where $\mathbf{a} = a\mathbf{i}$. Resolving in the $\mathbf{i}$- and $\mathbf{j}$-directions gives

$$-|\mathbf{F}| + mg\sin\theta = ma, \quad -mg\cos\theta + |\mathbf{N}| = 0.$$

Using $|\mathbf{F}| = \mu'|\mathbf{N}|$, since the car is slipping, we obtain

$$ma = mg\sin\theta - \mu' mg\cos\theta,$$

so

$$a = (\sin\theta - \mu'\cos\theta)g.$$

Hence if the slope is constant (so that θ is constant) and the road and tyre conditions are constant (so that μ' is constant), then a is constant and is the same for both cars. So the model does apply to a skid down a constant slope.

(We must assume here that $\tan\theta < \mu'$, so that $a < 0$. Otherwise the car will never stop!)

(c) (i) The model will apply to a Rolls-Royce provided that the test car has similar tyres. The size and weight of the crashed car (let alone the price) are not relevant, so long as the coefficient of sliding friction can be duplicated in the test.

(ii) A crash into a headwind will be covered by the model provided that we continue to ignore the effects of air resistance (this assumption was referred to in part (b) of Exercise 5). Otherwise, the headwind will increase the air resistance force, which depends on the speed of the car relative to the air.

If we include air resistance in the model, then the assumption of constant deceleration will no longer apply.

(iii) If the road is wet at the time of the crash, but dries before the test, then the model will not apply. The assumption that the acceleration in the test is the same as that in the crash will not be valid. The coefficient of sliding friction for a wet road is different from (less than) that for a dry road.

Solution to Exercise 7

(a) The formula $u_{\text{car}} = \sqrt{-2a_{\text{car}}x_{\text{car}}}$ does predict that u_{car} is an increasing function of x_{car}, as expected (see the figure in the margin).

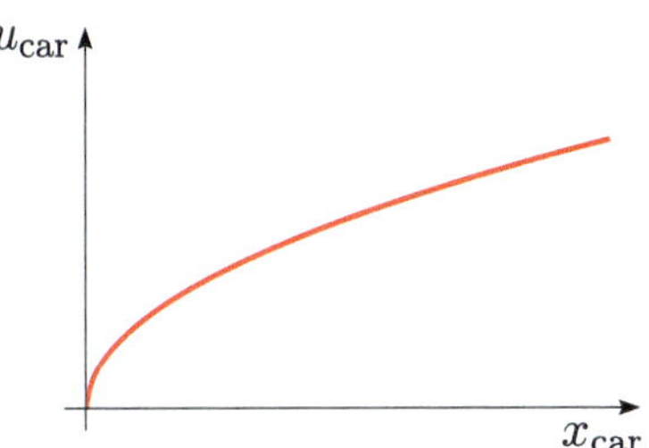

However, by itself this is not enough to solve the problem, because there is no direct way of finding the value of a_{car}. The test overcomes this difficulty, and yields $a_{\text{car}} = a_{\text{test}}$, provided that the assumptions are satisfied.

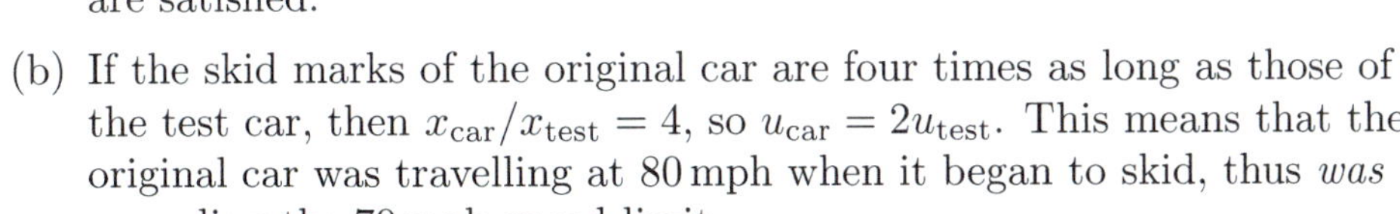

(b) If the skid marks of the original car are four times as long as those of the test car, then $x_{\text{car}}/x_{\text{test}} = 4$, so $u_{\text{car}} = 2u_{\text{test}}$. This means that the original car was travelling at 80 mph when it began to skid, thus *was* exceeding the 70 mph speed limit.

Solution to Exercise 8

(a) The instructions given to a police officer who is going to the scene of an accident for the first time might take the following form.

- If there has been a skid that has left a mark on the road, measure the length of this mark.
- Check that the tyres on the car involved in the accident are in a comparable state of wear to those on the test car, and that the slipperiness of the road has not altered significantly since the accident (e.g. due to a change in the weather).
- With a clear road, drive the test car (at a safe speed) towards the spot where the accident took place, and induce a skid. Note the initial speed of the test car and measure the length of its skid.
- To calculate the speed of the original car when it began to skid, divide the length of the original skid by the length of the test skid. Then take the square root of the result, and multiply this by the initial speed of the test car (in mph). The answer will be an estimate of the speed, in mph, of the original car when it started to skid.

(b) The model has been based on the assumption that the final speed of the original car was $v_{\text{car}} = 0$. If this was not the case, then provided that the test still gives the correct value for the deceleration, we have

$$u_{\text{car}} = \sqrt{v_{\text{car}}^2 - 2a_{\text{car}}x_{\text{car}}} > \sqrt{-2a_{\text{car}}x_{\text{car}}}.$$

Hence the actual speed of the crashed car at the start of its skid would have been greater than that estimated using the model.

Solution to Exercise 9

(a) Compound symbols for variables, such as u_{test} and u_{car}, immediately indicate what they represent. It is also useful to have the same base symbol for all the variables of one type, as in this case where u stands for speed. However, compound symbols are more cumbersome to write and to manipulate.

Single symbols for variables, such as u and U, are easier to work with, but it may be difficult to remember what they represent.
A compromise would be to use single-letter suffices, for example u_{t} and u_{c}; there is some danger of forgetting what these represent, but less so than with symbols such as u and U.

(The use of compound symbols without suffices, such as uc or $ucar$, is not recommended, because it is unclear whether this is intended to represent a single variable or whether it stands for a product of variables, each of which is represented by a single symbol.)

(b) Parameters do not change their value during the process. Mathematical models often involve differential equations, and knowing which symbols are constant in such equations is important when considering their solution.

It *is* possible to have a coefficient of friction greater than 1; this is the case for racing tyres on tarmac.

(c) Although in practice there seems to be an upper limit of 1 for the coefficient of friction between tyre and road surface, there is no point in stating this as an assumption. One of the reasons for the model is to avoid determining the coefficient of friction. This is in any case a statement about a data value and not an assumption.

(d) There are four variables in the final formula: u_{car}, u_{test}, x_{car} and x_{test}. The last three are measured, and their values are used to estimate u_{car}.

To validate the model, it would be necessary to measure corresponding values for all four variables, and to check that the predicted value for u_{car} is a reasonable approximation to the measured value.

There are no parameters in the final formula. However, data values for u_{test} and x_{test} are required in order to obtain a value for the parameter $a = -u_{\text{test}}^2/(2x_{\text{test}})$, which is constant for the given road conditions. This parameter value and the measured value of x_{car} are then used to find $u_{\text{car}} = \sqrt{-2ax_{\text{car}}}$.

(e) This is a subjective assessment and depends on the reader. The report could have been improved by the inclusion of a diagram, a table of variables and parameters (including units of measurements), and a list of the assumptions on which the model is based. The final sentence is intuitively correct, but is it confirmed by the mathematics? On the whole, this report sets out what it wants to do fairly well, but requires a better format.

Solution to Exercise 10

(a) Features such as convenience of handling and stacking must, presumably, be taken into account. However, the most important thing, from the manufacturer's point of view, is surely to minimise the cost of making each can.

In terms of the shape of a cylindrical can, this means that the 'best' shape will be the one that uses as little tin plate as possible, while meeting the requirement that the can is to hold a specified quantity of baked beans.

The objective may be different for different parties, and the purpose of the objective needs to be included.

(b) A possible statement is as follows.

> The problem is first to find a formula for the area of tin plate needed to make a cylindrical can for containing a specified volume of food, where the formula is in terms of the height and radius of the can. This formula will then be used to find the dimensions of the can for which the area of tin plate required is a minimum.

Note that volume, rather than weight (or, more properly, mass) is used to specify the quantity of food here because, as mentioned in Example 2, there is a degree of uniformity in the sizes of cans containing a range of different weights of foodstuff. The '400 gram' cans mentioned, on the evidence of supermarket shelves, may contain anything between 385 and 425 grams of food.

Solution to Exercise 11

We suggest the following assumptions.

- The circular ends fit exactly at the top and bottom of the cylinder – there is no lip.
- The seams use a negligible amount of tin plate (as it will be difficult to estimate the amount of tin plate used in the seams – at least for a first model).
- There is no wastage of tin plate: any 'leftover bits', resulting from cutting circular pieces for the ends of a can from a sheet of tin plate, are assumed to be recycled at negligible cost.
- The cans are perfectly smooth cylinders; there are no corrugations on the surface, and no ring pulls.

You might have thought of assuming that the cost of a tin can is proportional to the area of tin plate used, and that the manufacturing cost is negligible. These would have been appropriate if the purpose of the model had been stated in terms of minimising a financial cost, which is the ultimate aim, but they are essentially built into the problem statement in Exercise 10(b).

Solution to Exercise 12

In SI units, the linear dimensions would be in metres. However, the metre is rather too large a unit here, while the millimetre is too small, so the centimetre seems to be the best choice.

The key variables are as follows.

Symbol	Definition	Units
A	Area of tin plate used in manufacture of a can	cm^2
h	Height of can	cm
r	Radius of each end	cm

The specified volume V (in cm^3) is a parameter of the model.

Solution to Exercise 13

A possible diagram is as follows.

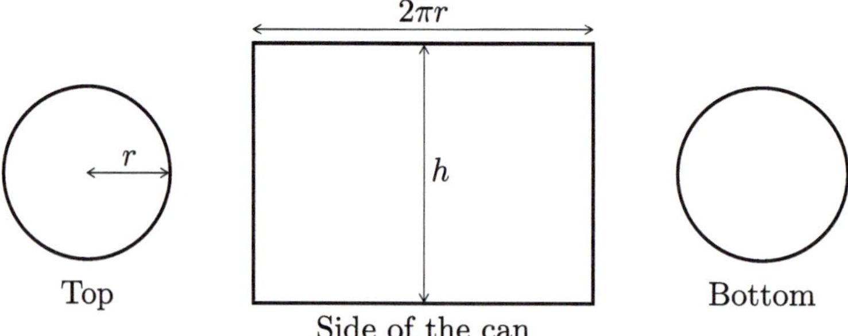

Solution to Exercise 14

(a) The can is made up of one rectangle (which forms the curved body of the can), of sides h and $2\pi r$, and two circular ends, each of area πr^2. Hence the area of tin plate used to make a can is

$$A = 2\pi rh + 2\pi r^2.$$

This formula relies on the assumptions that no tin plate is used for the seams, that there is no wastage of tin plate, and that the surfaces of the can are perfectly smooth.

(b) The volume of the can is given by $V = \pi r^2 h$. To express A in terms of r (and V, but not h), write $h = V/(\pi r^2)$ then eliminate h from the equation in part (a), to obtain

$$A = 2\pi r\left(\frac{V}{\pi r^2}\right) + 2\pi r^2 = 2\left(\frac{V}{r} + \pi r^2\right).$$

(Note that the area will always be positive.)

(c) All of the assumptions have been used in the formulation. This is a useful check to perform; if any assumptions are not used in the derivation of the model, then their presence is questionable.

(d) All of the variables and parameters that were defined have been used in the derivation. This is a useful check to perform; if any defined parameters or variables are not used in the derivation of the model, then their presence is questionable.

Solution to Exercise 15

It is easy to sketch the graph of A against r by noting that it is the sum of a quadratic function and a multiple of $1/r$; the graph is given in the margin.

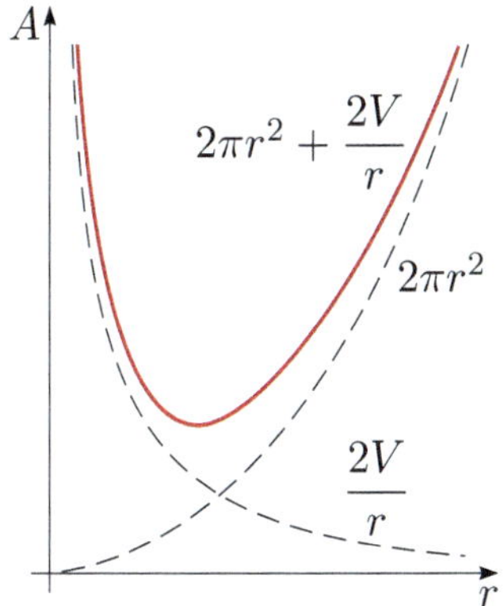

There appears to be just one local minimum, which will provide the solution to the problem.

From equation (4), we have

$$A = 2\left(\frac{V}{r} + \pi r^2\right),$$

where V is a constant. Thus

$$\frac{dA}{dr} = 2\left(-\frac{V}{r^2} + 2\pi r\right),$$

so $dA/dr = 0$ where

$$r^3 = \frac{V}{2\pi}.$$

(It is clear from the graph that this gives a minimum; but to confirm this, note that

$$\frac{d^2A}{dr^2} = 4\left(\frac{V}{r^3} + \pi\right),$$

which is positive for all $r > 0$.)

Thus the minimum area, for a specified volume V, is obtained when

$$r = \left(\frac{V}{2\pi}\right)^{1/3} \quad \text{and} \quad h = \frac{V}{\pi r^2} = \left(\frac{4V}{\pi}\right)^{1/3}.$$

The corresponding expression for the minimum area is

$$A_{\text{min}} = 2V\left(\frac{2\pi}{V}\right)^{1/3} + 2\pi\left(\frac{V}{2\pi}\right)^{2/3} = 3(2\pi V^2)^{1/3}.$$

Solution to Exercise 16

(a) With V fixed, we expect that $A \to \infty$ as $r \to 0$ (i.e. the can becomes long and thin). The model (4) predicts this.

(b) We expect that $A \to \infty$ as $r \to \infty$ (i.e. the can becomes short and fat). This also is predicted by the model.

(c) The solution to the model predicts that the minimum area increases as the volume is increased, which agrees with common sense.

Solution to Exercise 17

(a) The minimum area of a cylindrical can, of specified volume V, is given by the solution to Exercise 15 as $3(2\pi V^2)^{1/3}$. In this case its height is $h = (4V/\pi)^{1/3}$ and the radius of either end is $r = (V/(2\pi))^{1/3}$, for which the shape of the can satisfies $h/r = 2$. This means that the height of the can is equal to the diameter of its ends, so if it is seen from the side, it will appear square.

(b) In practice, cans are rarely the shape indicated in part (a). Most of them are taller than they are wide.

Solution to Exercise 18

Possible key phrases are: cylindrical can of fixed volume; area of tin plate; only cost is the tin plate used; minimise the area.

A possible description might read as follows.

> This model considers a cylindrical can of fixed volume and the area of tin plate used in its construction. This area, assumed to be proportional to the total cost of manufacture, is expressed as a function of the radius. This function is differentiated to find the radius for which the area is a minimum.

Solution to Exercise 19

(a) $[\text{area}] = \text{L}^2$.

(b) $[\text{density}] = \text{M}\,\text{L}^{-3}$.

(c) $[\theta] = [l/r] = [l]\,[r]^{-1} = \text{L} \times \text{L}^{-1} = 1$.

So angle is a dimensionless quantity, and radians are dimensionless units.

Solution to Exercise 20

The symbol v represents a speed, so

$$[v^2] = [v]^2 = (\text{L}\,\text{T}^{-1})^2 = \text{L}^2\,\text{T}^{-2}.$$

Similarly, v_0 is a speed, so v_0^2 has dimensions $\text{L}^2\,\text{T}^{-2}$.

The final term in the equation has dimensions

$$[2a_0x] = [2]\,[a_0]\,[x] = 1 \times (\text{L}\,\text{T}^{-2}) \times \text{L} = \text{L}^2\,\text{T}^{-2}$$

also.

Hence the equation is dimensionally consistent.

Solution to Exercise 21

The left-hand side of the equation, dm/dt, has the same dimensions as m/t, namely $\mathrm{M\,T^{-1}}$. The right-hand side of the equation has dimensions

$$\begin{aligned}[-km] &= \left[-\frac{r}{V}m\right] \\ &= [-1]\,[r]\,[V]^{-1}\,[m] \\ &= 1 \times (\mathrm{L^3\,T^{-1}}) \times (\mathrm{L^3})^{-1} \times \mathrm{M} = \mathrm{M\,T^{-1}}.\end{aligned}$$

Hence the equation is dimensionally consistent.

Solution to Exercise 22

Consider the dimensions of the terms on the left-hand side of the equation one by one:

$$\begin{aligned}&\left[\frac{p}{\rho}\right] = \frac{[p]}{[\rho]} = \frac{\mathrm{M\,L^{-1}\,T^{-2}}}{\mathrm{M\,L^{-3}}} = \mathrm{L^2\,T^{-2}}, \\ &\left[\tfrac{1}{2}u\right] = [u] = \mathrm{L\,T^{-1}}, \\ &[gz] = [g]\,[z] = (\mathrm{L\,T^{-2}}) \times \mathrm{L} = \mathrm{L^2\,T^{-2}}.\end{aligned}$$

The equation is dimensionally inconsistent because the second term has different dimensions from the other two. It therefore seems likely that the second term is incorrect, and that its dimensions should be $\mathrm{L^2\,T^{-2}}$. One form that would be consistent is $\frac{1}{2}u^2$ rather than $\frac{1}{2}u$.

Note that the 'constant' on the right-hand side of this equation cannot be dimensionless. It must have the same dimensions as the terms on the left-hand side.

Solution to Exercise 23

We require the dimensions of

$$c_2 = \frac{R}{D^2 v^2}.$$

R is the magnitude of a force thus has dimensions $\mathrm{M\,L\,T^{-2}}$. D is a length and has dimensions L. v is a speed and has dimensions $\mathrm{L\,T^{-1}}$. For dimensional consistency we must have

$$\begin{aligned}[c_2] &= \left[\frac{R}{D^2 v^2}\right] \\ &= [R]\,[D]^{-2}\,[v]^{-2} \\ &= (\mathrm{M\,L\,T^{-2}}) \times \mathrm{L^{-2}} \times (\mathrm{L\,T^{-1}})^{-2} = \mathrm{M\,L^{-3}}.\end{aligned}$$

Hence in SI units, c_2 would be measured in $\mathrm{kg\,m^{-3}}$ (the same units as density).

Solution to Exercise 24

First consider the argument $\omega t + \phi$ of the cosine function. Since ϕ is an argument of a cosine, $[\phi] = 1$. For dimensional consistency, $[\omega t] = 1$. Since $[t] = \mathrm{T}$, we must have $[\omega] = \mathrm{T}^{-1}$.

The result of the cosine function is dimensionless, so $[\cos(\omega t + \phi)] = 1$. Now $[x] = \mathrm{L}$, so the equation is dimensionally consistent if $[A] = \mathrm{L}$.

Solution to Exercise 25

We have

$$1\,\mathrm{kg} = 10^3\,\mathrm{g} \quad \text{and} \quad 1\,\mathrm{cm} = 10^{-2}\,\mathrm{m}.$$

Hence

$$\begin{aligned}13.546\,\mathrm{g\,cm^{-3}} &= 13.546\,\mathrm{g\,cm^{-3}} \times \left(\frac{1\,\mathrm{kg}}{10^3\,\mathrm{g}}\right) \times \left(\frac{10^{-2}\,\mathrm{m}}{1\,\mathrm{cm}}\right)^{-3}\\ &= 13.546 \times 10^{-3} \times 10^{6}\,\mathrm{kg\,m^{-3}}\\ &= 1.3546 \times 10^{4}\,\mathrm{kg\,m^{-3}}.\end{aligned}$$

So the density of the metal is $1.3546 \times 10^4\,\mathrm{kg\,m^{-3}}$.

Solution to Exercise 26

We have

$$1\,\mathrm{lb} = 0.453\,592\,\mathrm{kg} \quad \text{and} \quad 1\,\mathrm{in} = 2.54 \times 10^{-2}\,\mathrm{m}.$$

Hence

$$\begin{aligned}61\,\mathrm{lb\,in^{-2}} &= 61\,\mathrm{lb\,in^{-2}} \times \left(\frac{0.453\,592\,\mathrm{kg}}{1\,\mathrm{lb}}\right) \times \left(\frac{2.54 \times 10^{-2}\,\mathrm{m}}{1\,\mathrm{in}}\right)^{-2}\\ &= \frac{61 \times 0.453\,592}{(2.54 \times 10^{-2})^2}\,\mathrm{kg\,m^{-2}}\\ &\simeq 4.2887 \times 10^{4}\,\mathrm{kg\,m^{-2}}.\end{aligned}$$

So the pressure on the piston is $42\,890\,\mathrm{kg\,m^{-2}}$ (to 4 s.f.).

As mentioned in the question, in the SI system, pressure is *force* per unit area, not *mass* per unit area – this arises because of the difference between mass and weight. So we should multiply the above answer by g, the magnitude of the acceleration due to gravity, to give a pressure of $4.289 \times 9.81 \times 10^4 = 420\,700\,\mathrm{kg\,m^{-1}\,s^{-2}}$.

Solution to Exercise 27

We have

$$1 \text{ boliviano} = 1 \text{ boliviano} \times \left(\frac{\$1}{7.75 \text{ boliviano}}\right) \times \left(\frac{£1}{\$1.75}\right) = £0.074.$$

I approximated this exchange rate as 10 boliviano = $£\frac{3}{4}$.

For example, I bought a memory card for my camera for 830 boliviano, and I approximated this cost in £ as $84 \times \frac{3}{4} = £63$ (using 84 because it is divisible by 4). (The cost on my credit card bill was £61.59.)

Unit 9

Oscillations and energy

Introduction

All around you there are mechanical systems that vibrate or oscillate (the two words can be used interchangeably). Each day you probably experience oscillations in a wide variety of forms: the buzzing of an alarm clock, the vibrations of an electric hair dryer or razor, the sideways movements of a train or boat, and so on. Vibrations vary in scale from the small motions of atoms within solid objects, to the swaying of bridges and tall buildings; and from the irregular bending of a tree in the wind, to the extremely regular oscillations involved in any device that is designed to measure accurately the passage of time.

A common feature of many of these vibrations or oscillations is *periodicity*, where the vibration exhibits a pattern of movement or displacement that repeats itself over and over again as time progresses. Not all oscillations repeat themselves exactly. They may also become more or less pronounced if the system is *forced* or *damped*, or the oscillations may even become chaotic. In this unit, however, attention will be directed towards systems for which the effects of damping or forcing are of secondary importance. We model such a system by assuming that such complicating effects may be ignored completely. As a result, the behaviour predicted by the model is periodic and, in particular, *sinusoidal* (having a pattern of motion in which the displacement can be represented by a sine or cosine function of time).

Forced and damped oscillations are considered in Unit 10.

Sinusoidal oscillations arise in a wide variety of electrical, thermal and mechanical systems, but we concentrate here on the last of these. In this context, the sinusoidal nature of the oscillations can often be seen as a consequence of a further modelling assumption: namely, that the force acting on a moving object in order to bring it back towards its equilibrium position is proportional to its displacement from that position. This 'force law' was first put forward by Robert Hooke (Figure 1) in 1678.

This force is often referred to as a *restoring force*.

Robert Hooke

Robert Hooke was a contemporary of Isaac Newton. There was a dispute between the two men as to which had first proposed an inverse square law to describe the effects of gravitation.

Figure 1 Robert Hooke (1635–1703)

As an aid to visualising a restoring force, we introduce the concept of a *model spring* to embody it. The sinusoidal behaviour that arises from a model spring is known as *simple harmonic motion*, and is the simplest effective model for an oscillating system.

A model spring is a hypothetical object that can be either stretched or compressed from its 'natural length', and for which the force exerted depends on its length. This is in contrast to the *model string* introduced in Unit 2. A model string is inelastic, and hence has a fixed length, though it is capable of sustaining a range of pulling forces. The model spring, on the other hand, can provide pushing as well as pulling forces.

Figure 2 shows, in diagrammatic form, some of the systems that we will discuss. In each case, part of the system moves up and down, or to and fro, about an average position. We will attempt to explain these oscillations by applying the laws of Newtonian mechanics.

The system in Figure 2(a) is a weight hanging from an elastic string. This is a simplified representation of a system that might be modelled using a model spring. Figure 2(b) shows an object connected to a spring, which is attached to a fixed point at its other end. This might be used as a model for a spring-loaded buffer, which reverses the motion of a vehicle making contact with it. Alternatively, it could model the release mechanism in a pinball machine while in contact with the ball. Figure 2(c) represents a floating pontoon, bobbing up and down in water.

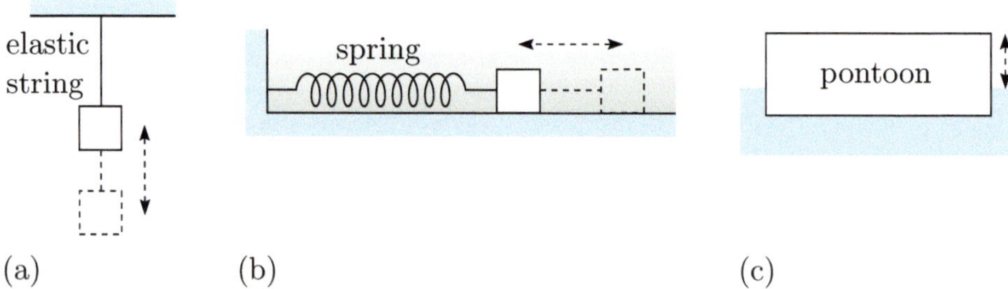

Figure 2 Examples of oscillating systems

Section 1 describes a simple experiment involving an oscillating system. It also introduces *Hooke's law* as a model for the force exerted by a spring, and goes on to consider how this law applies in various situations where no movement takes place.

Unit 11 takes up this story where more than one particle is involved.

Section 2 uses Newton's second law to model the oscillations of the simplest oscillating system, which consists of a single particle attached to a single horizontal spring. Variations on extended systems, such as vertical springs and systems with two springs, are also analysed.

Energy is an important concept in physics. What we mean by mechanical energy will be formally defined in Section 3, but for now we will use the everyday notion. It is a relatively new concept when compared to the ideas of mass and force. The concept of energy that can be converted from one form to another came to the fore when James Joule (see Figure 3) did experiments converting mechanical energy into heat in 1843.

Figure 3 James Joule (1818–1889)

James Joule

James Joule was an amateur physicist who did many experiments over a forty-year period that established laws about conversion of many types of energy, including mechanical energy, heat and electrical energy. It is for this reason that the SI unit of energy is named after him.

Joule found that energy can be neither created nor destroyed: it can merely be converted from one form to another. A particular instance of this fact will be given in Section 4 as the *law of conservation of mechanical energy*.

In Section 4 you will see that the use of conservation of mechanical energy to solve mechanics problems is entirely equivalent to using Newton's second law of motion. In certain cases, it is possible to derive the law of conservation of mechanical energy from Newton's second law, and vice versa. So in a sense, no new problems can be solved using conservation of energy – every problem could equally well be solved by appealing to Newton's second law.

So why should one learn about energy? There are many answers to this. One answer is that in some situations it is easier to solve problems by considering energy, as you will see in this unit.

A second reason to introduce energy is because it is a physical concept that is important in many fields other than mechanics. By introducing other forms of energy, such as electromagnetic, chemical, heat, and so on, conservation of energy can be extended beyond the field of mechanics, and becomes one of the fundamental principles of physics. In this module, however, we concentrate on mechanical energy only.

A third reason to introduce energy is because it leads the way from Newtonian mechanics to Lagrangian and Hamiltonian mechanics. These other methods for solving mechanics problems have richer mathematical structures than Newtonian mechanics, making the generalisation to relativistic and quantum mechanics easier.

Lagrangian and Hamiltonian mechanics are Level 3 applied mathematics topics.

In Section 4 the law of conservation of mechanical energy is applied to systems involving gravity and springs. For this first look at energy, the discussion will mostly be limited to one-dimensional problems.

Two- and three-dimensional problems are studied in Unit 16.

1 Characteristics of a model spring

In this section the important concept of a *model spring* is introduced. Subsection 1.2 shows how the idealised concept of a model spring leads to a simple type of force law. In Subsection 1.3 this model is applied to some statics problems.

The behaviour of a simple mechanical system that can be set up at home is used first to obtain a feel for the concept of a model spring, before it is defined mathematically.

1.1 An experiment

Figure 4 Experimental arrangement

The mechanical system to be considered is illustrated in Figure 4. It is constructed by clamping a metal ruler of length 30 cm to a heavy table so that the ruler is horizontal. A hanger is attached to the free end of the ruler, and 'particles' of known mass (metal nuts) are hung from the hanger. This may not seem like a spring, but it has all the characteristics of a spring – applying a force to the end of the ruler causes a displacement. An accurate measuring device is used to measure the displacement d of the top of the end of the ruler from the horizontal.

A member of the module team carried out the experiment described below. Six of the nuts used as 'particles' have total mass 0.055 kg, so we assume that each nut has a mass of about 9.2×10^{-3} kg. (For machine-produced simple articles, it is a good assumption that each nut has the same mass.)

With no nuts on the hanger, the displacement was 0.22 mm. We call this value of d the zero displacement, and denote it by the symbol d_0, where the suffix indicates that there are no nuts hanging. Similarly, d_n represents the displacement when there are n nuts hanging.

Table 1 shows the displacements when 1 to 6 nuts were hung from the end of the ruler. It also shows the magnitudes of the weights corresponding to n nuts, to three decimal places. These were found (to two decimal places) by multiplying the total mass of the nuts by $g = 9.81\,\mathrm{m\,s^{-2}}$.

Table 1 Experimental data for displacement of ruler and number of nuts

Number of nuts, n	0	1	2	3	4	5	6
Displacement of end, d_n (millimetres)	0.22	1.15	2.07	2.99	3.85	4.78	5.68
Magnitude of weight of nuts, $w(d_n)$ (newtons)	0	0.09	0.18	0.27	0.36	0.45	0.54

Figure 5 gives a graph of the some of the results in Table 1. We work in SI units, so the millimetres used in the table have been converted to metres for use in the graph.

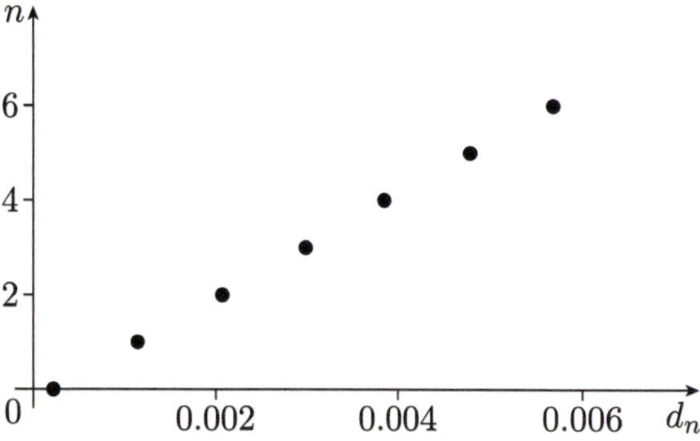

Figure 5 Plot of number of nuts n against displacement d_n (with d_n in metres)

Figure 5 shows that the number of nuts n and the displacement d_n are related in an approximately linear way when up to six nuts are hanging from the end of the ruler.

Figure 6 shows the magnitude of the weight of the nuts plotted against the displacement as well as the best straight line that goes through these points. Whereas Figure 5 considered discrete values of displacement, Figure 6 shows continuous values, hence d_n is replaced by d.

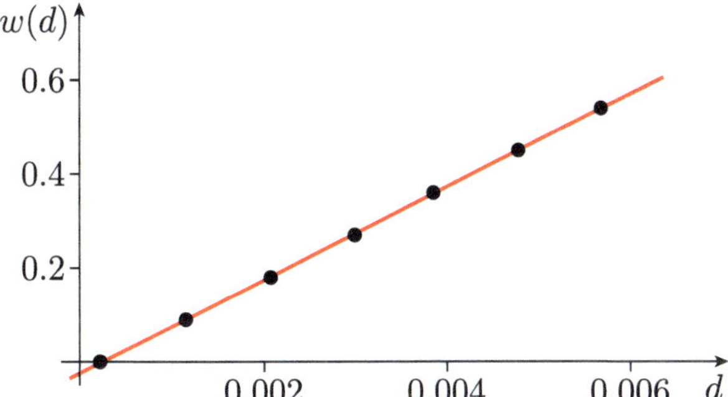

Figure 6 Plot of magnitude of weight against displacement, plus the best straight line through the points

A straight line that fits the experimental data points well is given by

$$w(d) \simeq -0.024 + 99.044d \quad \text{to 3 d.p.} \tag{1}$$

Later reference will be made to this equation, which is a relationship between the displacement of the ruler and the magnitude of the vertical force applied to it. Figure 7 shows the forces acting on the hanger. Since the hanger with n nuts is in equilibrium, its weight $\mathbf{W}$ must be exactly counteracted by the upward force provided by the ruler on the hanger. The force provided by the ruler is therefore $\mathbf{F} = -\mathbf{W}$, and the magnitude of this force is $|-\mathbf{W}| = |\mathbf{W}|$. Hence equation (1) tells us that the displacement of the ruler is related almost linearly to the magnitude of the force applied to the ruler. Another way of putting this is that as nuts are added to the hanger, the *increase* in the displacement of the ruler is approximately proportional to the *increase* in the magnitude of the force on it (or vice versa).

Figure 7 Forces acting on the hanger

From now on this unit considers only model springs, not real springs. The word 'spring' suggests a helical coil made of metal wire, and in representing a model spring diagrammatically, it is a stylised form of this physical realisation that is portrayed. However, a 'model spring' is in fact, as it says, nothing more than a model for a particular simple, linear, variation of a force with displacement. While it may be used as an idealised representation of a real coil spring, it can also be made to stand for part of a mechanical system that contains no spring at all, but behaves in the appropriate way, as in the experiment described above. In our idealised model, the model spring has zero mass.

The concept of a model spring may be compared with the concepts of a model string and a model pulley, which were introduced in Unit 2.

Note that the relationship between the length of the spring and the magnitude of the force in a model spring is represented by a straight line, and a straight line has two parameters – the slope (or gradient) and the intercept. For a spring, these two parameters are called the spring *stiffness* and the *natural length* of the spring, usually denoted by the symbols k and l_0, respectively. These parameters are often given next to a spring in a diagram, as shown in Figure 8. They will be explained further in the next subsection.

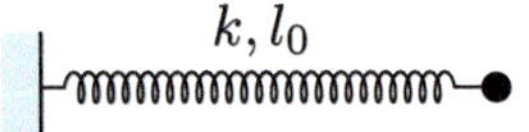

Figure 8 Spring representation

Figure 8 shows the diagrammatic convention that this module adopts. The bar on the left indicates a fixed and immovable wall to which the left-hand end of the spring is attached. The spring itself is coiled, and usually an end is attached to a particle, which is represented by a black filled circle; if no particle is attached, then the free end of the spring often ends in an open circle.

Model spring

A **model spring** is characterised by two positive constants, its **natural length** l_0 and its **stiffness** k. It has zero mass.

Exercise 1

Use equation (1) to estimate answers to the following questions.

(a) If the displacement is 2.5 mm, what is the magnitude of the weight on the ruler?

(b) If the magnitude of the weight on the hanger is 0.67 N, what will be the displacement?

(c) If the magnitude of the weight on the hanger is 60 N, what will be the displacement?

(d) If the displacement is −2.5 mm, what is the magnitude of the weight on the ruler?

1.2 Force law for a model spring

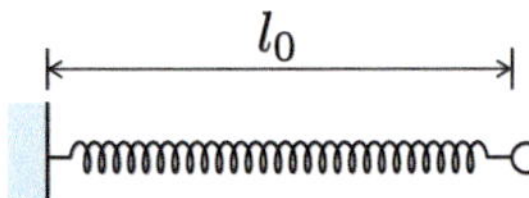

Figure 9 Free spring

Suppose that Figure 9 represents a model spring that is in its undisturbed state. One end is attached to a fixed object such as a wall, and the other end is free to move on a frictionless horizontal surface. The spring has no weight, and there is no normal reaction on it from the surface; there are no horizontal forces. So although the free end of the spring is free to move, it does not do so because the resultant force is zero. The length of the spring under these circumstances is its **natural length**, which we denote by l_0. If its length is made longer or shorter than l_0, then the spring will always try to return to its natural length.

Suppose now that the spring is *extended* (see Figure 10), so that its length is greater than its natural length; such a spring is said to be *in extension*, or alternatively *in tension*. You probably know from experience that if you hold the free end of an extended spring, then you will be pulled towards the fixed end. The magnitude of this pulling force is called the **tension** in the spring (just as, in Unit 2, we refer to the tension in a taut string). The more the spring is extended, the more it pulls; in other words, the tension in the spring increases with its extension.

Figure 10 Spring extended by e

Now suppose that the spring is *compressed* (see Figure 11), so that its length is less than its natural length; such a spring is said to be *in compression*. In this case, if you hold the free end of the spring, then the spring will push you away from its fixed end. Again, the more the spring is compressed, the more it pushes; the magnitude of the pushing force increases with compression.

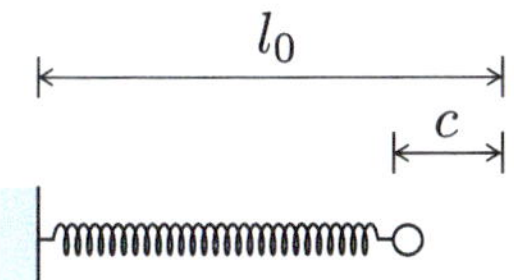

Figure 11 Spring compressed by c

A *model spring* is a special model of a real spring. This model is based on the assumption that *the magnitude of the force exerted by the spring is proportional to the amount of its deformation* (extension or compression). The constant of proportionality k, which relates the force magnitude to the deformation for a model spring, is known as the **stiffness** of the spring. The SI units for stiffness are newtons per metre ($\mathrm{N\,m^{-1}}$).

Force law

A model spring exerts a force on any object attached to either of its ends, as follows.

- When the length of the spring equals l_0, the magnitude of the force exerted by the spring is zero.
- When the spring is extended by an amount e (where $e > 0$), so that its length is $l_0 + e$, the force is directed along the axis of the spring towards the centre of the spring and has magnitude ke (stiffness × extension).
- When the spring is compressed by an amount c (where $c > 0$), so that its length is $l_0 - c$, the force is directed along the axis of the spring away from the centre of the spring and has magnitude kc (stiffness × compression).

The force law for a model spring is often known as **Hooke's law**, in honour of Robert Hooke. In 1678 Hooke declared, in *De Potentia Restitutiva*:

> It is very evident that the Rule or Law of Nature in every springing body is, that the force or power thereof to restore itself to its natural position is always proportionate to the Distance or space it is removed therefrom.

Actually, this rather overstates the case. The 'model spring' is only a model. Real springs exert forces that have a non-linear dependence on extension or compression and that may also depend on other factors such as the velocities of the ends or the mass of the spring. Also, as pointed out earlier, the concept of natural length may not carry over to an actual physical length associated with a real spring. Nevertheless, many real springs conform fairly closely to the idealised behaviour of a model spring for certain ranges of extension and compression. We will therefore use model springs to represent all the springs discussed in this unit. A bonus of this model is that it leads to a differential equation that can be solved. However, you should remember that this is only an *approximation* when comparing predictions with the real world. As in Exercise 1(c), if the modelling results in a value that is outside the validity of the model, then it must be revised.

Exercise 2

A model spring has natural length $0.3\,\text{m}$ and stiffness $200\,\text{N}\,\text{m}^{-1}$. It is attached to a fixed bracket at one end, while the other end is attached to an object (lying on a frictionless horizontal surface) whose distance l from the fixed bracket is variable (see Figure 12).

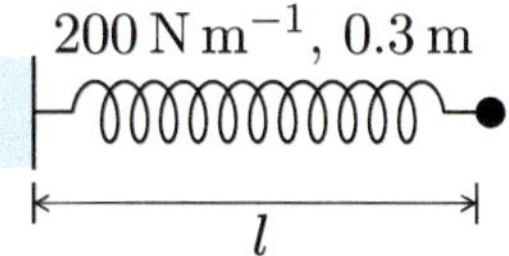

Figure 12

Determine the magnitude and direction, relative to the centre of the spring, of the force on the object when $l = 0.35$ and when $l = 0.2$.

So far we have discussed a spring with one end attached to a fixed point and considered the force at the free end of the spring. However, if you take a spring with an end in each hand and extend it, you will be made aware that there are pulling forces exerted by the spring at *both* ends.

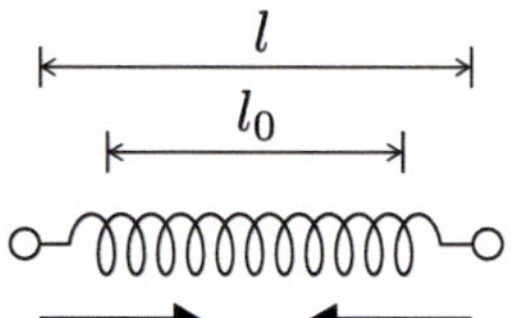

Figure 13 Spring extended at both ends

For a model spring in tension, these two forces have equal magnitudes and are both directed towards the centre of the spring (see Figure 13). The magnitude of each of these forces is given, as before, by stiffness $\times$ extension. If we denote the length of the extended spring by l, and its natural length by l_0, then its extension is $e = l - l_0$ and so, by the force law for a model spring, the tension in the spring is equal to $k(l - l_0)$.

If the same spring is compressed by being pushed towards its centre at both ends, then its length l is less than its natural length l_0 by the amount $c = l_0 - l$, and the magnitude of the force provided by the spring is equal to $k(l_0 - l)$.

We need to look at two different cases whenever we consider the force that may be exerted by a model spring on a particle or object attached to an end of the spring, depending on whether the spring is extended or compressed. However, it is possible to write a single *vector* formula for the force, which will apply correctly whatever the configuration of the spring.

The force exerted by a model spring of length l, natural length l_0 and stiffness k, on any object attached to one of its ends, is:

- directed from the object towards the spring, with magnitude $k(l - l_0)$, if the spring is extended from its natural length ($l > l_0$)
- directed away from the spring towards the object, with magnitude $k(l_0 - l)$, if the spring is compressed ($l < l_0$).

Suppose that we are concerned with the forces on an object attached to one end of a spring, and that $\widehat{\mathbf{s}}$ is a unit vector in the direction from the object towards the spring (see Figure 14). The force $\mathbf{H}$ that the spring exerts on the object can be written in terms of $\widehat{\mathbf{s}}$ as follows.

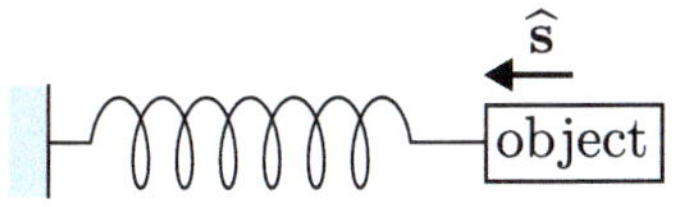

Figure 14 Direction of $\widehat{\mathbf{s}}$

- If the spring is extended ($l > l_0$), then it provides a force in the same direction as $\widehat{\mathbf{s}}$, namely $\mathbf{H} = k(l - l_0)\,\widehat{\mathbf{s}}$.
- If the spring is compressed ($l < l_0$), then it provides a force in the opposite direction to $\widehat{\mathbf{s}}$, namely $\mathbf{H} = k(l_0 - l)(-\widehat{\mathbf{s}})$.

Since $-k(l - l_0) = k(l_0 - l)$, we actually have the same algebraic relationship in each case. It follows that this vector equation provides a complete description of the model spring force law, whether the spring is extended or compressed, and this is a consequence of there being a linear relationship between the component of force and the change in length of the spring.

The vector equation also gives the correct outcome of zero force when the spring has its natural length ($l = l_0$).

Vector form of Hooke's law

Suppose that a model spring has natural length l_0 and stiffness k. Then on an object attached to a particular end, the spring exerts a force

$$\mathbf{H} = k(l - l_0)\,\widehat{\mathbf{s}}, \tag{2}$$

where l is the length of the spring and $\widehat{\mathbf{s}}$ is a unit vector in the direction from the end where the object is attached towards the centre of the spring.

Exercise 3

Use the vector form of Hooke's law to answer once more the questions posed in Exercise 2. The diagram of the spring is repeated as Figure 15.

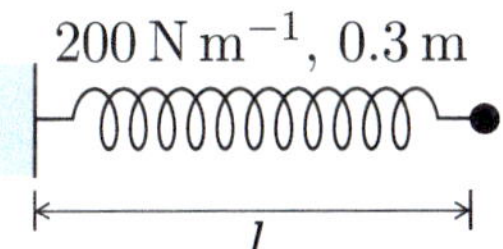

Figure 15

1.3 Model springs at rest

The chief purpose behind introducing the concept of a model spring is to provide a simple mathematical model that describes the oscillatory behaviour of certain systems, such as those shown in the Introduction. However, before considering systems in motion, it is instructive to look at some static configurations that feature model springs.

The approach to analysing these situations is similar to that which you studied in Unit 2, with the addition of Hooke's law to describe the force provided by a model spring.

See Procedure 2 in Unit 2.

Example 1

A particle of mass 5 kg is suspended from a model spring that is attached at its top end to a fixed point. The natural length of the spring is 0.3 m. The length of the spring is 0.5 m when the particle is suspended from it.

What is the stiffness k of the spring?

Solution

◀ Draw picture ▶

◀ Choose axes ▶

◀ Draw force diagram(s) ▶

Figure 16 shows the spring and the force diagram. Take the **i**-direction to be vertically downwards, as shown.

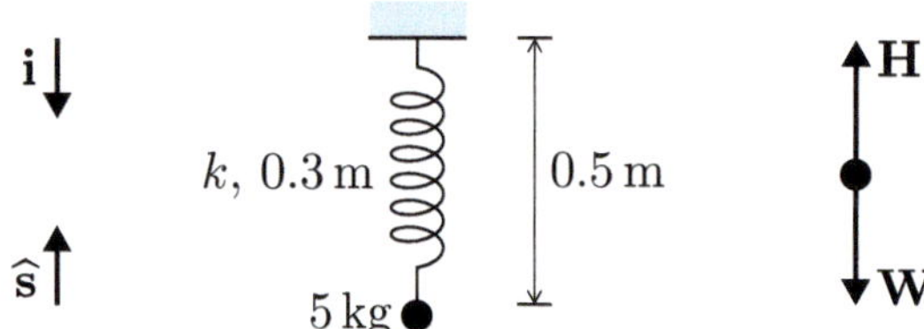

Figure 16 Mechanical system and force diagram

The mass of the particle is 5 kg, so its weight is $\mathbf{W} = 5g\mathbf{i}$.

The spring force $\mathbf{H}$ will depend on the values of k, l, l_0 and $\widehat{\mathbf{s}}$. We are given that $l = 0.5$ and $l_0 = 0.3$, and k is unknown. The unit vector $\widehat{\mathbf{s}}$ in the direction from the particle towards the centre of the spring points upwards, so $\widehat{\mathbf{s}} = -\mathbf{i}$.

◀ Apply law(s) ▶

By Hooke's law (equation (2)), the model spring provides a force

$$\mathbf{H} = k(l - l_0)\,\widehat{\mathbf{s}} = k(0.5 - 0.3)(-\mathbf{i}) = -0.2k\mathbf{i}.$$

The equilibrium condition for particles was introduced in Unit 2.

Since the particle is in equilibrium, we have

$$\mathbf{W} + \mathbf{H} = \mathbf{0}, \quad \text{that is,} \quad 5g\mathbf{i} - 0.2k\mathbf{i} = \mathbf{0}.$$

◀ Solve equation(s) ▶

Resolving in the **i**-direction gives

$$5g - 0.2k = 0,$$

so

$$k = 25g \simeq 245 \quad \text{to 3 s.f., using } g = 9.81\,\mathrm{m\,s^{-2}}.$$

◀ Interpret solution ▶

We conclude that the stiffness of the spring is $245\,\mathrm{N\,m^{-1}}$.

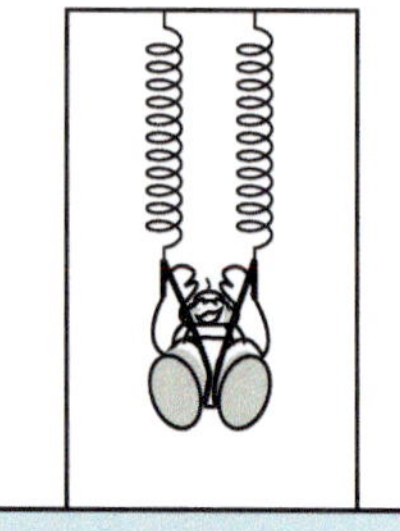

Figure 17 A baby bouncer

Exercise 4

A 'baby bouncer' (see Figure 17) consists of a seat and straps, whose combined mass is 0.5 kg, suspended from two identical springs that are attached to a door lintel. When a baby of mass 7.5 kg is strapped into the seat, the seat descends by 3 cm.

What is the stiffness of each of the springs?

The procedure to find the force due to a spring can be made more systematic.

Procedure 1 Finding the force on a particle due to a spring

1. Determine the length l of the spring; this is best done by drawing a diagram to show the various lengths in the system. Note that l may be in terms of unknown displacements such as x.
2. Use the given natural length l_0 of the spring to find the extension $l - l_0$.
3. From the diagram, determine $\widehat{\mathbf{s}}$, the unit vector from the particle towards the centre of the spring.
4. The spring tension force is $\mathbf{H} = k(l - l_0)\,\widehat{\mathbf{s}}$, where k is the stiffness of the spring.

If there is more than one spring, then consider a tabular approach (see Example 2).

Note that l and $\widehat{\mathbf{s}}$ are deduced from the position of the spring in relation to the particle, or the point at which the force due to the spring is required; k and l_0 are characteristics of the spring. The converse, finding the length of a spring when the spring tension force is known, is also possible using an approach like Procedure 1.

Example 2

An object of mass m is attached to two model springs, whose other ends are attached to fixed points that are 0.9 m apart. The object can be considered as a particle of negligible size. The left-hand spring has stiffness $40\,\mathrm{N\,m^{-1}}$ and natural length 0.3 m, while the right-hand spring has stiffness $60\,\mathrm{N\,m^{-1}}$ and natural length 0.2 m. The object is supported by a table and is free to move on the table without friction. This set-up is shown in Figure 18.

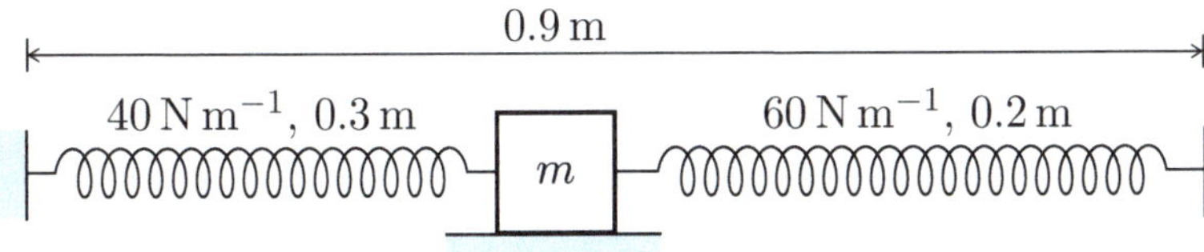

Figure 18

Determine the equilibrium position of the object, and find the tension in the springs.

Solution

All the forces acting on a particle should be drawn on a force diagram even if it is anticipated that some forces may not play a part in the subsequent derivation. The application of the equilibrium law or Newton's second law sorts out what is required.

The forces acting on the particle are shown in Figure 19, where $\mathbf{W}$ is the weight, $\mathbf{N}$ is the normal reaction, and $\mathbf{H}_1$ and $\mathbf{H}_2$ are the forces exerted by the left-hand and right-hand springs, respectively.

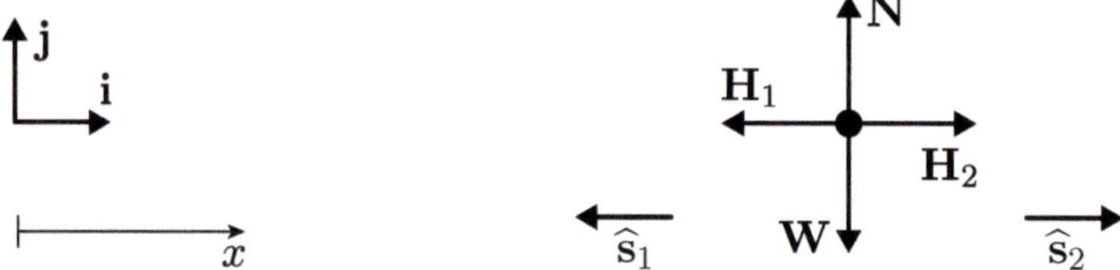

Figure 19 Axis and force diagram

Take $\mathbf{i}$ to be horizontal and positive to the right, and take $\mathbf{j}$ to be positive vertically upwards. Hence $\mathbf{W} = mg(-\mathbf{j})$ and $\mathbf{N} = |\mathbf{N}|\,\mathbf{j}$.

For each spring, the force exerted is given by Hooke's law as

$$\mathbf{H} = k(l - l_0)\,\widehat{\mathbf{s}},$$

where $\widehat{\mathbf{s}}$ is a unit vector directed from the particle towards the centre of the spring. The x-axis is chosen to point from left to right, in the same direction as $\mathbf{i}$, with origin at the left-hand fixed point, as shown in Figure 19. Then the lengths of the two springs are x (left spring) and $0.9 - x$ (right spring).

Choosing x to be in the contrary direction to $\mathbf{i}$ is not recommended; similarly for y and $\mathbf{j}$.

For problems where there is more than one spring, you might like to use a tabular approach to find the spring forces $\mathbf{H}$, and identify the values of k, l, l_0 and $\widehat{\mathbf{s}}$ either from the problem statement or from the diagram that accompanies it, as is done in Table 2.

Here $\widehat{\mathbf{s}}_1 = -\mathbf{i}$ since the direction from the particle towards the spring is to the left, that is, in the $-\mathbf{i}$-direction. Similarly, $\widehat{\mathbf{s}}_2 = \mathbf{i}$ since the direction from the particle towards the spring is to the right.

Table 2 Tabular approach to finding spring forces

Spring	l	l_0	$l - l_0$	k	$\widehat{\mathbf{s}}$	$\mathbf{H}$
Left	x	0.3	$x - 0.3$	40	$-\mathbf{i}$	$40(x - 0.3)(-\mathbf{i})$
Right	$0.9 - x$	0.2	$0.7 - x$	60	$\mathbf{i}$	$60(0.7 - x)\mathbf{i}$

From Table 2,

$$\mathbf{H}_1 = 40(x - 0.3)(-\mathbf{i}) = 40(0.3 - x)\mathbf{i},$$
$$\mathbf{H}_2 = 60(0.7 - x)\mathbf{i}.$$

Since the particle is held in equilibrium,

$$\mathbf{W} + \mathbf{N} + \mathbf{H}_1 + \mathbf{H}_2 = mg(-\mathbf{j}) + |\mathbf{N}|\,\mathbf{j} + 40(0.3 - x)\mathbf{i} + 60(0.7 - x)\mathbf{i} = \mathbf{0}.$$

Resolving in the **i**-direction gives

$$60(0.7 - x) + 40(0.3 - x) = 0.$$

The solution of this equation is $x = 0.54$, so the particle is 0.54 m from the left-hand fixed point (and hence 0.36 m from the right-hand fixed point).

Resolving in the **j**-direction gives $|\mathbf{N}| - mg = 0$, but this is of no consequence in this solution as it just tells us that the weight is balanced by the normal reaction.

The magnitudes of the spring forces can be found by substituting $x = 0.54$ into the above expressions for the forces to obtain

$$\mathbf{H}_1 = 40(0.3 - 0.54)\mathbf{i} = -9.6\mathbf{i},$$
$$\mathbf{H}_2 = 60(0.7 - 0.54)\mathbf{i} = 9.6\mathbf{i}.$$

So, as expected, since the particle is not moving, the forces exerted by the springs are of equal magnitude and in opposite directions (which provides a useful quick check). The tension in the springs is the magnitude of the above forces, that is, 9.6 N.

Exercise 5

Consider a modified version of the problem stated in Example 2, in which the distance between the two fixed ends is reduced to 0.3 m, but all other aspects of the problem remain unchanged (see Figure 20).

Solve this modified problem (finding, at the end, the magnitude of the forces exerted by the springs).

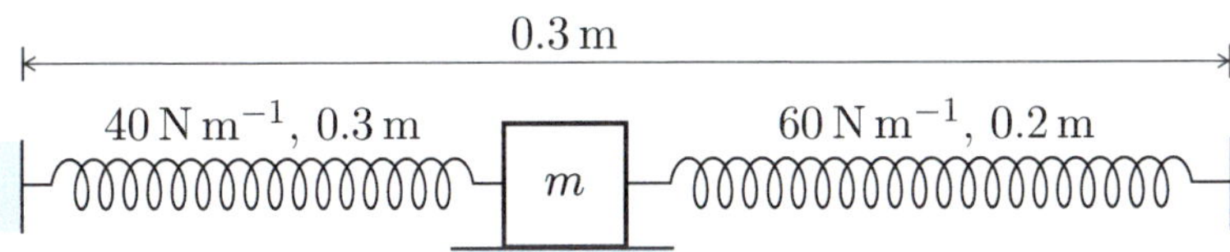

Figure 20

Example 3

A model spring hangs vertically from a fixed point, with a particle of mass m suspended from its lower end. The particle is in equilibrium.

Express the length l of the spring in terms of its stiffness k and natural length l_0, the mass m, and the magnitude of the acceleration due to gravity g.

Solution

Choose the **i**-direction to be vertically downwards, as shown in Figure 21, which also shows the force diagram.

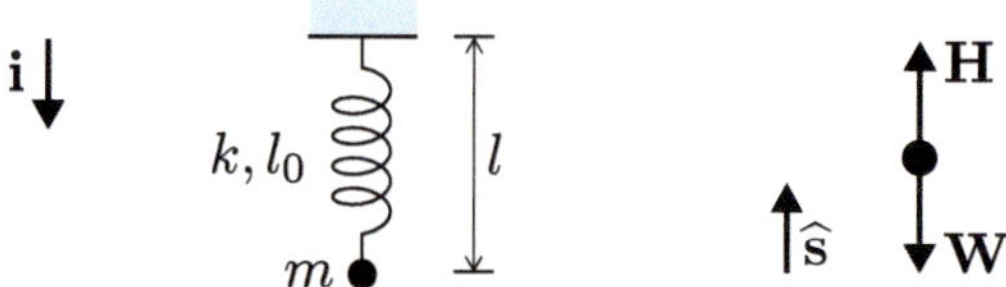

Figure 21 Set-up and force diagram

The weight of the particle is $\mathbf{W} = mg\mathbf{i}$, and the force exerted on it by the spring is

$$\mathbf{H} = k(l - l_0)\,\widehat{\mathbf{s}} = k(l - l_0)(-\mathbf{i}).$$

Since $\mathbf{H} + \mathbf{W} = \mathbf{0}$ (because the particle is in equilibrium), we have

$$-k(l - l_0) + mg = 0,$$

that is,

$$l = l_0 + \frac{mg}{k}.$$

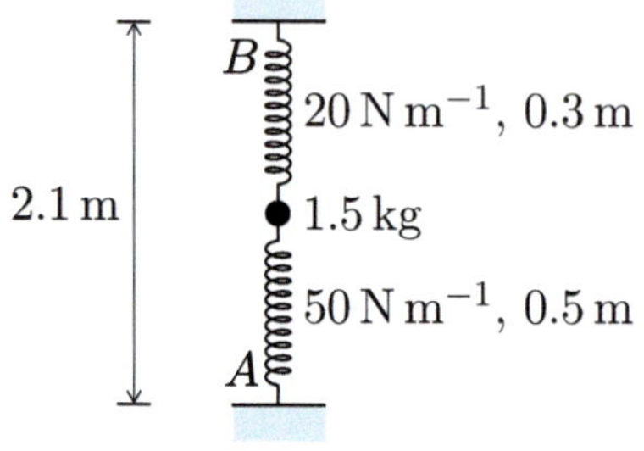

Figure 22

Exercise 6

An object of mass 1.5 kg is attached to the floor at A by a model spring of stiffness $50\,\mathrm{N\,m^{-1}}$ and natural length 0.5 m. It is also attached to the ceiling at B, a point 2.1 m vertically above A, by a model spring of stiffness $20\,\mathrm{N\,m^{-1}}$ and natural length 0.3 m (see Figure 22).

When the object is in equilibrium, what is the height of the object above the floor?

Exercise 7

A battery of length 5 cm is put into a battery holder. The holder has an internal length of 5.5 cm, and has a spring at each end to hold the battery in place (see Figure 23). The left-hand spring has stiffness $30\,\mathrm{N\,m^{-1}}$ and natural length 0.4 cm, while the right-hand spring has stiffness $10\,\mathrm{N\,m^{-1}}$ and natural length 0.3 cm.

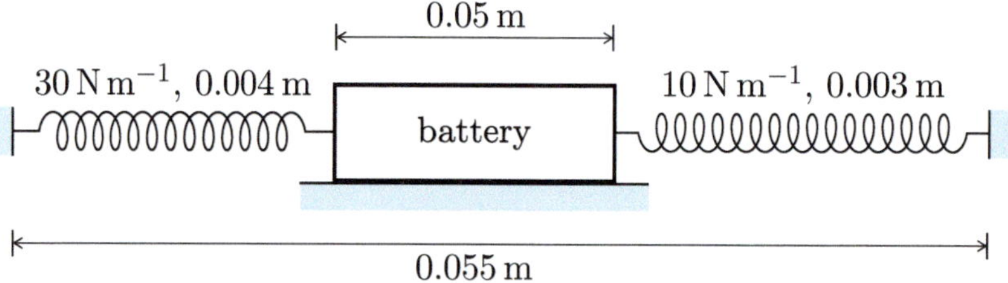

Figure 23 Schematic arrangement of the battery holder (not to scale)

What is the magnitude of the spring forces acting on the battery, assuming that there are no friction forces?

2 Oscillating systems

In Unit 3 you saw how to analyse the one-dimensional motion of a particle under the influence of forces such as gravity and air resistance. Here we are again concerned solely with one-dimensional motion, but now we concentrate on how a particle behaves when connected to the end of a model spring. In this unit friction and air resistance will be considered to be negligible, and the only forces that will be considered will be gravity, normal reactions and spring forces. There is no fundamental difference in approach; it is just that the force due to a spring depends on the length of the spring, and this results in a different outcome.

2.1 Setting up an equation of motion

As before, the analysis is based on Newton's second law, which states that if a particle has mass m and experiences a resultant force $\mathbf{F}$, then its acceleration $\mathbf{a}$ is given by

$$\mathbf{F} = m\mathbf{a}.$$

You saw in Unit 3 that the acceleration $\mathbf{a}(t)$ of a particle at time t is defined as the derivative of its velocity $\mathbf{v}(t)$, which in turn is defined as the derivative of its position vector $\mathbf{r}(t)$. Using Newton's 'dot' notation for derivatives with respect to time, we therefore have $\mathbf{a} = \ddot{\mathbf{r}}$, so Newton's second law may also be expressed as

Recall that $\dot{\mathbf{r}}$ means $d\mathbf{r}/dt$ and $\ddot{\mathbf{r}}$ means $d\dot{\mathbf{r}}/dt = d^2\mathbf{r}/dt^2$. Each dot stands for '$d/dt$'.

$$m\ddot{\mathbf{r}} = \mathbf{F}. \qquad (3)$$

Example 4 illustrates the type of approach that is required where a model spring is involved. As indicated in the margin alongside the solution, it is still appropriate to adopt here the procedural steps for applying Newton's second law that were used in Unit 3. The last two stages of this procedure, solving the differential equation and interpreting the solution, will be dealt with in the following subsection.

See Procedure 1 of Unit 3 for the listing of these steps.

Example 4

A spring is attached to a fixed wall at one end. The other end of the spring is attached to a glider that moves along a track, such that the motion of the glider is friction-free. Model the glider as a particle of mass m and the spring as a model spring of natural length l_0 and stiffness k.

As mentioned above, air resistance is deemed to be negligible.

Find the equation of motion for the particle when its position x is measured in the direction away from the wall, with the origin at the fixed end.

(This model system is similar to that of Figure 2(b) in the Introduction, and as mentioned there, it might provide a first model for a spring-loaded buffer.)

Solution

◀ Draw picture ▶
◀ Choose axes ▶
◀ State assumptions ▶

Figure 24 sketches the physical situation, including the x-axis and unit vectors **i** and **j**. We model the glider as a particle of mass m and the spring as a model spring. All friction and air resistance forces are ignored.

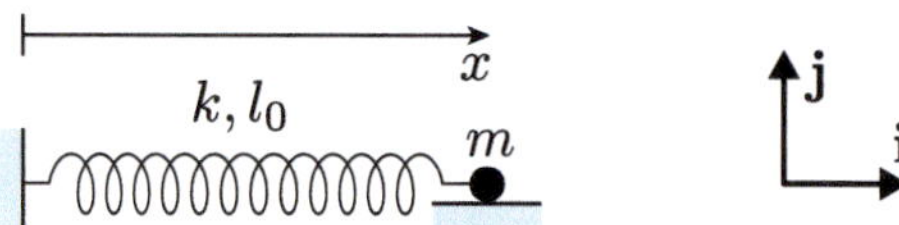

Figure 24 Simple oscillating system

The unit vector **i** is directed away from the fixed end. The length of the spring is given by x, in the same direction as **i**.

◀ Draw force diagram ▶

In addition to its weight **W** and the normal reaction **N**, the only force acting on the particle is that provided by the model spring, but how should this be portrayed on a force diagram? As you saw in Subsection 1.2, this force will pull the particle if the spring is extended and push the particle if the spring is compressed, so we can expect the force exerted by the spring on the particle to act in the direction of **i** at some times and in the direction of $-$**i** at others. We are therefore unable to make a single 'correct' choice for the direction of this force **H**, but *either choice will do* for the purposes of the diagram. (The force diagram shown in Figure 25 corresponds to the situation when the spring is extended, and this is the usual convention.)

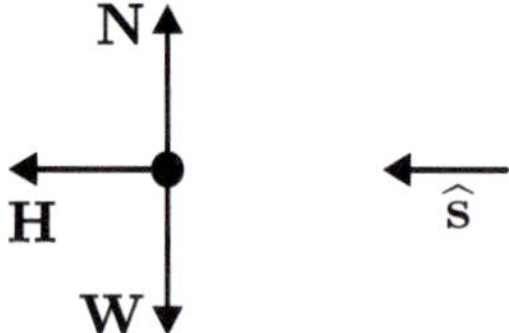

Figure 25 Force diagram

◀ Apply Newton's 2nd law ▶

Applying Newton's second law to the particle gives

$$m\ddot{\mathbf{r}} = \mathbf{H} + \mathbf{W} + \mathbf{N}. \tag{4}$$

To proceed further we need to model all the forces. The force **H** exerted by the model spring on the particle is described by Hooke's law, that is,

$$\mathbf{H} = k(l - l_0)\,\widehat{\mathbf{s}},$$

where l is the length of the spring and $\widehat{\mathbf{s}}$ is a unit vector in the direction from the end where the particle is attached towards the centre of the spring.

For the current situation, $l = x$ and $\widehat{\mathbf{s}} = -\mathbf{i}$, so the force in the spring that is exerted on the particle is

$$\mathbf{H} = k(x - l_0)(-\mathbf{i}).$$

The other two forces are $\mathbf{W} = mg(-\mathbf{j})$ and $\mathbf{N} = |\mathbf{N}|\,\mathbf{j}$. Substituting into equation (4) gives

$$m\ddot{\mathbf{r}} = \mathbf{H} + \mathbf{W} + \mathbf{N} = -k(x - l_0)\mathbf{i} - mg\mathbf{j} + |\mathbf{N}|\,\mathbf{j}. \tag{5}$$

Since the motion is only in the x-direction, the position vector of the particle is $\mathbf{r}(t) = x(t)\mathbf{i}$, so $\dot{\mathbf{r}}(t) = \dot{x}(t)\mathbf{i}$ and $\ddot{\mathbf{r}}(t) = \ddot{x}(t)\mathbf{i}$. Hence resolving equation (5) in the **i**-direction gives $m\ddot{x} = -k(x - l_0)$, that is,

$$m\ddot{x} + kx = kl_0. \tag{6}$$

This is the equation of motion.

Equation (6) is a second-order inhomogeneous linear constant-coefficient differential equation. This is exactly the type of differential equation that was studied in Unit 1 (this is no coincidence), and it can be solved using the methods described in that unit. This will be done in the next subsection, but to conclude this subsection, we ask you to look at the derivation of another equation of motion.

As in Unit 1, we collect all the terms involving the unknown function (x in this case) and its derivatives onto the left-hand side, and any remaining terms onto the right-hand side.

Exercise 8

Figure 26 shows the same spring and mass as described in Example 4, but now it is arranged so that it moves vertically, hanging from a fixed support at the top. Find the equation of motion for the particle in this case.

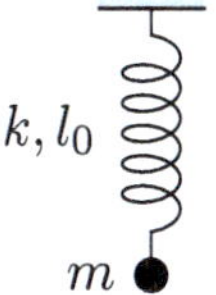

Figure 26 Vertical spring and mass

Exercise 9

What are the similarities, and the differences, between the equation of motion derived in Exercise 8 and equation (6) derived in Example 4 for a system that moves horizontally?

2.2 Solving the equation of motion

In the previous subsection we derived

$$m\ddot{x} + kx = kl_0. \tag{7}$$

as the equation of motion for a particle attached to one end of a horizontal model spring, whose other end is fixed. The solution of this equation will be found in two parts: first the complementary function, and then the particular integral.

We now complete Procedure 1 of Unit 3 for the problem stated in Example 4, by solving the differential equation and interpreting the solution.

The method for solving the associated homogeneous equation

$$m\ddot{x} + kx = 0 \tag{8}$$

to obtain the complementary function was described in Unit 1, but quite a lot can be deduced simply by looking at the form of the equation.

Equation (8), written as $\ddot{x} = (-k/m)x$, must be satisfied by a function $x(t)$ whose second derivative $\ddot{x}(t)$ is a negative constant multiple $(-k/m)$ of the function $x(t)$ itself. What functions satisfy this specification? There are at least two that come to mind, namely the sine and cosine functions. Hence functions of the form $x(t) = B\cos\omega t$ and $x(t) = C\sin\omega t$ (where B, C and ω are constants) are contenders to be solutions of equation (8). Since we are dealing with a homogeneous linear differential equation, it follows from the principle of superposition that the sum of two such functions is also a contender.

See Unit 1, Theorem 2.

Exercise 10

Show by substitution that $x(t) = B\cos\omega t + C\sin\omega t$ is a solution of the differential equation $m\ddot{x} + kx = 0$, where $\omega^2 = k/m$.

Exercise 11

This method of solution is described in Unit 1.

By using the auxiliary equation, find the general solution of the differential equation $m\ddot{x} + kx = 0$.

Exercises 10 and 11 confirm that our reasoning and the methods of Unit 1 give the same solution.

Exercise 12

Find the general solutions of the equations of motion found in Example 4 and Exercise 8, namely

$$m\ddot{x} + kx = kl_0,$$
$$m\ddot{x} + kx = kl_0 + mg.$$

Exercise 13

The equilibrium length of a horizontal spring is its natural length, l_0. Example 3 found that the equilibrium length of a hanging spring is $l_0 + mg/k$.

Compare each of these values with the value of the constant in the appropriate general solution found in Exercise 12.

Exercise 13 shows that the constant term in the solution of the differential equation for a horizontal or hanging spring is exactly the same as the equilibrium position of the particle. From now on this will be denoted by x_{eq}. In many systems that oscillate, the complementary function, which is associated with the oscillatory behaviour of the system, is of more consequence than the equilibrium position, which is associated with the particular integral.

Using the results of Exercises 12 and 13, the general solution of

$$m\ddot{x} + kx = kx_{\text{eq}}$$

is

$$x(t) = B\cos\omega t + C\sin\omega t + x_{\text{eq}}, \tag{9}$$

where $\omega = \sqrt{k/m}$, the constant term x_{eq} is the equilibrium position of the particle, and B and C are arbitrary constants. The values of these arbitrary constants can be found if initial conditions are provided for the function $x(t)$, usually $x(0)$ and $\dot{x}(0)$.

Motion of the type described by equation (9), arising from a differential equation of the form $\ddot{x} + \omega^2 x = 0$, is known as **simple harmonic motion** (sometimes abbreviated to **SHM**).

It is sometimes more convenient to replace $B\cos\omega t + C\sin\omega t$ with $A\cos(\omega t + \phi)$, which is directly equivalent. To see this, use a compound-angle formula to write

$$A\cos(\omega t + \phi) = A(\cos\omega t\cos\phi - \sin\omega t\sin\phi).$$

Equating terms in $\cos\omega t$ and $\sin\omega t$ gives $B = A\cos\phi$ and $C = -A\sin\phi$. Then

$$B^2 + C^2 = A^2\cos^2\phi + (-A)^2\sin^2\phi = A^2(\cos^2\phi + \sin^2\phi) = A^2.$$

So

$$A = \sqrt{B^2 + C^2}, \quad \cos\phi = \frac{B}{A}, \quad \sin\phi = -\frac{C}{A}. \tag{10}$$

We take the positive root for A so that $A > 0$. The value of ϕ that satisfies the equations for $\cos\phi$ and $\sin\phi$ lies in the interval $[-\pi, \pi]$.

The equivalent form of equation (9) is

$$x(t) = B\cos\omega t + C\sin\omega t + x_{\text{eq}} = A\cos(\omega t + \phi) + x_{\text{eq}}. \tag{11}$$

Note that the addition of a cosine term and a sine term with the same argument leads to a cosine (or sine) term that is shifted.

Exercise 14

Given that

$$x(t) = 10\cos\left(7t + \arctan\left(\tfrac{3}{4}\right)\right),$$

find constants B, C and ω in the equivalent form

$$x(t) = B\cos\omega t + C\sin\omega t.$$

When applying initial conditions, it is usually easier to use the form $B\cos\omega t + C\sin\omega t$, whereas the expression $A\cos(\omega t + \phi)$ gives a better feel for the features of motion.

The quantity ϕ, which is called the **phase angle** (or **phase**) of the motion, has the interpretation that $A\cos\phi$ is the particle's position at time $t = 0$. Also, if ϕ is positive, then $-\phi/\omega$ gives the last time, prior to $t = 0$, at which the position x reaches its maximum value (see Figure 27). If ϕ is negative, then $-\phi/\omega$ is the first time after $t = 0$ at which the position x reaches its maximum value. The graph of $A\cos(\omega t + \phi)$ may be obtained from the graph of $A\cos\omega t$ by a horizontal shift of ϕ/ω to the left.

The parameters in the general solution of a spring–mass system have special terminology, and at this point it is useful to define the terms used for describing oscillations. First, consider a plot of the solution

$$x(t) = A\cos(\omega t + \phi) + x_{\text{eq}} = A\cos\left(\omega\left(t + \frac{\phi}{\omega}\right)\right) + x_{\text{eq}}.$$

$A > 0$ and $-\pi < \phi \leq \pi$.

The form of the solution is a sinusoid added to a constant; the cosine has amplitude A and is displaced by an amount $t = \phi/\omega$ to the left. The period is the distance between successive peaks, and represents the time taken for the oscillation to repeat itself. These features are shown in Figure 27. (The graph has been drawn for $\phi > 0$.)

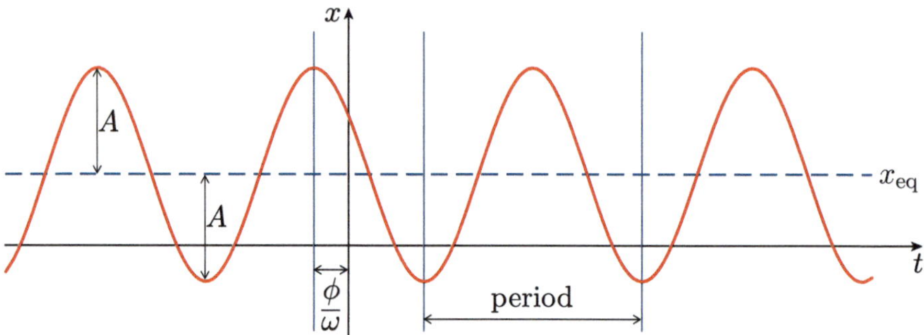

Figure 27 Solution for a general oscillating system

As the angle ωt is measured in radians, the SI units of the angular frequency ω are radians per second ($\text{rad}\,\text{s}^{-1}$).

The variable ω is called the **angular frequency** of the motion. Simple harmonic motion is periodic with a **period** τ given by $\tau = 2\pi/\omega$, that is, the motion repeats itself every $\tau = 2\pi/\omega$ seconds.

The SI unit for frequency is the *hertz* (Hz), where 1 hertz is 1 cycle per second. Hence $1\,\text{Hz} = 1\,\text{s}^{-1}$.

Since τ is the time for one cycle, the number f of cycles per unit time is equal to $1/\tau$. This is called the **frequency** of the motion and is given by

$$f = \frac{1}{\tau} = \frac{\omega}{2\pi},$$

so the angular frequency ω (in $\text{rad}\,\text{s}^{-1}$) is equal to 2π times the frequency f (in Hz).

We now demonstrate how particular values may be found for the arbitrary constants and when specific initial conditions are given.

Example 5

In a model of a vibrating system, the position $x(t)$, with origin at zero, of a particle at time t is given by

$$x(t) = B\cos 10t + C\sin 10t,$$

where B and C are constants. The particle is set in motion at time $t = 0$, when its position is $-0.2\,\text{m}$ and its speed is $1\,\text{m}\,\text{s}^{-1}$ in the direction of the positive x-axis (both measured relative to the given x-axis and origin).

(a) Find the particular position function for the particle that corresponds to these initial conditions. Where is the particle after 0.5 seconds?

(b) What are the angular frequency, frequency and period of the oscillations?

(c) Write the solution in the form $A\cos(\omega t + \phi)$, and hence state the amplitude and phase angle to three decimal places. Sketch the graph of $x(t)$ against t for $0 \leq t \leq 2$.

Solution

(a) The initial conditions may be written as

$$x(0) = -0.2, \quad \dot{x}(0) = 1.$$

Since $x(t) = B\cos 10t + C\sin 10t$, we have

$$\dot{x}(t) = -10B\sin 10t + 10C\cos 10t.$$

Substituting $t = 0$ into these equations and using the given initial conditions, we have

$$x(0) = B = -0.2, \quad \dot{x}(0) = 10C = 1.$$

Hence $B = -0.2$ and $C = 0.1$, giving the particular position function

$$x(t) = -0.2\cos 10t + 0.1\sin 10t.$$

After 0.5 seconds, the particle is at

$$x(0.5) = -0.2\cos 5 + 0.1\sin 5 \simeq -0.15 \quad \text{to 2 d.p.},$$

that is, 0.15 m along the x-axis in the negative direction from the origin.

(b) The angular frequency is 10, so the frequency is $10/2\pi = 5/\pi$, and the period is $2\pi/10 = \pi/5$.

(c) Using equations (10), we have

$$A = \sqrt{B^2 + C^2} = \sqrt{(-0.2)^2 + 0.1^2} = \sqrt{0.05}$$

and

$$\cos\phi = \frac{B}{A} = -\frac{0.2}{\sqrt{0.05}}, \quad \sin\phi = -\frac{C}{A} = -\frac{0.1}{\sqrt{0.05}}.$$

Thus

$$\tan\phi = \frac{0.1}{0.2} = \tfrac{1}{2},$$

and since both $\sin\phi$ and $\cos\phi$ are negative, this gives $\phi = -\pi + \arctan\left(\frac{1}{2}\right)$. Hence

$$x(t) = \sqrt{0.05}\cos\left(10t - \pi + \arctan\left(\tfrac{1}{2}\right)\right).$$

The amplitude A is $\sqrt{0.05} \simeq 0.224$ and the phase angle ϕ is $-\pi + 0.464 \simeq -2.678$, both to three decimal places, so $\phi/\omega = -2.678/10 \simeq -0.27$.

A sketch of x against t is shown in Figure 28.

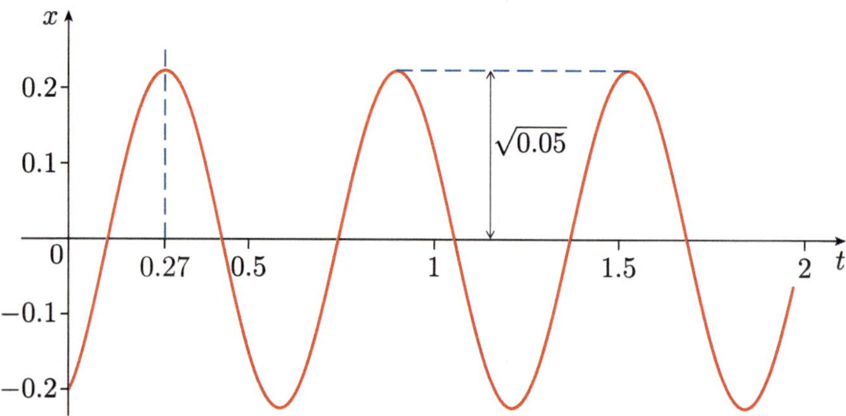

Figure 28

Exercise 15

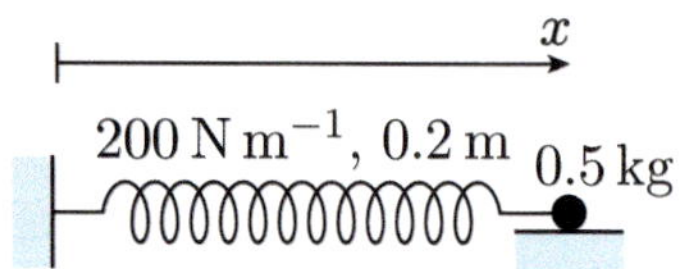

Figure 29 Oscillating system

For a particular case of the system shown in Figure 24, and reproduced here as Figure 29, the model spring's stiffness k is $200\,\mathrm{N\,m^{-1}}$, the natural length l_0 is 0.2 m, and the particle's mass m is 0.5 kg. The system is set in motion when $t = 0$, at which time the position is 0.3 m from the left-hand end and the speed is $2\,\mathrm{m\,s^{-1}}$ to the right.

The solution of the equation of motion is

$$x(t) = B\cos\omega t + C\sin\omega t + l_0,$$

where $\omega = \sqrt{k/m}$, and B and C are constants.

(a) Find the angular frequency ω and the period τ of the motion.

(b) Find the frequency f of the motion.

(c) Find the position function for the particle at time t.

(d) Find the minimum distance of the particle from the left-hand end.

You saw in Exercise 14 that the equivalence of the expressions for $x(t)$ in equations (11) is obtained by showing how the constants B and C in the first form are related to the constants A and ϕ in the second form. It is often helpful to draw a diagram such as the figure shown in the solution to Exercise 14.

Exercise 16

In Exercise 15 you showed that the particular oscillation described there is represented by the equation

$$x(t) = 0.1\cos 20t + 0.1\sin 20t + 0.2.$$

(a) Find the phase angle of this oscillation.

(b) Sketch the graph of x against t.

(c) Determine the times when the spring takes its natural length.

We summarise the results about simple harmonic motion in the following box.

Simple harmonic motion

The simple harmonic motion equation

$$\ddot{x} + \omega^2 x = \omega^2 x_{\text{eq}} \qquad (12)$$

has a general solution that can be written in the form

$$x(t) = B\cos\omega t + C\sin\omega t + x_{\text{eq}}, \qquad (13)$$

where B and C are arbitrary constants, or alternatively as

$$x(t) = A\cos(\omega t + \phi) + x_{\text{eq}}, \qquad (14)$$

where

$$A = \sqrt{B^2 + C^2}, \quad \cos\phi = B/A, \quad \sin\phi = C/A, \qquad (15)$$

with $A > 0$ and $-\pi < \phi \leq \pi$.

The following names are given to quantities that are characteristic of simple harmonic motion:

- the *angular frequency* of the oscillations is ω
- the *period* (time for one complete cycle) is $\tau = 2\pi/\omega$
- the *frequency* (number of cycles per second) is $f = 1/\tau = \omega/(2\pi)$
- the *amplitude* of the oscillations is A
- the *phase angle* of the oscillations is ϕ
- the equilibrium position about which the particle oscillates is x_{eq}.

This equation is equivalent to

$$m\ddot{x} + kx = kx_{\text{eq}},$$

where $\omega^2 = k/m$.

We now extend the experiment that was described in Subsection 1.1.

We can consider the system of the ruler and weights as a particle on the end of a model spring, as shown in Figure 30.

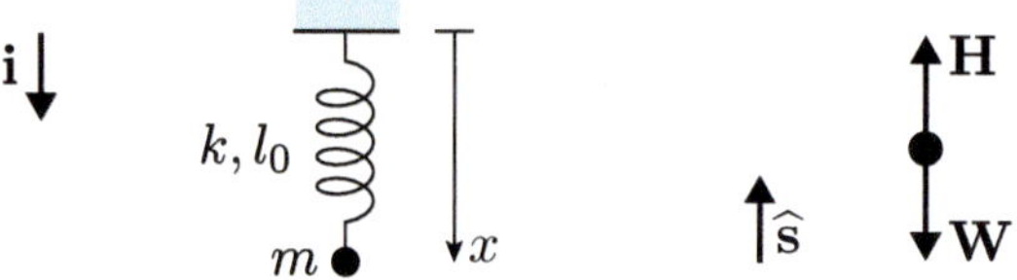

Figure 30 Model for the ruler experiment

Equation (1) of Subsection 1.1 can be rewritten in the vector form of Hooke's law, using x instead of d (the displacement of the end of the ruler). The spring force is in the upward direction, so

$$\mathbf{H} = -99.044\left(x - \frac{0.024}{99.044}\right)\mathbf{i} = -99.044(x - 0.000\,24)\,\mathbf{i},$$

which gives the spring stiffness as about $99\,\mathrm{N\,m^{-1}}$.

To test the mathematical model for oscillations, a mass of 0.03 kg was firmly attached to the ruler, then the ruler was deflected downwards by about a centimetre and released. The resulting oscillations were measured by a recording device, and some of them are shown in Figure 31, where t is the time in seconds and x is the distance of the end of the ruler below a fixed point.

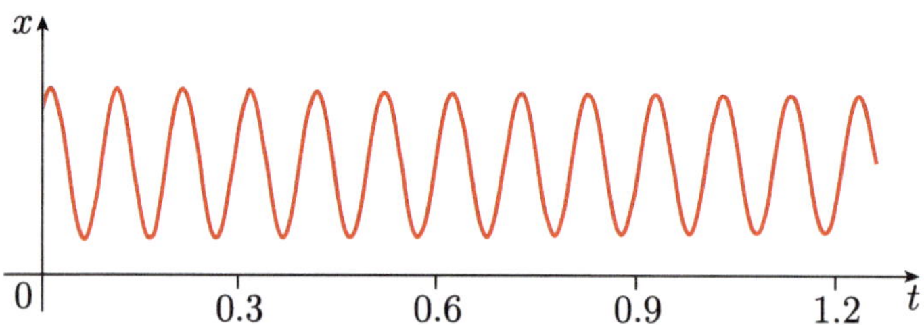

Figure 31 Measurements of an oscillating ruler

Example 6

Consider the oscillations of the end of the ruler as shown in Figure 31.

(a) Is the motion periodic?

(b) From the graph, estimate the period of the motion.

(c) Compare the measured period with the period obtained from considering the stiffness of the model spring and the mass of the particle on the end.

Solution

(a) The motion does look periodic.

(b) There are thirteen peaks in all, so twelve oscillations, and the time interval for these twelve oscillations is approximately 1.25 seconds, so the period is $1.25/12 \simeq 0.104$ seconds to three significant figures.

(c) For a mass m and spring stiffness k, the period is given by $\tau = 2\pi\sqrt{m/k}$. From above, the spring stiffness was estimated to be $99\,\mathrm{N\,m^{-1}}$, and the mass is $0.03\,\mathrm{kg}$. Therefore an estimate of the period is $2\pi\sqrt{0.03/99} \simeq 0.109$ seconds, which compares well with the measured value.

This simple model of a ruler that is oscillating with a mass on its end gives a very good comparison between the observed period and that deduced from considering the equivalent spring stiffness of the ruler. One may detect a slight decrease in the amplitude of the oscillations as time goes on, and this will be the subject of the next unit.

This section concludes with some exercises involving springs and masses in different set-ups.

Note that a useful check when setting up equations of motion for oscillating systems is that all the individual terms in x and $\ddot{x}$ when taken onto one side have the same sign; this confirms that each spring does what is required of it to restore the system to equilibrium.

Exercise 17

A block of mass m lying on a frictionless surface oscillates horizontally and is attached to two springs: the one on the left has stiffness k and natural length $2l_0$; the one on the right has stiffness $3k$ and natural length l_0 (see Figure 32). The other ends of the springs are attached to fixed points aligned horizontally, a distance $5l_0$ apart. Assume that the block may be modelled as a particle.

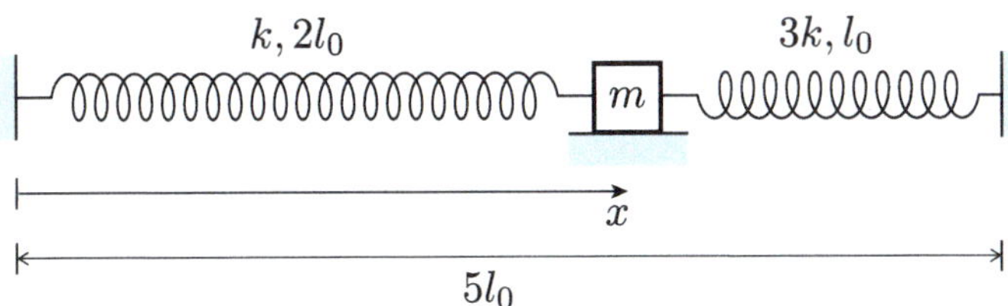

Figure 32

(a) Show that the distance x of the block from the left-hand end at time t during the oscillations satisfies the differential equation

$$m\ddot{x} + 4kx = 14kl_0.$$

(b) Hence find the distance from the left-hand end at which the block could remain in equilibrium.

(c) Find the period τ of the block's oscillations.

(d) Suppose that the block is initially (at $t = 0$) released from rest halfway between the two ends. Find an expression for $x(t)$.

(e) Does the block hit one of the ends in its subsequent motion?

Exercise 18

A block P of mass m is attached to three springs whose other ends are attached to fixed points Q, R and S. The point R is a distance $\frac{1}{2}l_0$ below Q, and the point S is a distance $4l_0$ below Q. The parameters of the three springs are given in Figure 33, which also illustrates the arrangement of the springs and the block.

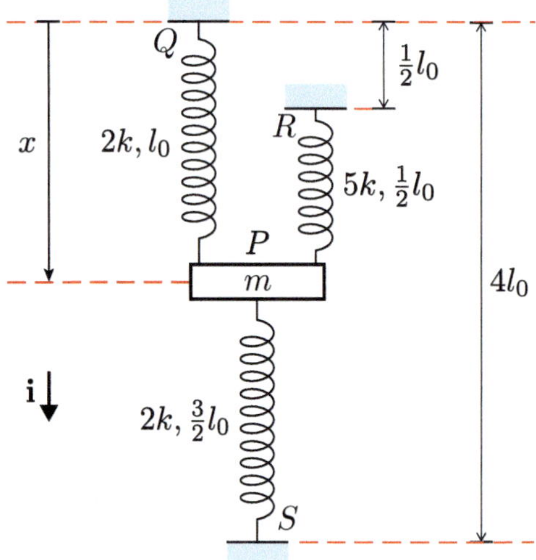

Figure 33

Model the block as a particle and the springs as model springs. Take the origin at Q, with the vertical distance of P from Q being x, so that the x-axis is as shown, and $\mathbf{i}$ is pointing downwards.

(a) Draw a force diagram indicating all the forces acting on the particle.

(b) Find the spring force for each spring.

(c) Derive the equation of motion of the particle.

(d) Find the position of equilibrium for the particle.

(e) Find the general solution of the differential equation that you found in part (c).

(f) The particle is initially released from rest at a distance $\frac{4}{3}l_0$ below Q. Determine the solution of the differential equation that satisfies these initial conditions.

(g) Write down the period and the amplitude of the oscillations of the particle during its subsequent motion.

(h) Draw a sketch of the graph of x against t for $t \geq 0$, clearly indicating the amplitude, period, starting position and average position of the block.

Figure 34 shows the uniform cross-section of a rectangular pontoon that floats on water. The position x of the upper surface of the pontoon is measured upwards from the water level as shown in Figure 34, which also shows the positive **i**-direction. The pontoon has height h, mass m and uniform horizontal cross-sectional area A; ρ is the density of water, and g is the magnitude of the acceleration due to gravity.

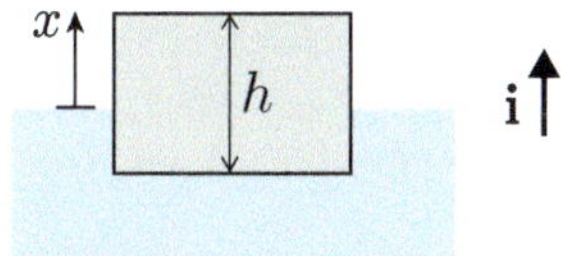

Figure 34 Pontoon on the water

Archimedes' principle states that if an object is wholly or partly immersed in a liquid, then the resulting buoyancy force on the object is directed vertically upwards and is equal in magnitude to the weight of liquid displaced by the object.

There are two forces acting on the pontoon: $\mathbf{W} = mg(-\mathbf{i})$, the weight of the pontoon, and $\mathbf{B}$, the buoyancy force (see Figure 35). In order to use Archimedes' principle to find the buoyancy force, first we need to find the volume of water displaced, which is $A(h - x)$. We then multiply this by ρg to obtain the magnitude of the buoyancy force; since the force is upwards,

Figure 35 Forces acting on the pontoon

$$\mathbf{B} = A\rho g(h - x)\mathbf{i}. \tag{16}$$

Note the similarity of the buoyancy force to the spring force $\mathbf{H} = k(l - l_0)\widehat{\mathbf{s}}$.

Applying Newton's second law gives

$$m\ddot{x}\mathbf{i} = \mathbf{W} + \mathbf{B} = -mg\mathbf{i} + A\rho g(h - x)\mathbf{i}.$$

Resolving in the **i**-direction and taking the x terms onto the left-hand side gives

$$m\ddot{x} + A\rho g x = A\rho g h - mg. \tag{17}$$

According to this model, the 'bobbing up and down' of the pontoon is therefore a simple harmonic motion.

Exercise 19

(a) What is the range of x for which model (16) for the buoyancy force is applicable?

(b) Show that for a pontoon of given dimensions, this model predicts that the period τ of the oscillations is proportional to $\sqrt{m}$, and hence increases with the mass of the pontoon. Do you think that there is any limit to the period that could be attained, for a suitably massive pontoon?

---

3 Introduction to energy

In this section an alternative approach to solving a certain class of mechanics problems is introduced. This approach is based on the concept of *mechanical energy*. The application of this concept to solving mechanics problems is based on the constancy or *conservation* of mechanical energy – we discuss this in Section 4.

Here we introduce a method for solving mechanics problems for systems where the forces *depend only on the position* of the particle. Such forces are very common. For example, the force acting on a steel pin due to the attraction of a magnet depends on the distance between the pin and the magnet. In fact, the magnitude of this force increases as the pin is brought closer to the magnet. On the other hand, a dog tethered to a tree by a length of elastic rope feels a restraining force, once the elastic is taut, whose magnitude increases as the dog moves away from the tree. In either case, we can consider a one-dimensional situation and define the **i**-direction to be the direction of the force, and then we have

$$\mathbf{F} = F\mathbf{i} = F(x)\,\mathbf{i},$$

where we have written $F(x)$ in the last expression to emphasise that the **i**-component of the force depends on x alone. We assume that the form of the function $F(x)$ is known, and use this knowledge to answer questions about the motion. It is possible here for $F(x)$ to be a constant function, so in saying that 'F depends only on position', we really mean that F depends *at most* on position, that is, it does not depend directly on other variables of the motion, such as velocity, acceleration or time.

For example, the weight and tension force in a spring are forces that depend on position only. It is tempting to think that friction is a constant, in the same way that weight is, but this is not so. Friction **F** opposes the motion in the direction opposite to the direction in which the particle is travelling, thus can be written in terms of the velocity **v** of the particle as

$$\mathbf{F} = -|\mathbf{F}|\,\frac{\mathbf{v}}{|\mathbf{v}|}.$$

Air resistance and friction are examples of dissipative forces that cause a loss of mechanical energy to other forms of energy such as heat or sound.

3.1 Forms of energy

Consider a particle of mass m acted on by a total force $\mathbf{F}(x)$ that is dependent on x alone. Choose **i** to be a unit vector in the direction of the force, so $\mathbf{F}(x) = F(x)\,\mathbf{i}$ (where the distinction between scalar and vector notation is important, because $F(x)$ can be positive or negative).

It is natural to begin by writing down Newton's second law,

$$\mathbf{F} = m\mathbf{a},$$

where as usual **a** is the acceleration of the particle. The motion is one-dimensional and we have $\mathbf{a} = \ddot{x}\mathbf{i}$, so this equation can be resolved in the **i**-direction to obtain

$$F(x) = m\ddot{x}. \tag{18}$$

This is a differential equation in which the independent variable is t. You have seen in Unit 3 equations (9) how $\ddot{x}$ can be transformed into a derivative with x as the independent variable, so we can write

$$\ddot{x} = v\,\frac{dv}{dx}, \tag{19}$$

where $v = \dot{x}$ is the component of velocity in the **i**-direction. A derivative with respect to t has been transformed into a derivative with respect to x. Using this transformation on equation (18), and integrating both sides with respect to x from the initial position with $x = x_0$, $v = v_0$ to a general position x, v gives

$$\int_{x_0}^{x} F(x)\,dx = m\int_{x_0}^{x} v\,\frac{dv}{dx}\,dx = m\int_{v_0}^{v} v\,dv = m\left[\tfrac{1}{2}v^2\right]_{v_0}^{v}, \tag{20}$$

where the limits for the right-hand integral have been changed to those of the variable being integrated.

Let $U(x) = -\int F(x)\,dx$, and substitute this into equation (20) to give

The reason for the negative sign will be explained soon.

$$-\left[U(x)\right]_{x_0}^{x} = m\left[\tfrac{1}{2}v^2\right]_{v_0}^{v},$$

which on rearrangement becomes

$$U(x_0) + \tfrac{1}{2}mv_0^2 = U(x) + \tfrac{1}{2}mv^2. \tag{21}$$

This equation has a useful symmetry: both sides of the equation involve the expression $\frac{1}{2}mv^2 + U(x)$; the only difference is that the right-hand side involves v and x at a general position x, and the left-hand side involves the initial values of those variables.

Each of these quantities, $\frac{1}{2}mv^2$ and $U(x)$, has a special significance, and we will consider them individually.

Kinetic energy

The term $\frac{1}{2}mv^2$ is dependent on speed (note that the sign of v is immaterial since it is squared – the direction of motion is not important), and it is given the name *kinetic energy*.

Kinetic energy

The **kinetic energy** T of a particle of mass m moving with component of velocity v (where $v = \dot{x}$) is given by

$$T = \tfrac{1}{2}mv^2 = \tfrac{1}{2}m\dot{x}^2. \tag{22}$$

It is conventional to use T to denote the kinetic energy because E is usually reserved for the total energy, as you will see later.

The word *kinetic* in the term *kinetic energy* emphasises that the energy is due to the motion of the object.

The SI unit for energy is the *joule* (J). So a particle of mass 4 kg moving with speed $5\,\mathrm{m\,s^{-1}}$ has kinetic energy $\frac{1}{2} \times 4 \times 5^2 = 50$ joules.

$1\,\mathrm{J} = 1\,\mathrm{kg\,m^2\,s^{-2}}$.

Potential energy

Now consider the term $U(x)$ in equation (21). We call this term the potential energy *function* to emphasise that it is dependent on the position of the particle.

Potential energy

For a particle under the influence of a force $\mathbf{F} = F(x)\,\mathbf{i}$ that is either constant or depends only on the position $\mathbf{r} = x\mathbf{i}$, the **potential energy function** $U(x)$ is

$$U(x) = -\int F(x)\,dx, \qquad (23)$$

or equivalently,

$$F(x) = -\frac{dU}{dx}. \qquad (24)$$

It is conventional to use U to denote the potential energy.

By definition, $U(x)$ will incorporate a constant of integration, so it will not be unique. However, note that it occurs on both sides of equation (21), so the same constant of integration will occur on both sides.

Exercise 20

Check that potential energy has the same dimensions as kinetic energy.

Total energy

Consider again equation (21), repeated here:

$$U(x_0) + \tfrac{1}{2}mv_0^2 = U(x) + \tfrac{1}{2}mv^2.$$

The left-hand side of this equation is a constant, which we call E, the *total mechanical energy* of the particle.

Total mechanical energy

The **total mechanical energy** of a particle is given by the sum of its kinetic and potential energies, that is,

$$E = \tfrac{1}{2}mv^2 + U(x). \qquad (25)$$

If there is more than one force, then the total potential energy $U(x)$ is the sum of the potential energy functions $U_1(x), U_2(x), \ldots, U_n(x)$ due to each of the n forces.

The potential energy function is defined in equation (23) as an indefinite integral, so it is defined only to within a constant of integration. In practice, however, the value of this constant is unimportant, because it can be absorbed into the constant E appearing in equation (25). The constant of integration in the definition of the potential energy function can therefore be chosen to have any convenient value. In other words, we can choose the value of the potential energy function $U(x)$ to be zero at any convenient point. This point is called the **datum** of the potential energy function. Thus if $x = x_0$ is the chosen datum, then $U(x_0) = 0$.

You saw a similar situation in Unit 1, where the integrating factor for a first-order linear differential equation was defined in terms of an indefinite integral.

3.2 Calculating potential energy functions

For the remainder of this section, we ask you to concentrate on the process of finding the potential energy function $U(x)$ for given forces.

Exercise 21

Using an x-axis pointing vertically upwards (see Figure 36) and equation (23), verify that the potential energy function for the force due to gravity acting on a particle of mass m is

$$U(x) = mgx,$$

where the datum O for this function is taken to be the origin $x = 0$.

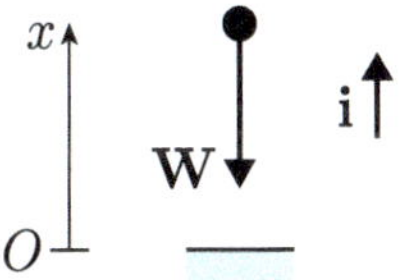

Figure 36

Exercise 22

Consider the motion of a particle of mass m along a straight frictionless track that is at an angle θ to the horizontal. Using an x-axis pointing down the slope, verify that the potential energy function for the total force acting on the particle is

$$U(x) = -mgx \sin\theta,$$

where the datum is taken to be the origin $x = 0$.

In Exercise 21, you showed that for a suitable choice of datum ($x = 0$), the potential energy function is $U(x) = mgx$ if the x-axis is directed vertically upwards. In Exercise 22, you showed that for a suitable choice of datum ($x = 0$), the potential energy function is $U(x) = -mgx \sin\theta$ if the x-axis is directed down the slope. It can be shown that it is always the case that for positions above the datum the potential energy function is positive, and for those below the datum the potential energy function is negative.

If the x-axis in Exercise 21 were pointing in the opposite direction, that is, vertically downwards (see Figure 37), then the potential energy function would be $U(x) = -mgx$. Now if the particle is above the datum, then the coordinate x is negative, so $U(x)$ is again positive. Conversely, if the particle is below the datum, then x is positive, so $U(x)$ is negative. These signs are the same as for the potential energy function with an upwards pointing x-axis. This leads to the following memorable result for the potential energy function due to gravity, henceforth called the *gravitational potential energy*.

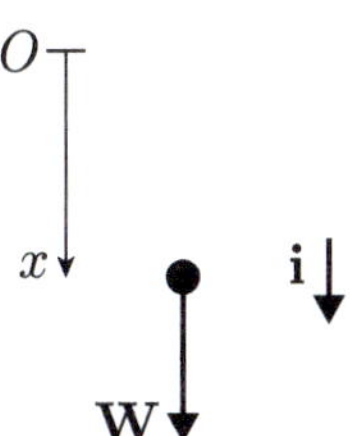

Figure 37 Reversing the direction of the x-axis

Gravitational potential energy

The **gravitational potential energy** of a particle of mass m is given by

$$U = mg \times \text{height above datum}, \qquad (26)$$

where the height is measured vertically from some chosen datum, and g is the magnitude of the acceleration due to gravity.

The result stated for gravitational potential energy applies even when the direction of motion of the particle is not vertical. For example, in Exercise 22 you showed that for a particle moving on a slope at an angle θ to the horizontal, with the x-axis pointing down the slope, the potential energy function is $U(x) = -mgx\sin\theta$. Since the vertical height of the particle above the datum is $-x\sin\theta$, we again have $U(x) = mg \times \text{height}$.

Exercise 23

In equation (16) the buoyancy force was derived to be

$$\mathbf{B} = A\rho g(h - x)\mathbf{i},$$

where x is the height of the top of a pontoon above the water level, h is the height of the pontoon, and $A\rho g$ is a constant.

Find the potential energy function for the buoyancy force, taking the datum to be the level when the pontoon is just out of the water, that is, $x = h$.

Exercise 24

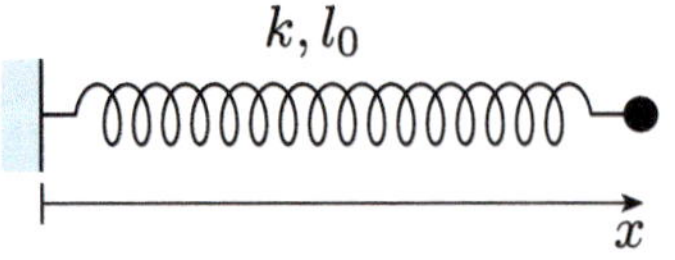

Figure 38

Figure 38 shows a model spring of stiffness k and natural length l_0 whose total length is x.

Verify that the potential energy function for the force exerted by the spring is

$$U(x) = \tfrac{1}{2}k(x - l_0)^2,$$

where the datum is taken to be the point of zero deformation, $x = l_0$.

As before, 'deformation' refers to either extension or compression from the natural length, as appropriate.

For the model spring force in Exercise 24, you showed the following, which we state for future reference.

Potential energy stored in a spring

The potential energy stored in a model spring is

$$U = \tfrac{1}{2} \times \text{stiffness} \times (\text{deformation})^2, \qquad (27)$$

where the datum is chosen to be at the natural length of the spring.

You could use a different datum for the potential energy of a spring, but this is the most natural choice.

We now investigate what happens to the potential energy function when the force acting on a particle is the sum of two forces.

Exercise 25

Find the potential energy function for each of the following forces (where a and b are constants). State, in each case, the datum that you use.

(a) $F(x)\,\mathbf{i} = -ax^2\mathbf{i}$

(b) $F(x)\,\mathbf{i} = bx^{-2}\mathbf{i}$ $(x > 0)$

Exercise 26

(a) Suppose that the potential energy functions $U_1(x)$ and $U_2(x)$ correspond to the forces $F_1(x)\,\mathbf{i}$ and $F_2(x)\,\mathbf{i}$, respectively. Show that a potential energy function corresponding to the force $F(x)\,\mathbf{i} = F_1(x)\,\mathbf{i} + F_2(x)\,\mathbf{i}$ is

$$U(x) = U_1(x) + U_2(x).$$

(b) Using the result of part (a) and your answers to Exercise 25, write down a potential energy function for the force

$$F(x)\,\mathbf{i} = (-ax^2 + bx^{-2})\mathbf{i} \quad (x > 0),$$

where a and b are constants.

Exercise 26 raises the question of what occurs when two potential energy functions that have different datum points are added together. In this case, it is possible to set up a fresh datum at any desired point for the combined potential energy function, by adding an appropriate constant to the expression for $U(x)$, since the potential energy is defined in terms of an indefinite integral (equation (23)). Specifying a datum amounts to pinning down a value for the arbitrary constant in the indefinite integral, but while this may sometimes be convenient, it is not essential to do it. Indeed, it is not necessary for the potential energy to take the value zero at any point.

Note that $U(x) + C$ is also a potential energy function, for any constant C.

4 Energy conservation

In the previous section we concentrated mainly on the relationship between the potential energy function $U(x)$ and the force component $F(x)$. Now we look at how to use energy conservation to solve problems.

The fact that the total mechanical energy E is constant throughout the motion of a particle that is subject to forces that depend only on position means that the particle *does not lose or gain mechanical energy.*

Law of conservation of mechanical energy

If the force on a particle depends only on the particle's position, then the total mechanical energy of the particle is constant. In other words,

$$E = \tfrac{1}{2}mv^2 + U(x), \tag{28}$$

is a constant.

To see the reasonableness of this, consider the familiar example of an object falling vertically downwards under gravity, ignoring air resistance. As the object speeds up (and gains kinetic energy), it loses height (and loses potential energy). On the other hand, if the object is rising (gaining potential energy), then it is also slowing down (losing kinetic energy). So conservation of mechanical energy says that a gain in kinetic energy must be accompanied by a loss of potential energy, and vice versa.

Energy conservation is useful in solving problems that involve only the speed and position of objects, rather than, say, the position of the object at specific times. This is illustrated in the following example and exercise.

Example 7

A stone of mass m is thrown vertically upwards from ground level with initial speed $15\,\mathrm{m\,s^{-1}}$. Assume that gravity is the only force acting on the stone.

(a) Find the speed of the stone when its height is $5\,\mathrm{m}$.

(b) Find the maximum height of the stone.

Solution

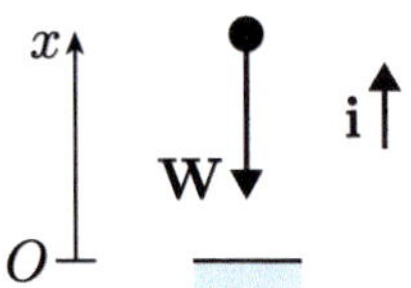

Figure 39 Stone falling under gravity

(a) Choose an x-axis pointing upwards, with the origin (and datum) at the point of projection (see Figure 39). The only force acting on the stone is the weight due to gravity, which is constant, so equation (25) applies. Using equation (26), the potential energy is mgx, thus the total energy is given by

$$E = \tfrac{1}{2}mv^2 + mgx.$$

Initially, $v = 15$ when $x = 0$, so

$$E = \tfrac{1}{2}m \times 15^2 + m \times 9.81 \times 0 = 112.5m.$$

Hence throughout the motion, since energy is conserved, we have

$$\tfrac{1}{2}v^2 + gx = 112.5. \tag{29}$$

Notice that the m terms have cancelled, so this equation applies to a stone of any mass; all stones will follow the same trajectory.

When $x = 5$, this gives

$$\tfrac{1}{2}v^2 + 9.81 \times 5 = 112.5,$$

hence

$$v^2 = 2 \times (112.5 - 9.81 \times 5) = 126.9.$$

So the speed of the stone at height $x = 5$ is

$$|v| = \sqrt{126.9} \simeq 11.3 \quad \text{to 3 s.f.}$$

This speed of about $11.3\,\text{m}\,\text{s}^{-1}$ is the same whether the stone is moving up or down. The sign of the velocity will depend on the direction of motion.

(b) At maximum height, we have $v = 0$. Substituting $v = 0$ into equation (29) produces

$$gx = 112.5$$

or

$$x = 112.5/9.81 \simeq 11.5 \quad \text{to 3 s.f.},$$

so the maximum height attained by the stone is approximately 11.5 m.

Exercise 27

(a) A marble, initially at rest, is dropped from the Clifton Suspension Bridge and falls into the River Avon, 77 m below (see Figure 40). Assuming that the only force acting on the marble is the force of gravity, use energy conservation to estimate the speed of the marble just before it hits the water. How far has it fallen when its speed reaches $20\,\text{m}\,\text{s}^{-1}$?

(Choose a downward-pointing x-axis with origin at the point of release, and take this origin as the datum of potential energy. Take care with the sign of the potential energy function.)

(b) Why can the law of conservation of mechanical energy not be used if air resistance is taken into account in the situation of part (a)?

You considered this situation in Unit 3.

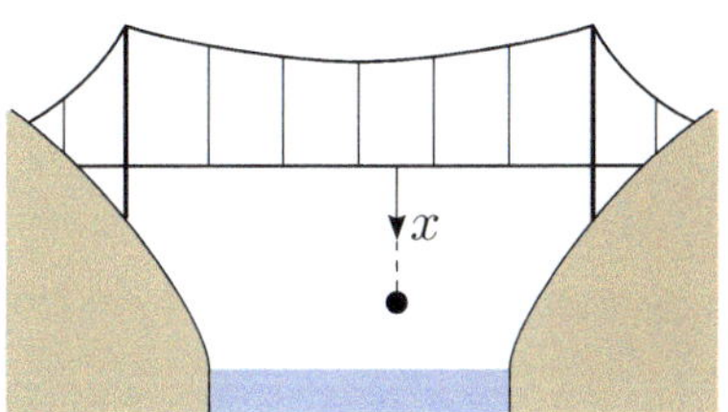

Figure 40 A marble dropped from the Clifton Suspension Bridge

---

Models involving friction or air resistance do not exhibit conservation of mechanical energy. In fact, energy is conserved, but the mechanical energy lost by a particle is converted to another form of energy such as heat or sound.

4.1 General method for solving problems using energy principles

Example 7 is typical of many problems that can be solved by using the law of conservation of mechanical energy. There are two reasons why this problem is particularly suitable. First, it involves a force, namely gravity, that depends only on position (in this case, of course, the force is a constant function of position). This is crucial, because the law of conservation of mechanical energy applies only to such forces. Second, the problem asks for a relationship between position and speed, which is ideal because the conservation law immediately gives the desired relationship.

Procedure 2 Applying conservation of energy

This procedure may be applied to a mechanics problem involving the one-dimensional motion of a particle in which:

- the total force depends only on the position of the particle (or is constant)
- the question to be answered refers to a relationship between the position and the speed of the particle.

Carry out the following steps.

◀ Draw picture ▶

1. Drawing a picture is always the first step in solving a mechanics problem.

◀ State assumptions ▶

2. State any assumptions used, such as ignoring friction or air resistance.

◀ Choose datum ▶

3. The next step is to choose a datum. For gravitational potential energy it is usual to choose the origin. For a spring, the datum is usually chosen to be the point of zero deformation. (For other forces this step could be deferred until the potential energy function is known; choose the datum to simplify the expression.)

◀ Find potential energy ▶

4. If gravity or model springs are present, simply add together the corresponding potential energy functions as given in equations (26) and (27). Otherwise, identify the x-component $F(x)$ of the total force acting on the particle, then apply the potential energy function definition in equation (23).

◀ Find E ▶

5. Use equation (28) and the initial conditions to calculate the value of the constant E, the total mechanical energy of the system.

◀ Find x or v ▶

6. Solve the resulting version of equation (28), either for x (at a specified value of v) or for v (at a specified value of x).

◀ Interpret solution ▶

7. Interpret the solution in terms of the original problem.

The steps outlined in Procedure 2 are used in the following example.

Example 8

A toy manufacturer is testing a trampoline and wants to estimate the largest mass of child that can safely use it. An experiment is devised in which a sandbag of mass m is dropped from a height h, and the maximum deformation d of the trampoline is measured.

(a) Model the trampoline as a model spring of stiffness k. Find an equation relating the given parameters and g. You may assume that no energy is lost in the impact.

(b) It is observed that when a 20 kg sandbag is dropped from a height of 1 m above the trampoline, the maximum deformation of the trampoline is 25 cm. Use these data to calculate the stiffness of the model spring.

A better approach would be to collect more data so that a better estimate could be made and the model could be tested.

(c) If the maximum safe deformation is 30 cm, estimate the maximum mass that may be dropped on the trampoline from a height of 1 m.

Solution

(a) Figure 41 shows pictures of the sandbag at the top and bottom of its motion. As shown, we choose an x-axis pointing upwards, with origin at the height of the undeformed trampoline. ◀ Draw picture ▶

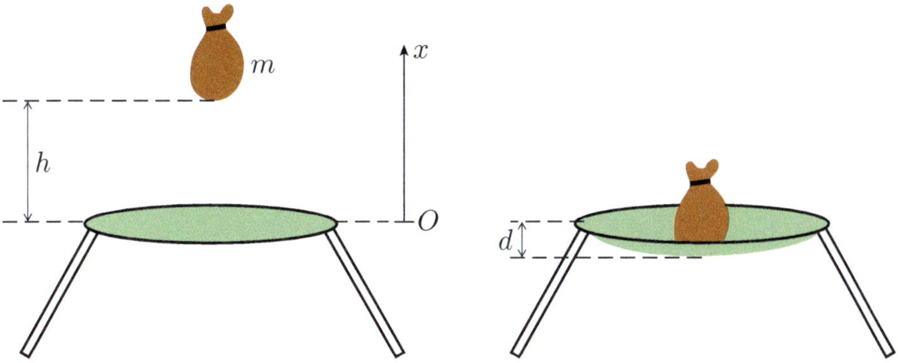

Figure 41 A sandbag being dropped onto a trampoline

We model the sandbag as a particle, which is subject only to the gravitational force in flight. We model the trampoline as a model spring. ◀ State assumptions ▶

Take the origin as the datum for the gravitational potential energy. Then the potential energy function due to the weight is ◀ Choose datum ▶ ◀ Find potential energy ▶

$$U_{\text{grav}}(x) = mgx.$$

Take the origin as the datum for the spring potential energy function (as this corresponds to the point of zero deformation of the spring). The trampoline exerts a force only when the sandbag is in contact with it, so the potential energy is stored in the spring only if $x < 0$. So the potential energy stored in the spring is

$$U_{\text{spring}}(x) = \begin{cases} 0, & x \geq 0, \\ \frac{1}{2}kx^2, & x < 0. \end{cases}$$

To find the total potential energy, we add together the two contributions:

$$U(x) = U_{\text{spring}}(x) + U_{\text{grav}}(x) = \begin{cases} mgx, & x \geq 0, \\ \frac{1}{2}kx^2 + mgx, & x < 0. \end{cases}$$

Initially, the sandbag is at rest ($v = 0$) at a height $x = h > 0$ above the trampoline, so using equation (28) gives ◀ Find E ▶

$$E = \tfrac{1}{2}m0^2 + mgh = mgh.$$

At the bottom of the motion, the sandbag is again at rest, at the point $x = -d$ (where $d > 0$ is the maximum deformation of the trampoline). At this point we have ◀ Find x or v ▶

$$E = \tfrac{1}{2}m0^2 + U(-d) = \tfrac{1}{2}kd^2 - mgd.$$

This gives the desired relationship

$$mgh = \tfrac{1}{2}kd^2 - mgd. \qquad (30)$$

◀ Interpret solution ▶

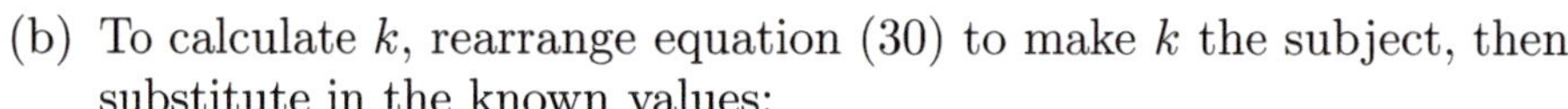

(b) To calculate k, rearrange equation (30) to make k the subject, then substitute in the known values:

$$k = \frac{2mg(h+d)}{d^2} = \frac{2 \times 20 \times 9.81 \times (1 + 0.25)}{0.25^2} = 7848.$$

So the spring has stiffness approximately $8000\,\mathrm{N\,m^{-1}}$.

(c) To calculate the mass that corresponds to the maximum deformation of the trampoline, rearrange equation (30) to make m the subject, then substitute in the known values:

$$m = \frac{kd^2}{2g(h+d)} = \frac{7848 \times 0.3^2}{2 \times 9.81 \times (1+0.3)} \simeq 27.7 \quad \text{to 3 s.f.}$$

So the maximum safe mass is approximately 27 kg.

Note that here we round *down* rather than to the nearest integer (because 28 kg would be *above* the safe maximum mass).

Now try applying the law of conservation of energy yourself by attempting the following exercises.

Exercise 28

A particle of mass m moves without friction down a slope inclined at an angle θ to the horizontal (see Figure 42). The particle starts from rest at the point O, and slides down to the point A, which is a vertical distance h below O. The displacement of the particle from O, measured down the slope, is denoted by x. The point A is to be taken as the datum for the potential energy function.

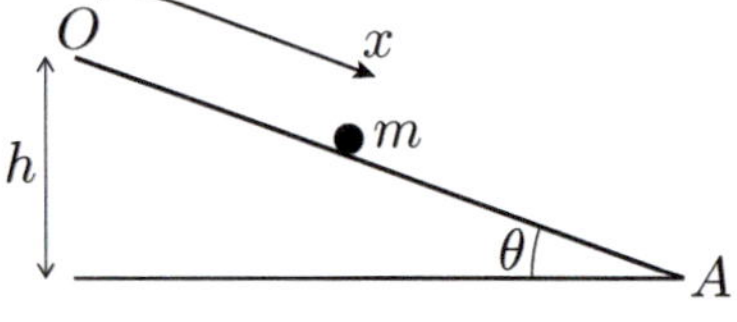

Figure 42 A mass moving down a slope

(a) Show that the total mechanical energy of the particle can be written as

$$E = \tfrac{1}{2}mv^2 + mg(h - x\sin\theta),$$

where v is the speed of the particle.

(b) Obtain, in terms of h, an expression for the speed of the particle at A, and comment on your result.

(c) Apply the result from part (b) to the specific case of a crate of empty bottles that slides 2 m down a smooth ramp set at an angle of $\pi/6$ radians to the horizontal, starting from rest.

This problem was considered in Example 5 of Unit 3.

Exercise 29

A particle of mass 0.5 kg moves along a horizontal straight frictionless track. The particle is attached to an end of a model spring of stiffness $2\,\mathrm{N\,m^{-1}}$, the other end of the spring being fixed (see Figure 43).

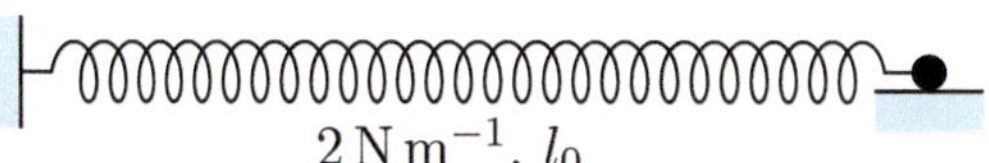

Figure 43

Initially, the extension of the spring is 2 m and the particle's speed is $3\,\mathrm{m\,s^{-1}}$, directed towards the centre of the spring. What is the particle's maximum speed during the subsequent motion, and at what positions is it momentarily at rest? (Assume that the natural length l_0 of the model spring is greater than 2.5 m.)

Exercise 30

A particle of mass m is attached to the lower end of a model spring of natural length l_0 and stiffness $k = 10mg/l_0$, where g is the magnitude of the acceleration due to gravity. The spring is hung vertically, with its upper end fixed (see Figure 44).

The particle is pulled vertically downwards until the spring's length is $\frac{5}{4}l_0$ and then released from rest. By using the law of conservation of energy, find the speed of the particle when the spring is of length l_0 in the subsequent motion.

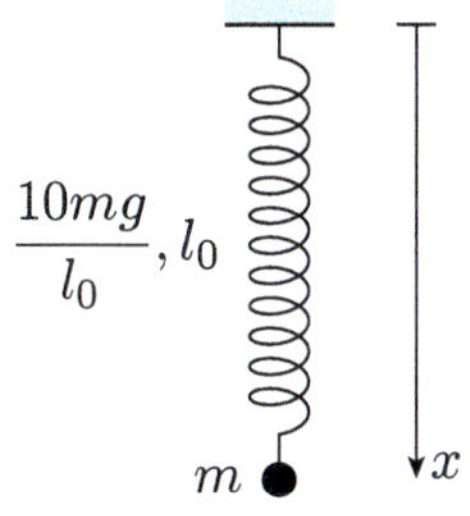

Figure 44

4.2 Energy in oscillating systems

As you have seen in the above example and exercises, the law of conservation of mechanical energy,

$$E = \tfrac{1}{2}mv^2 + U(x),$$

provides a convenient method for finding the relationship between the velocity $v\mathbf{i}$ and the position $x\mathbf{i}$ whenever the force acting on a particle is a function of position. Since the term $\frac{1}{2}mv^2$ can never be negative, the same must be true of the quantity $E - U(x)$, which means that motion is possible only in regions for which $E - U(x) \geq 0$. Also, the particle will be momentarily at rest ($v = 0$) at points for which $E - U(x) = 0$. At such points, called **turning points**, the particle changes its direction of motion; this is illustrated in Figure 45. A particle moving under the action of a force that gives rise to the illustrated potential energy function $U(x)$, and with total mechanical energy E, will oscillate backwards and forwards between the two turning points A and B, changing its direction of motion each time it reaches one of these points.

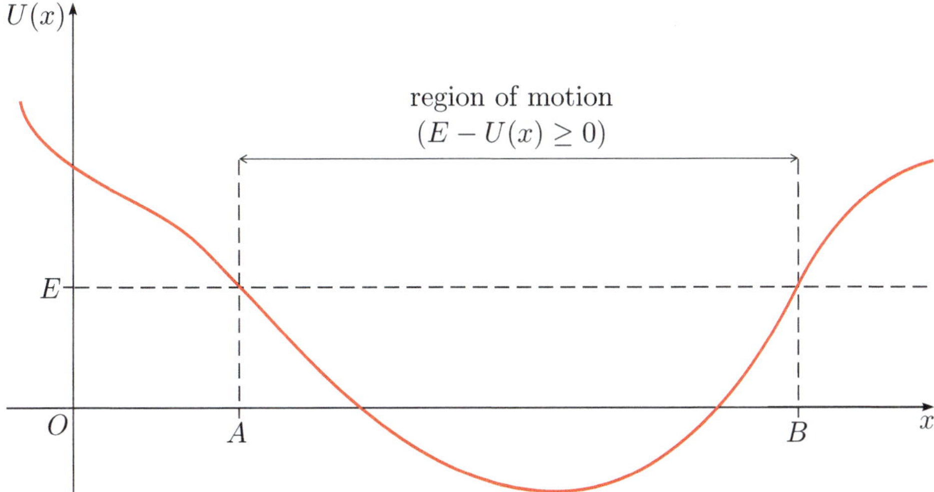

Figure 45 Turning points

Exercise 31

Throughout this question, all quantities are measured in the appropriate SI units.

A particle of mass $m = 2$ moves under the influence of a force that has x-component $F(x) = 2 - 2x$.

(a) Find a potential energy function $U(x)$ for this force.

(b) If the particle is released from rest at $x = -1$, find the total mechanical energy of the particle in the subsequent motion.

(c) Use the law of conservation of mechanical energy to find an expression for the velocity $v\mathbf{i}$ of the particle at position $x\mathbf{i}$.

(d) Find the region of motion of the particle, and its speed at the midpoint of this region.

We now show how energy methods can be used to find solutions for oscillating systems. The process is the reverse of that used in Section 3 to introduce the concept of energy. This may seem like a circular argument, but it is sometimes easier to solve problems with oscillations using energy methods.

Start with the equation for total energy,

$$E = \tfrac{1}{2}mv^2 + U(x).$$

Differentiate this with respect to t to obtain

$$\frac{dE}{dt} = \tfrac{1}{2}m\,\frac{d}{dt}(v^2) + \frac{d}{dt}(U(x)).$$

Use the chain rule on the derivatives on the right-hand side:

$$\frac{dE}{dt} = \tfrac{1}{2}m\,\frac{d}{dv}(v^2)\,\frac{dv}{dt} + \frac{d}{dx}(U(x))\,\frac{dx}{dt} = \tfrac{1}{2}m2v\,\frac{dv}{dt} + U'(x)\,v.$$

If the system is such that energy is conserved (no dissipative forces), then E is a constant, so $dE/dt = 0$. Divide by v to obtain

We may assume that $v \neq 0$ in general.

$$0 = m\,\frac{dv}{dt} + U'(x)$$

or

$$0 = m\ddot{x} + U'(x). \tag{31}$$

One consequence of equation (31) is that it makes it easy to find positions of equilibrium. (The system is in equilibrium if the particle remains at rest.) At equilibrium $\ddot{x} = 0$, so the solution of $U'(x) = 0$ will give the values of x at which a system is in equilibrium.

Exercise 32

Find the positions of equilibrium for the potential energy function $U(x) = x^3 - 3x$.

The plot of the potential energy function in Figure 46 shows that the positions of equilibrium found in Exercise 32 occur at a local maximum or local minimum of $U(x)$. The nature of the equilibrium at each of these points is different.

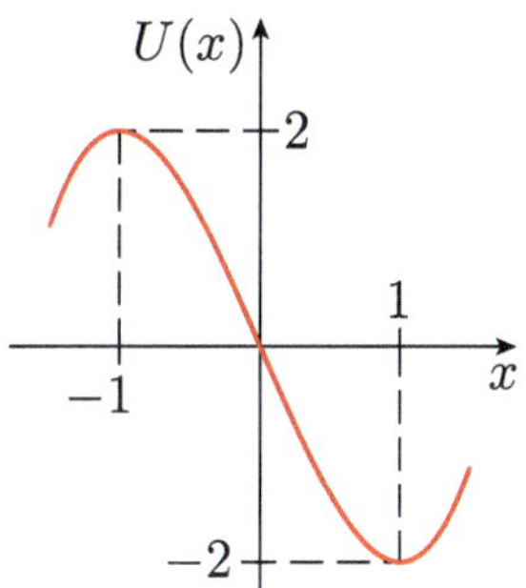

Figure 46 Positions of equilibrium

Consider an equilibrium point x_{eq} of $U(x)$ so that $U'(x_{\text{eq}}) = 0$. Expand $U(x)$ as a Taylor series about x_{eq} to give

$$U(x) = U(x_{\text{eq}}) + U'(x_{\text{eq}})(x - x_{\text{eq}}) + \frac{1}{2!}U''(x_{\text{eq}})(x - x_{\text{eq}})^2 + \frac{1}{3!}U'''(x_{\text{eq}})(x - x_{\text{eq}})^3 + \cdots .$$

Since $U'(x_{\text{eq}}) = 0$, the Taylor series simplifies to

$$U(x) = U(x_{\text{eq}}) + \frac{1}{2!}U''(x_{\text{eq}})(x - x_{\text{eq}})^2 + \frac{1}{3!}U'''(x_{\text{eq}})(x - x_{\text{eq}})^3 + \cdots .$$

Now we make the approximation that $x - x_{\text{eq}}$ is small, so the second term of the above series will dominate the remainder if $U''(x_{\text{eq}}) \neq 0$; if $U''(x_{\text{eq}}) = 0$, then the quadratic term disappears and the following discussion does not hold.

The case where $U''(x_{\text{eq}}) = 0$ is not considered further.

So the potential energy is approximately a quadratic:

$$U(x) \simeq U(x_{\text{eq}}) + \tfrac{1}{2}U''(x_{\text{eq}})\,(x - x_{\text{eq}})^2.$$

Since we require $U'(x)$ in equation (31), we differentiate to obtain

$$U'(x) \simeq U''(x_{\text{eq}})\,(x - x_{\text{eq}}).$$

From the above arguments it is sensible to construct a model of the behaviour of the small oscillations where the potential energy is exactly quadratic, and then

$$U'(x) = U''(x_{\text{eq}})\,(x - x_{\text{eq}}).$$

Using this in equation (31) gives

$$0 = m\ddot{x} + U''(x_{\text{eq}})\,(x - x_{\text{eq}}),$$

which on rearrangement becomes

$$m\ddot{x} + U''(x_{\text{eq}})\,x = U''(x_{\text{eq}})\,x_{\text{eq}}.$$

If $U''(x_{\text{eq}}) > 0$, then the motion is simple harmonic around x_{eq}, so the motion is simple harmonic with angular frequency $\omega = \sqrt{U''(x_{\text{eq}})/m}$. You will appreciate from this general argument that this kind of motion occurs frequently. Hence we give a mechanical system of this kind a name: it is called a **harmonic oscillator**.

Note that for the angular frequency to be real, we must have $U''(x_{\text{eq}}) > 0$, which *is* the case if $U(x)$ has a local minimum at x_{eq}.

If $U''(x_{\text{eq}}) < 0$, then the solution is

$$x = Ae^{\lambda t} + Be^{-\lambda t} + x_{\text{eq}}$$

where A and B are arbitrary constants, and $\lambda = \sqrt{U''(x_{\text{eq}})/m}$. Unless $A = 0$, which is extremely unlikely in practice, x will increase exponentially.

An example of the two types of equilibrium is provided by a ball in a concave bowl, which will oscillate about the bottom point, and a ball on top of a convex bowl, where any slight departure from equilibrium leads to catastrophic departure from equilibrium.

Equilibrium for a potential function

For a system with potential energy function $U(x)$, positions of equilibrium x_{eq} occur whenever $U'(x_{\text{eq}}) = 0$.

If $U''(x_{\text{eq}}) > 0$, then for small oscillations the motion is simple harmonic with angular frequency $\omega = \sqrt{U''(x_{\text{eq}})/m}$.

If $U''(x_{\text{eq}}) < 0$, then the equilibrium point is unstable.

Of course, when $\mathbf{F}$ is a non-zero constant force, $U'(x) = -F(x)$ is never zero, and there are no equilibrium points.

For the potential energy function defined in Exercise 32, the equilibrium point at $x = 1$ (a minimum) will have oscillations around it, and the one at $x = -1$ (a maximum) will not. For this case $U''(-1) = -6$ and the equation of motion is $m\ddot{x} - 6x = -6x_{\text{eq}}$, for which the general solution is $x = Be^{\sqrt{6/m}} + Ce^{-\sqrt{6/m}} + x_{\text{eq}}$, showing that the point is unstable.

Exercise 33

All quantities here are measured in the appropriate SI units.

A particle of mass $m = 3$ moves along a straight line and experiences a force that repels it from the origin at position $x = 0$. This force has x-component $F(x) = 2/x^2$ for $x > 0$. Initially, the particle is moving towards the origin, being at position $x\mathbf{i} = 10\mathbf{i}$ with velocity $v\mathbf{i} = -\mathbf{i}$.

(a) Find a potential energy function for this force for $x > 0$, stating the datum that you use.

(b) Find the total mechanical energy of the particle.

(c) What is the closest point to the origin that is reached by the particle?

(d) Sketch the graph of the potential energy function, and indicate the region of motion. Describe the motion of the particle.

(e) Use your graph to determine whether there are any equilibrium points for this potential energy function.

Exercise 34

This is not a realistic model, since the truck would bounce off such buffers.

The system in Figure 47 can be used to model the motion of a railway truck while in contact with buffers. The mass of the truck is 2000 kg, and the stiffness of the (model) buffer spring is $10^5\,\mathrm{N\,m^{-1}}$. If the truck comes to rest at a position where the spring is compressed by 0.1 m, find the speed of the truck when it first came into contact with the buffers. (You should neglect friction, and assume that the spring has its natural length when the truck first touches the buffers.)

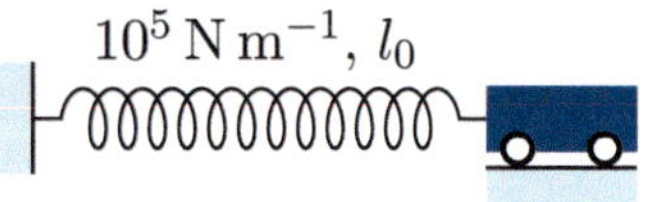

Figure 47

Exercise 35

This question analyses the same system of springs and a mass that was used in Exercise 18, but now using energy methods.

A block P of mass m is attached to three springs whose other ends are attached to fixed points Q, R and S. The point R is a distance $\frac{1}{2}l_0$ below Q, and the point S is a distance $4l_0$ below Q. The parameters of the three springs are given in Figure 48, which also illustrates the arrangement of the springs and the block.

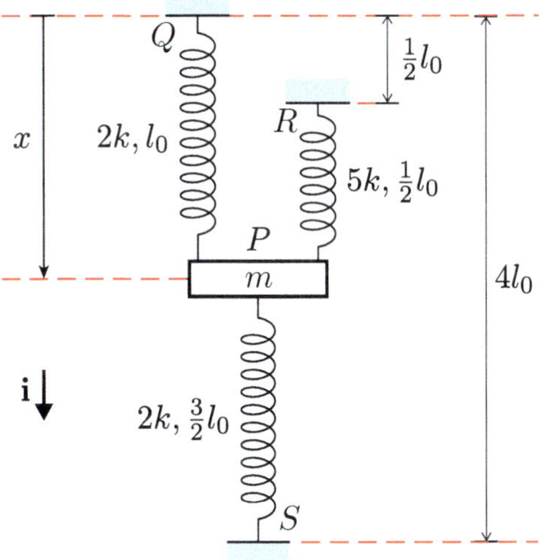

Figure 48

Model the block as a particle and the springs as model springs. Take the origin at Q, with the vertical distance of P from Q being x, so that the x-axis is as shown, and $\mathbf{i}$ is pointing downwards.

(a) Choose a datum and write down the gravitational potential energy of particle P at a general point of its motion.

(b) Write down the kinetic energy of particle P at a general point of its motion.

(c) Determine the potential energy stored in each spring at a general point of its motion.

(d) Write down an equation representing the total mechanical energy for the system at a general point of its motion. Use equation (31) to verify that your answer is equivalent to the equation of motion derived in Exercise 18(c).

Exercise 36

This question concerns a system with potential energy function given by

$$U(x) = \frac{x^4}{4} - \frac{x^3}{3} - x^2$$

whose graph is sketched in Figure 49.

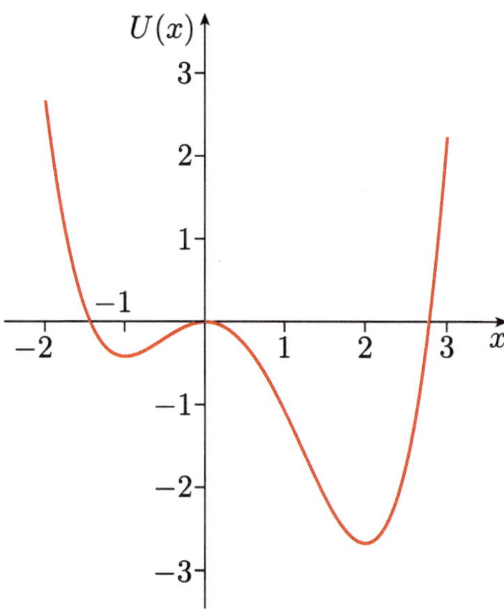

Figure 49

(a) Write down, in the form $F(x)\,\mathbf{i}$, the force that gives rise to this potential function.

(b) The total energy of the system is a constant E. For each of the following values of E, state, with a reason, a range (or ranges) accurate to one decimal place of x-values (if any) that could represent a motion of the system.

- $E = -3$
- $E = -2$
- $E = -0.25$
- $E = 1$

(c) Find the equilibrium points for this potential energy function, and state when motion around the equilibrium point will be oscillatory.

Learning outcomes

After studying this unit, you should be able to:

- apply the force law (Hooke's law) for a model spring in statics and dynamics problems
- find the equilibrium position of a particle that is acted on by model springs and gravity
- derive an equation of motion for a system involving model springs and gravity
- understand the basic features of simple harmonic motion, and use the terminology associated with it
- find the amplitude, phase angle, angular frequency and period of a system undergoing simple harmonic motion
- understand and explain the concepts of kinetic energy, potential energy and total mechanical energy
- find the kinetic energy of a particle
- find the gravitational potential energy of a particle
- find the energy stored in a spring
- apply, in appropriate circumstances, the law of conservation of mechanical energy to solve simple problems in mechanics
- identify the turning points and region of motion for a particle with given potential energy and total mechanical energy.

Solutions to exercises

Solution to Exercise 1

(a) If $d = 0.0025$, then equation (1) gives

$$w(d) = -0.024 + 99.044 \times 0.0025 = 0.224 \quad \text{to 3 d.p.}$$

The magnitude of the weight on the ruler is approximately 0.22 N when the displacement is 2.5 mm.

(b) If $w(d) = 0.67$, then equation (1) gives

$$0.67 = -0.024 + 99.044 \times d.$$

Solving for d, we obtain

$$d = \frac{0.67 + 0.024}{99.044} \simeq 0.007 \quad \text{to 3 d.p.,}$$

so the displacement is 7 mm.

(c) If $w(d) = 60$, then equation (1) gives

$$60 = -0.024 + 99.044 \times d.$$

Solving for d, we obtain

$$d = \frac{60 + 0.024}{99.044} \simeq 0.606 \quad \text{to 3 d.p.,}$$

so the mathematical model predicts a displacement of 0.61 m to two decimal places. (In reality this is impossible, since the ruler itself is only 0.3 m long. Any solution of the mathematical model must be reviewed in the light of its validity to represent the actual situation.)

(d) If $d = -0.0025$, then equation (1) gives

$$w(d) = -0.024 + 99.044 \times (-0.0025) = -0.272 \quad \text{to 3 d.p.}$$

The magnitude of the weight on the ruler is about 0.27 N when the displacement is −2.5 mm; however, the negative sign indicates that the force applied to the ruler is upwards. In this particular case the ruler can be bent upwards as well as downwards, so it is practically possible; in other situations it may not be possible to have a negative displacement.

Solution to Exercise 2

When $l = 0.35$, the spring is extended by an amount $e = 0.05$, so the magnitude of the spring force is $0.05 \times 200 = 10$ (newtons). Since the spring is in tension, the force on the object at the end is directed towards the centre of the spring.

When $l = 0.2$, the spring is compressed by an amount $c = 0.1$, so the magnitude of the spring force is $0.1 \times 200 = 20$ (newtons). Since the spring is in compression, the force on the object at the end is directed away from the centre of the spring.

Solution to Exercise 3

In each case we have $k = 200$ and $l_0 = 0.3$. Take $\widehat{\mathbf{s}}$ to be a unit vector in the direction from the object towards the centre of the spring.

With $l = 0.35$, the force acting on the object is

$$\mathbf{H} = 200(0.35 - 0.3)\,\widehat{\mathbf{s}} = 10\,\widehat{\mathbf{s}},$$

with magnitude 10 N and direction given by $\widehat{\mathbf{s}}$, that is, towards the centre of the spring.

When $l = 0.2$, the force acting on the object is

$$\mathbf{H} = 200(0.2 - 0.3)\,\widehat{\mathbf{s}} = -20\,\widehat{\mathbf{s}} = 20(-\widehat{\mathbf{s}}),$$

with magnitude 20 N and direction given by $-\widehat{\mathbf{s}}$, that is, away from the centre of the spring.

These results agree with what we found in Exercise 2.

Solution to Exercise 4

Let the length of each spring be l_1 when there is no baby in the baby bouncer. Assume that the two springs may be modelled satisfactorily by model springs, each with the same stiffness k and natural length l_0. Choose the $\mathbf{i}$-direction to be vertically downwards. Thus $\widehat{\mathbf{s}}$ for this problem is $-\mathbf{i}$ for both springs. The situation and force diagram are shown below.

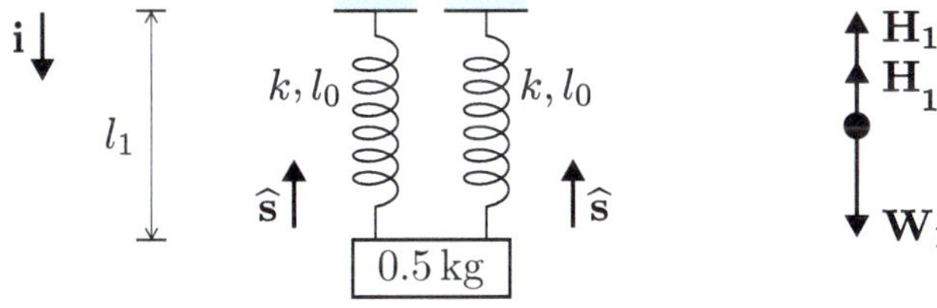

Each spring exerts a force

$$\mathbf{H}_1 = k(l_1 - l_0)\,\widehat{\mathbf{s}} = k(l_1 - l_0)(-\mathbf{i}).$$

Since the two springs are identical, and they are both vertical, the spring force in each can be represented by the same vector.

The weight of the seat plus straps is $\mathbf{W}_1 = 0.5g\mathbf{i}$.

The system is in equilibrium, so

$$\mathbf{W}_1 + 2\mathbf{H}_1 = \mathbf{0}.$$

Resolving in the $\mathbf{i}$-direction gives

$$2k(l_1 - l_0) - 0.5g = 0.$$

With the baby strapped into the seat, the length of each spring becomes $l_1 + 0.03$ (in metres). The new situation and force diagram are shown below.

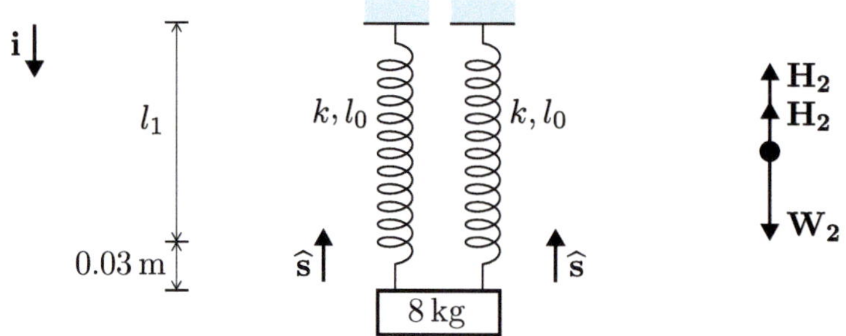

Now each spring exerts a force

$$\mathbf{H}_2 = k(l_1 + 0.03 - l_0)(-\mathbf{i}),$$

while the total weight of the baby, seat and straps is $\mathbf{W}_2 = 8g\mathbf{i}$.

The system is again in equilibrium, so

$$\mathbf{W}_2 + 2\mathbf{H}_2 = \mathbf{0}.$$

Resolving in the **i**-direction gives

$$2k(l_1 + 0.03 - l_0) - 8g = 0.$$

We have $2k(l_1 - l_0) = 0.5g$ from the earlier calculation, and subtracting this from the equation above gives $0.06k = 7.5g$. Hence $k = 125g \simeq 1200$, and the stiffness of each spring is $1200\,\mathrm{N\,m^{-1}}$.

Solution to Exercise 5

As before, the object is modelled as a particle. In Example 2, the distance between the fixed points was greater than the sum of the natural lengths of the two springs, hence in equilibrium each spring was extended. Here the distance between the fixed points is less than the sum of the natural lengths of the two springs, hence in equilibrium each spring is compressed. However, this does not affect the approach to the solution. (This shows the usefulness of the vector formulation for the force in a spring, be it in extension or compression.)

◀ Choose axes ▶

◀ Draw force diagram(s) ▶

We again choose the x-axis to point from left to right, with origin at the left-hand fixed point. The unit vectors will be the same as before, as shown in the diagram below.

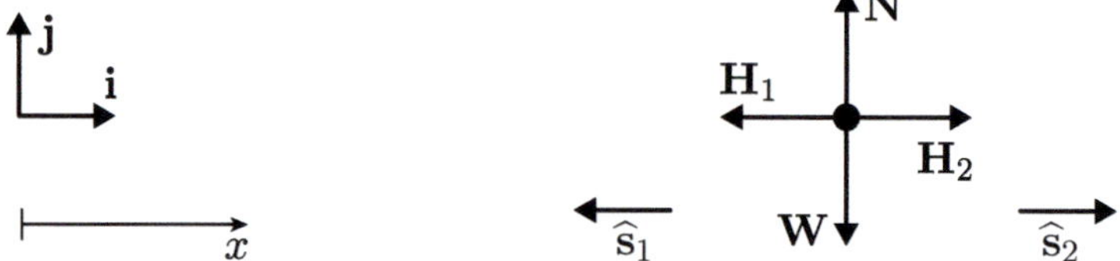

Strictly speaking, as we know that both springs will be in compression, we should have $\mathbf{H}_1$ and $\mathbf{H}_2$ in the opposite directions. We address this issue at the end of the solution.

The lengths of the two springs are x (left spring) and $0.3 - x$ (right spring), and this will alter the spring forces. As can be seen from the diagram above, everything else is the same as in the solution to Example 2.

◀ Apply law(s) ▶

Spring	l	l_0	$l - l_0$	k	$\widehat{\mathbf{s}}$	$\mathbf{H}$
Left	x	0.3	$x - 0.3$	40	$-\mathbf{i}$	$40(x - 0.3)(-\mathbf{i})$
Right	$0.3 - x$	0.2	$0.1 - x$	60	$\mathbf{i}$	$60(0.1 - x)\mathbf{i}$

Thus we have

$$\mathbf{H}_1 = 40(x - 0.3)(-\mathbf{i}) = 40(0.3 - x)\mathbf{i},$$
$$\mathbf{H}_2 = 60(0.1 - x)\,\mathbf{i}.$$

Since the particle is held in equilibrium by these two forces plus the weight $\mathbf{W}$ of the particle and the normal reaction $\mathbf{N}$, we have

$$\begin{aligned}\mathbf{W} + \mathbf{N} + \mathbf{H}_1 + \mathbf{H}_2 &= mg(-\mathbf{j}) + |\mathbf{N}|\,\mathbf{j} + 40(0.3 - x)\mathbf{i} + 60(0.1 - x)\mathbf{i} \\ &= \mathbf{0}.\end{aligned}$$

Resolving in the $\mathbf{i}$-direction gives

◀ Solve equation(s) ▶

$$60(0.1 - x) + 40(0.3 - x) = 0.$$

The solution of this equation is $x = 0.18$, so the particle is 0.18 m from the left-hand fixed point (and hence 0.12 m from the right-hand fixed point). Using this value of x in the equation for $\mathbf{H}_1$ gives

$$\mathbf{H}_1 = 40(0.3 - 0.18)\mathbf{i} = 4.8\mathbf{i},$$

that is, the left-hand spring force has a magnitude of 4.8 N that acts to the right, so it is pushing the particle away – it is a compressive force. The spring force in the right-hand spring will be compressive as well, pushing the particle to the left. Also, as expected,

$$\mathbf{H}_2 = 60(0.1 - 0.18)\mathbf{i} = -4.8\mathbf{i}.$$

Note that the spring force $\mathbf{H}_1$ was shown as a tension, in the direction to the left, in the force diagram; but this analysis has shown it to be compressive and acting to the right. However, this does not affect the analysis.

Solution to Exercise 6

Model the object as a particle. The force diagram is shown below, where $\mathbf{W}$ is the weight of the particle, $\mathbf{H}_1$ is the spring force in the upper spring, and $\mathbf{H}_2$ is the spring force in the lower spring.

◀ Draw force diagram(s) ▶

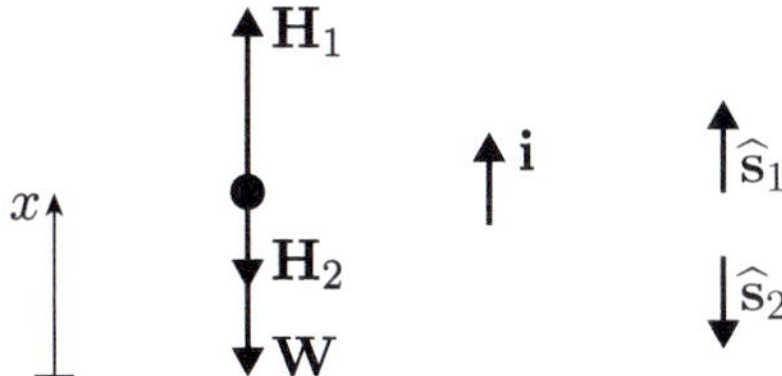

◀ Choose axes ▶

Choose the x-axis to point vertically upwards, with origin at A, so $\mathbf{i}$ is also upwards. Then the weight of the particle is

$$\mathbf{W} = 1.5g(-\mathbf{i}).$$

◀ Apply law(s) ▶

The springs have lengths x (lower spring) and $2.1 - x$ (upper spring), and the forces exerted by them are shown in the table below.

Spring	l	l_0	$l - l_0$	k	$\widehat{\mathbf{s}}$	$\mathbf{H}$
Upper	$2.1 - x$	0.3	$1.8 - x$	20	$\mathbf{i}$	$20(1.8 - x)\mathbf{i}$
Lower	x	0.5	$x - 0.5$	50	$-\mathbf{i}$	$50(x - 0.5)(-\mathbf{i})$

So

$$\mathbf{H}_1 = 20(1.8 - x)\mathbf{i},$$
$$\mathbf{H}_2 = 50(x - 0.5)(-\mathbf{i}) = 50(0.5 - x)\mathbf{i}.$$

In equilibrium,

$$\mathbf{H}_1 + \mathbf{H}_2 + \mathbf{W} = \mathbf{0}.$$

◀ Solve equation(s) ▶

Resolving in the $\mathbf{i}$-direction gives

$$20(1.8 - x) + 50(0.5 - x) - 1.5g = 0,$$

so $70x = 61 - 1.5g$. The solution of this equation is $x \simeq 0.66$.

◀ Interpret solution ▶

So the object is about 0.66 m above the floor.

You may like to reformulate the solution to the problem using the length y measured downwards from point B (and $\mathbf{i}$ now pointing downwards as well). Determine the value of y and hence confirm the value for x given above.

Solution to Exercise 7

Once the battery has been placed in its holder, the sum of the spring lengths is $(5.5 - 5) = 0.5$ cm, which in SI units is 0.005 m.

Let $\mathbf{W}$ be the weight of the battery, let $\mathbf{N}$ be the normal reaction from the battery holder acting on the battery, let $\mathbf{H}_1$ be the spring force in the left-hand spring, and let $\mathbf{H}_2$ be the spring force in the right-hand spring.

◀ Choose axes ▶

Take the x-axis, and the $\mathbf{i}$-direction, to be from left to right, with the origin for x at the left-hand end of the battery holder; x is used as the distance from the left-hand end of the battery holder to the left-hand end of the battery.

◀ Draw force diagram(s) ▶

The axis, unit vectors and force diagram are shown below.

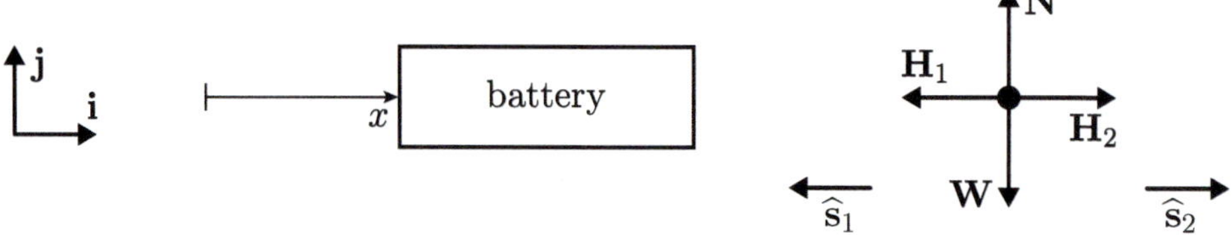

The spring at the left-hand side of the battery has length x. The length of the right-hand spring is $0.055 - 0.05 - x$ (allowing for the length of the battery). From the battery to the centre of the left-hand spring is from right to left, so $\widehat{\mathbf{s}}_1 = -\mathbf{i}$ (for the left-hand spring); we look in the opposite direction for the right-hand spring, so $\widehat{\mathbf{s}}_2 = \mathbf{i}$.

The spring forces are derived in the table below.

◀ Apply law(s) ▶

Spring	l	l_0	$l - l_0$	k	$\widehat{\mathbf{s}}$	$\mathbf{H}$
Left	x	0.004	$x - 0.004$	30	$-\mathbf{i}$	$30(x - 0.004)(-\mathbf{i})$
Right	$0.005 - x$	0.003	$0.002 - x$	10	$\mathbf{i}$	$10(0.002 - x)\,\mathbf{i}$

So

$$\mathbf{H}_1 = 30(x - 0.004)(-\mathbf{i}) = 30(0.004 - x)\mathbf{i},$$
$$\mathbf{H}_2 = 10(0.002 - x)\mathbf{i}.$$

Since the battery is held in equilibrium by these two forces plus the weight $\mathbf{W} = mg(-\mathbf{j})$ of the battery and the normal reaction $\mathbf{N} = |\mathbf{N}|\,\mathbf{j}$ on it (both acting vertically), we have

$$\mathbf{W} + \mathbf{N} + \mathbf{H}_1 + \mathbf{H}_2 = \mathbf{0}.$$

Resolving in the (horizontal) **i**-direction gives

◀ Solve equation(s) ▶

$$30(0.004 - x) + 10(0.002 - x) = 0,$$

which gives

$$0.14 - 40x,$$

with solution $x = 0.0035$ (metres). Resolving in the **i**-direction also tells us that the magnitudes of the (compressive) spring forces are equal. This magnitude is

$$\begin{aligned}|\mathbf{H}_1| &= |30(0.004 - 0.0035)| \\ &= 30 \times 0.0005 \\ &= 0.015,\end{aligned}$$

that is, 0.015 N.

Solution to Exercise 8

Define the length of the model spring to be x, with the origin at the support at the top. Define **i** to be in the direction of increasing x. (Only one unit vector is needed as all the forces are vertical.) We model the system as a particle of mass m and a model spring; air resistance forces are ignored. The set-up and force diagram are shown below.

◀ Draw picture ▶

◀ Choose axes ▶

◀ State assumptions ▶

◀ Draw force diagram ▶

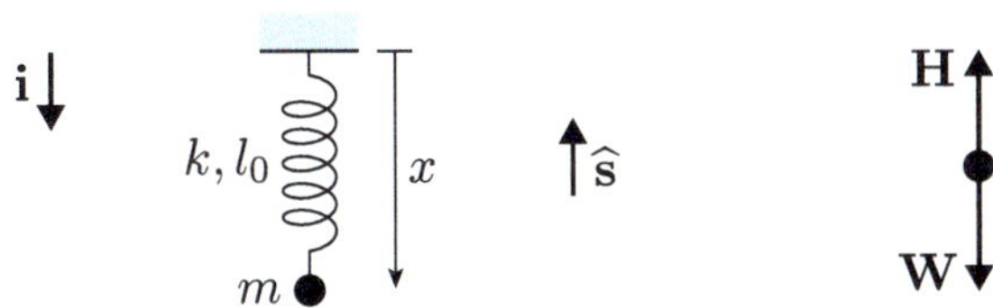

There are two forces acting on the particle, the weight and that due to the spring. Using the unit vector given, $\mathbf{W} = mg\mathbf{i}$.

The force $\mathbf{H}$ exerted by the spring on the particle is given by equation (2), that is,

$$\mathbf{H} = k(l - l_0)\,\widehat{\mathbf{s}},$$

where l is the length of the spring and $\widehat{\mathbf{s}}$ is a unit vector directed from the particle towards the centre of the spring. Here $l = x$ and $\widehat{\mathbf{s}} = -\mathbf{i}$, which gives

$$\mathbf{H} = k(x - l_0)(-\mathbf{i}).$$

◀ Apply Newton's 2nd law ▶

Apply Newton's second law (3) to obtain

$$m\ddot{\mathbf{r}} = m\ddot{x}\mathbf{i} = -k(x - l_0)\mathbf{i} + mg\mathbf{i}.$$

Resolving in the $\mathbf{i}$-direction gives

$$m\ddot{x} = -k(x - l_0) + mg,$$

that is,

$$m\ddot{x} + kx = kl_0 + mg.$$

Solution to Exercise 9

The associated homogeneous differential equation, $m\ddot{x} + kx = 0$, is the same in both cases. However, the constant term on the right-hand side is different. So the complementary function will be the same, but the particular integral will be different.

Solution to Exercise 10

If $x(t) = B\cos\omega t + C\sin\omega t$, then

$$\dot{x}(t) = -B\omega\sin\omega t + C\omega\cos\omega t,$$
$$\ddot{x}(t) = -B\omega^2\cos\omega t - C\omega^2\sin\omega t = -\omega^2\,x(t).$$

Putting $\omega^2 = k/m$ and writing x for $x(t)$, we have

$$\ddot{x} = -kx/m,$$

that is,

$$m\ddot{x} + kx = 0.$$

Hence $x(t) = B\cos\omega t + C\sin\omega t$ is a solution of this differential equation.

Solution to Exercise 11

Write the differential equation in the form

$$\ddot{x} + \omega^2 x = 0,$$

where $\omega^2 = k/m$. Then the auxiliary equation is

$$\lambda^2 + \omega^2 = 0,$$

with solutions $\lambda = 0 \pm i\omega$.

So the general solution is

$$\begin{aligned} x(t) &= e^0(B\cos\omega t + C\sin\omega t) \\ &= B\cos\omega t + C\sin\omega t, \end{aligned}$$

where B and C are arbitrary constants.

Solution to Exercise 12

Both differential equations have a constant on the right-hand side, so consider the solution of

$$m\ddot{x} + kx = K,$$

where $K = kl_0$ for Example 4 and $K = kl_0 + mg$ for Exercise 8.

Based on the methods in Unit 1, since the function on the right-hand side is a constant, try a solution of the form $x_\text{p} = c$, where c is a constant. Then $\dot{x}_\text{p} = \ddot{x}_\text{p} = 0$, and substituting this into the differential equation gives

$$m \times 0 + kc = K, \quad \text{thus} \quad c = \frac{K}{k}.$$

The general solution is the sum of the complementary function and the particular integral. So for Example 4, the general solution is

$$x(t) = B\cos\omega t + C\sin\omega t + l_0,$$

and for Exercise 8, the general solution is

$$x(t) = B\cos\omega t + C\sin\omega t + l_0 + \frac{mg}{k},$$

where B and C are arbitrary constants.

Solution to Exercise 13

For both the horizontal spring and the vertical spring, the equilibrium length is the same as the value of the constant.

Each general solution is a sinusoid plus a constant term. The sinusoid represents the oscillatory motion, and the constant term gives the position about which the oscillations take place. It is sometimes called the average position of the oscillations.

Solution to Exercise 14

The coefficient of t is 7, so $\omega = 7$.

Using a compound-angle formula gives

$$\cos\left(7t + \arctan\left(\tfrac{3}{4}\right)\right) = \cos 7t\cos\left(\arctan\left(\tfrac{3}{4}\right)\right) - \sin 7t\sin\left(\arctan\left(\tfrac{3}{4}\right)\right).$$

From the figure in the margin, it can be seen that if $\arctan\left(\frac{3}{4}\right) = \phi$, then $\cos\phi = \frac{4}{5}$ and $\sin\phi = \frac{3}{5}$, so

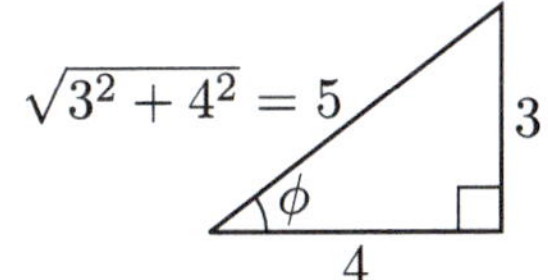

$$10\cos\left(7t + \arctan\left(\tfrac{3}{4}\right)\right) = 10\cos 7t \times \tfrac{4}{5} - 10\sin 7t \times \tfrac{3}{5}.$$

Thus $B = 8$ and $C = -6$, and we have the solution

$$x(t) = 8\cos\omega t - 6\sin\omega t.$$

Solution to Exercise 15

(a) The angular frequency is

$$\omega = \sqrt{k/m} = \sqrt{200/0.5} = 20,$$

that is, $20\,\text{rad}\,\text{s}^{-1}$.

The period is

$$\tau = 2\pi/\omega = 2\pi/20 \simeq 0.314 \quad \text{to 3 d.p.},$$

that is, about 0.31 s.

(b) The frequency is

$$f = \omega/(2\pi) = 20/(2\pi) \simeq 3.18 \quad \text{to 2 d.p.},$$

that is, about 3.18 Hz.

(c) With $\omega = 20$ and $l_0 = 0.2$, we have

$$x(t) = B\cos 20t + C\sin 20t + 0.2,$$

whose derivative is

$$\dot{x}(t) = -20B\sin 20t + 20C\cos 20t.$$

Now $x = 0.3$ and $\dot{x} = 2$ when $t = 0$, so

$$0.3 = B + 0.2, \quad 2 = 20C.$$

Hence $B = 0.1$ and $C = 0.1$. Thus the position function for the particle is

$$x(t) = 0.1\cos 20t + 0.1\sin 20t + 0.2.$$

(d) The amplitude of the oscillation is $\sqrt{0.1^2 + 0.1^2} = 0.141$ to three decimal places. So the minimum distance is $0.2 - 0.141 = 0.059$ m, or approximately 6 cm.

Solution to Exercise 16

(a) Using equations (10), we have $A = \sqrt{0.1^2 + 0.1^2} = 0.1\sqrt{2}$ and

$$\cos\phi = 0.1/A, \quad \sin\phi = -0.1/A.$$

Since $\cos\phi > 0$ and $\sin\phi < 0$, the angle ϕ lies in the fourth quadrant (where $-\frac{\pi}{2} < \phi < 0$), so

$$\phi = \arcsin\left(\frac{-0.1}{0.1\sqrt{2}}\right) = -\frac{\pi}{4}.$$

(b) Using the result of part (a), the oscillation may be described by the equation

$$x(t) = 0.1\sqrt{2}\cos\left(20t - \tfrac{\pi}{4}\right) + 0.2.$$

This is sketched below (where $\phi/\omega = -\frac{\pi}{4}/20 \simeq 0.04$).

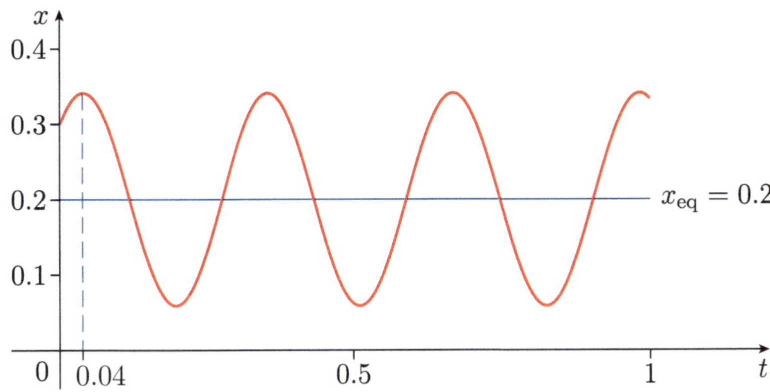

(c) The spring takes its natural length when $x = l_0$. From part (b), this occurs when

$$0.2 = 0.1\sqrt{2}\cos\left(20t - \tfrac{\pi}{4}\right) + 0.2,$$

that is, when $20t - \pi/4 = (2n+1)\pi/2$, with n any integer. Thus the times are given by

$$t = \frac{1}{20}\left(\frac{\pi}{4} + \frac{(2n+1)\pi}{2}\right) = \frac{\pi}{20}\left(\tfrac{3}{4} + n\right), \quad n \in \mathbb{Z}.$$

Solution to Exercise 17

(a) A picture of the set-up is shown below, along with unit vectors in the horizontal and vertical directions.

◂ Draw picture ▸

◂ Choose axes ▸

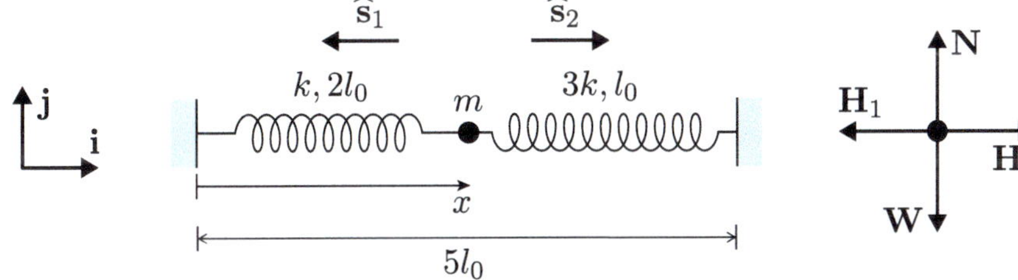

We model the block as a particle of mass m and the springs as model springs. All friction and air resistance forces are ignored.

◂ State assumptions ▸

The force diagram is drawn (in the picture above) for the case where both springs are extended. There are four forces acting on the particle: the weight $\mathbf{W}$ of the particle, the normal reaction $\mathbf{N}$ from the table, the spring force $\mathbf{H}_1$ from the left-hand spring, and the spring force $\mathbf{H}_2$ from the right-hand spring.

◂ Draw force diagram ▸

The spring forces are derived in the following table (where the headings refer to general notation and the rows contain the parameters for this particular set-up).

◂ Apply Newton's 2nd law ▸

Spring	l	l_0	$l - l_0$	k	$\widehat{\mathbf{s}}$	$\mathbf{H}$
Left	x	$2l_0$	$x - 2l_0$	k	$-\mathbf{i}$	$k(x - 2l_0)(-\mathbf{i})$
Right	$5l_0 - x$	l_0	$4l_0 - x$	$3k$	$\mathbf{i}$	$3k(4l_0 - x)\mathbf{i}$

So

$$\mathbf{H}_1 = k(2l_0 - x)\mathbf{i},$$
$$\mathbf{H}_2 = 3k(4l_0 - x)\mathbf{i}.$$

In addition, $\mathbf{W} = mg(-\mathbf{j})$ and $\mathbf{N} = |\mathbf{N}|\,\mathbf{j}$.

From Newton's second law, we have

$$\begin{aligned} m\ddot{x}\mathbf{i} &= \mathbf{H}_1 + \mathbf{H}_2 + \mathbf{W} + \mathbf{N} \\ &= k(2l_0 - x)\mathbf{i} + 3k(4l_0 - x)\mathbf{i} - mg\mathbf{j} + |\mathbf{N}|\,\mathbf{j}. \end{aligned}$$

Resolving in the **i**-direction and rearranging, we obtain the equation

$$m\ddot{x} + 4kx = 14kl_0,$$

as required.

◀ Solve differential equation ▶

(b) In equilibrium, with $x = x_{\text{eq}}$ say, we have $\dot{x} = 0$ and $\ddot{x} = 0$. Substituting into the equation of motion gives

$$4kx_{\text{eq}} = 14kl_0,$$

so the equilibrium distance from the left-hand end is

$$x_{\text{eq}} = \tfrac{7}{2}l_0.$$

(c) The general solution of the equation of motion is

$$x(t) = B\cos\omega t + C\sin\omega t + x_{\text{eq}},$$

where B and C are constants, and the angular frequency ω is $\sqrt{4k/m}$. So the period of the oscillations is

$$\tau = \frac{2\pi}{\omega} = \pi\sqrt{\frac{m}{k}}.$$

(d) The initial conditions are $x(0) = \frac{5}{2}l_0$ (the halfway point) and $\dot{x}(0) = 0$. So we have

$$\tfrac{5}{2}l_0 = B + \tfrac{7}{2}l_0, \quad 0 = C\omega.$$

Hence $B = -l_0$ and $C = 0$, giving the particular solution

$$x(t) = -l_0\cos\left(2\sqrt{\frac{k}{m}}\,t\right) + \tfrac{7}{2}l_0.$$

(e) The amplitude of the motion is $A = \sqrt{(-l_0)^2 + 0^2} = l_0$.

Note that one of the extrema is the point at which the particle was released from rest.

The minimum distance is $x_{\text{eq}} - A = \frac{7}{2}l_0 - l_0 = \frac{5}{2}l_0$.

The maximum distance is $x_{\text{eq}} + A = \frac{7}{2}l_0 + l_0 = \frac{9}{2}l_0$.

The range of motion is within the fixed ends at $x = 0$ and $x = 5l_0$, so the particle never hits the ends.

Solution to Exercise 18

(a) Since air resistance can be ignored, there are just four forces acting, namely the weight $\mathbf{W}$, and three spring forces: $\mathbf{H}_1$ from spring QP, $\mathbf{H}_2$ from spring RP, and $\mathbf{H}_3$ from spring SP. The block is modelled as a particle, so the forces act at the same point.

The force diagram in the margin also shows the unit vectors $\widehat{\mathbf{s}}_1$, $\widehat{\mathbf{s}}_2$ and $\widehat{\mathbf{s}}_3$ associated with each spring.

$\mathbf{H}_1$ $\mathbf{i}$ $\mathbf{H}_2$ $\mathbf{H}_3$ $\mathbf{W}$ $\widehat{\mathbf{s}}_1$ $\widehat{\mathbf{s}}_2$ $\widehat{\mathbf{s}}_3$

(b) The spring forces can be obtained using a table (where once again the headings refer to general notation and the rows contain the parameters for this particular set-up).

Spring	l	l_0	$l - l_0$	k	$\widehat{\mathbf{s}}$	$\mathbf{H}$
QP	x	l_0	$x - l_0$	$2k$	$-\mathbf{i}$	$2k(x - l_0)(-\mathbf{i})$
RP	$x - \frac{1}{2}l_0$	$\frac{1}{2}l_0$	$x - l_0$	$5k$	$-\mathbf{i}$	$5k(x - l_0)(-\mathbf{i})$
SP	$4l_0 - x$	$\frac{3}{2}l_0$	$\frac{5}{2}l_0 - x$	$2k$	$\mathbf{i}$	$2k\left(\frac{5}{2}l_0 - x\right)\mathbf{i}$

So

$$\mathbf{H}_1 = 2k(l_0 - x)\mathbf{i},$$
$$\mathbf{H}_2 = 5k(l_0 - x)\mathbf{i},$$
$$\mathbf{H}_3 = 2k\left(\tfrac{5}{2}l_0 - x\right)\mathbf{i}.$$

(c) Applying Newton's second law, the equation of motion is

$$m\ddot{x}\mathbf{i} = \mathbf{W} + \mathbf{H}_1 + \mathbf{H}_2 + \mathbf{H}_3.$$

The weight is given by $\mathbf{W} = mg\mathbf{i}$, so using the results from part (b), we have

$$m\ddot{x}\mathbf{i} = mg\mathbf{i} + 2k(l_0 - x)\mathbf{i} + 5k(l_0 - x)\mathbf{i} + 2k\left(\tfrac{5}{2}l_0 - x\right)\mathbf{i}.$$

Resolving in the $\mathbf{i}$-direction, the equation of motion is

$$m\ddot{x} + 9kx = mg + 12kl_0.$$

(d) In equilibrium, $x = x_{\text{eq}}$ and the acceleration is zero, so

$$9kx_{\text{eq}} = mg + 12kl_0.$$

Thus the equilibrium position is given by

$$x_{\text{eq}} = \tfrac{4}{3}l_0 + \frac{mg}{9k}.$$

(e) Starting with the equation of motion from part (c), the associated homogeneous differential equation is

$$\ddot{x} + \frac{9k}{m}x = 0.$$

Take $\omega^2 = 9k/m$, so that the complementary function is

$$x_c = A\cos\omega t + B\sin\omega t,$$

where A and B are constants, that is,

$$x_c = A\cos\left(3\sqrt{\frac{k}{m}}\,t\right) + B\sin\left(3\sqrt{\frac{k}{m}}\,t\right).$$

The particular integral x_p is the equilibrium position, thus the general solution is

$$\begin{aligned} x(t) &= x_c + x_p \\ &= A\cos\left(3\sqrt{\frac{k}{m}}\,t\right) + B\sin\left(3\sqrt{\frac{k}{m}}\,t\right) + \frac{mg}{9k} + \tfrac{4}{3}l_0. \end{aligned}$$

(f) If the particle is released from rest ($\dot{x}(0) = 0$) at time $t = 0$ at a position $\frac{4}{3}l_0$ below Q, then there are two equations to solve.

Using the initial condition on the displacement gives

$$\tfrac{4}{3}l_0 = A + \frac{mg}{9k} + \tfrac{4}{3}l_0,$$

so

$$A = -\frac{mg}{9k}.$$

After differentiating the general solution to obtain

$$\dot{x}(t) = -3A\sqrt{\frac{k}{m}}\sin\left(3\sqrt{\frac{k}{m}}\,t\right) + 3B\sqrt{\frac{k}{m}}\cos\left(3\sqrt{\frac{k}{m}}\,t\right),$$

insertion of the initial condition for the speed gives

$$0 = 3B\sqrt{\frac{k}{m}},$$

so $B = 0$.

Thus the equation of motion in this case is

$$x(t) = \tfrac{4}{3}l_0 + \frac{mg}{9k}\left(1 - \cos\left(3\sqrt{\frac{k}{m}}\,t\right)\right).$$

(g) From the form of the solution, the frequency is $\omega = 3\sqrt{k/m}$, so the period of the oscillations is

$$\tau = \frac{2\pi}{\omega} = \frac{2\pi}{3}\sqrt{\frac{m}{k}}.$$

The amplitude is $mg/9k$.

(h) A sketch graph is shown below.

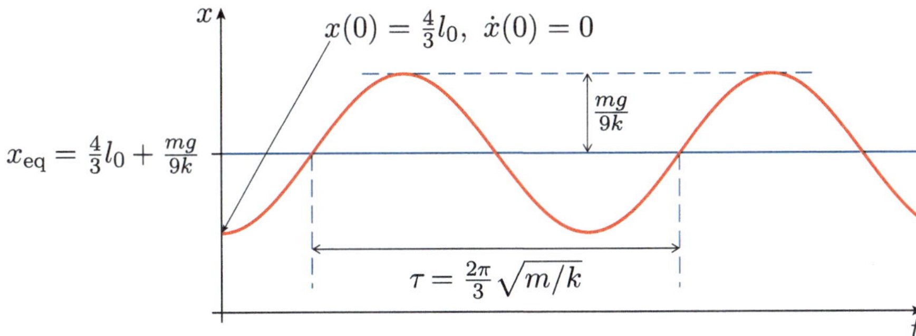

Solution to Exercise 19

(a) Once the pontoon is out of the water, the buoyancy force is zero, so the upper limit for x is h. When the pontoon is fully submerged, the buoyancy force is constant, and not dependent on x, so the lower limit for x is 0. The expression for the buoyancy force is valid for $0 < x < h$.

(b) The given equation of motion (17) is equivalent to

$$\ddot{x} + \omega^2 x = \omega^2 x_{eq},$$

where $\omega^2 = A\rho g/m$ and $x_{eq} = h - m/(\rho A)$. The period τ is therefore given by

$$\tau = \frac{2\pi}{\omega} = 2\pi\sqrt{\frac{m}{A\rho g}} = C\sqrt{m},$$

where $C = 2\pi/\sqrt{A\rho g}$ is constant if A is fixed. Thus τ is proportional to $\sqrt{m}$.

If the mass m is increased progressively, for given dimensions of the pontoon, then eventually the pontoon will sink. The quoted equation of motion then no longer applies.

From part (a), for the pontoon to float we must have $x_{eq} > 0$, which occurs if $m < Ah\rho$. This gives an upper limit $2\pi\sqrt{h/g}$ for the period of oscillations.

Solution to Exercise 20

Using the method of dimensions from Unit 8,

$$[T(x)] = \left[\tfrac{1}{2}\right][m]\left[v^2\right] = 1 \times \mathrm{M} \times \left(\mathrm{L\,T^{-1}}\right)^2 = \mathrm{M\,L^2\,T^{-2}}$$

Take care to distinguish T denoting kinetic energy from T used as the dimensions of time.

and

$$[U(x)] = \left[-\int F(x)\,dx\right] = [F(x)]\,[x] = \mathrm{M\,L\,T^{-2}} \times \mathrm{L} = \mathrm{M\,L^2\,T^{-2}}.$$

So, not surprisingly, kinetic energy and potential energy have the same dimensions.

Solution to Exercise 21

With the given x-axis, the force due to gravity has $\mathbf{i}$-component $F = -mg$. Using the definition of the potential energy function (23), we have

$$U(x) = -\int F(x)\,dx = \int mg\,dx = mgx + C,$$

where C is a constant. The given datum is the origin, so $U(0) = 0$, thus $C = 0$, and we obtain

This is a justification for the negative sign in the definition of $U(x)$; the energy is more positive the higher the particle is above the datum.

$$U(x) = mgx.$$

So for the force due to gravity, the potential energy function is mg times the height above the chosen datum.

Solution to Exercise 22

Choose the unit vectors $\mathbf{i}$ and $\mathbf{j}$ as shown below.

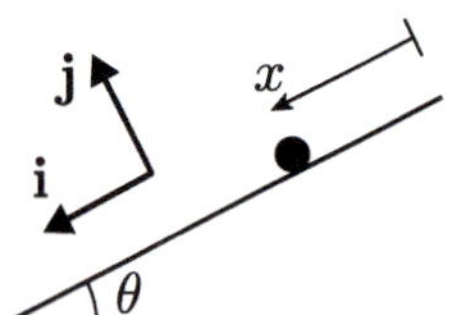

Then the forces acting on the particle are the normal reaction $\mathbf{N}$ (in the $\mathbf{j}$-direction) and the weight

$$\mathbf{W} = mg\,(\sin\theta\,\mathbf{i} + \cos\theta\,(-\mathbf{j})) = mg\sin\theta\,\mathbf{i} - mg\cos\theta\,\mathbf{j}.$$

The motion (and hence the resultant force) is along the slope. The total force $F(x)\,\mathbf{i}$ acting on the particle along the slope is given by

$$F(x) = (\mathbf{N} + \mathbf{W}) \cdot \mathbf{i} = mg\sin\theta.$$

Using the definition of the potential energy function (23), we have

$$U(x) = -\int F(x)\,dx = \int -mg\sin\theta\,dx = -mgx\sin\theta + C,$$

where C is a constant. The given datum is the origin, so $U(0) = 0$, thus $C = 0$, and we obtain

$$U(x) = -mgx\sin\theta.$$

Since $x\sin\theta$ is the vertical height of the particle as measured *below* the datum, the potential energy function is again mg times the height above the chosen datum (since this is now negative, because the x-axis is down the slope).

Solution to Exercise 23

The **i**-component of the buoyancy force is $A\rho g(h-x)$, so

$$U(x) = -\int A\rho g(h-x)\,dx = -A\rho g\left(hx - \tfrac{1}{2}x^2\right) + C,$$

Exercise 19(a) shows that the range of validity for the equation for **B** is $0 < x < h$.

where C is a constant. Using the given datum, $U(h) = 0$, so

$$0 = -A\rho g\left(h^2 - \tfrac{1}{2}h^2\right) + C,$$

therefore

$$C = \tfrac{1}{2}A\rho g h^2.$$

Thus

$$U(x) = -A\rho g\left(hx - \tfrac{1}{2}x^2 + \tfrac{1}{2}A\rho g h^2\right) = \tfrac{1}{2}A\rho g(h-x)^2.$$

This makes sense because when the pontoon is fully submerged ($x = 0$), the potential energy is a maximum. This can be demonstrated by holding a light ball (such as a table tennis ball) underwater; on being released, it shoots upwards. Potential energy is converted to kinetic energy.

Solution to Exercise 24

With the given choice of origin, the force due to the spring is

$$\mathbf{H} = k(x - l_0)(-\mathbf{i}) = -k(x - l_0)\mathbf{i},$$

whose x-component is $F(x) = -k(x - l_0)$. Hence the potential energy function is

$$U(x) = -\int F(x)\,dx = \int k(x - l_0)\,dx = \tfrac{1}{2}k(x - l_0)^2 + C,$$

where C is a constant. The given datum is $x = l_0$, so $U(0) = 0$, thus $C = 0$, and we have

$$U(x) = \tfrac{1}{2}k(x - l_0)^2.$$

So for the force due to the spring, the potential energy function is

$$\tfrac{1}{2} \times \text{stiffness} \times (\text{deformation})^2.$$

Solution to Exercise 25

(a) For $F(x) = -ax^2$, we have

$$U(x) = -\int F(x)\,dx = \int ax^2\,dx = \tfrac{1}{3}ax^3 + C,$$

where C is a constant. Choosing $x = 0$ as the datum, so $U(0) = 0$, gives $C = 0$. Hence the potential energy function is

$$U(x) = \tfrac{1}{3}ax^3.$$

(b) For $F(x) = bx^{-2}$ $(x > 0)$, we have

$$U(x) = -\int F(x)\,dx = -\int bx^{-2}\,dx = bx^{-1} + C,$$

It is not possible to use $x = 0$ as the datum, since $U(x)$ is not defined at $x = 0$.

where C is a constant. Choosing $x = \infty$ as the datum, so $U(\infty) = 0$, gives $C = 0$. Hence the potential energy function is

$$U(x) = bx^{-1} \quad (x > 0).$$

Choosing a different datum, say x_0 such that $U(x_0) = 0$, leads to

The motion of a comet can be represented by this model.

$$U(x) = \frac{b}{x} - \frac{b}{x_0} \quad (x > 0).$$

Solution to Exercise 26

(a) From the definition of potential energy, if $U(x)$ is a potential energy function corresponding to the force $F(x)\,\mathbf{i}$, then

$$\begin{aligned} U(x) &= -\int F(x)\,dx \\ &= -\int \left[F_1(x) + F_2(x)\right] dx \\ &= -\int F_1(x)\,dx - \int F_2(x)\,dx \\ &= U_1(x) + U_2(x). \end{aligned}$$

(b) If $F_1(x) = -ax^2$ and $F_2(x) = bx^{-2}$ $(x > 0)$, then from Exercise 25, we have

$$U_1(x) = \tfrac{1}{3}ax^3, \quad U_2(x) = bx^{-1} \ (x > 0).$$

Now the given force is $F(x)\,\mathbf{i} = F_1(x)\,\mathbf{i} + F_2(x)\,\mathbf{i}$, so its potential energy function is

$$\begin{aligned} U(x) &= U_1(x) + U_2(x) \\ &= \tfrac{1}{3}ax^3 + bx^{-1} \quad (x > 0). \end{aligned}$$

(Your answer could differ from the above by a constant if you chose a different datum for either F_1 or F_2; see the text that follows this exercise.)

Solution to Exercise 27

(a) Using the origin as the datum with a downward-pointing x-axis, the height above the datum is $-x$. So using equation (26), the potential energy is $mg(-x)$, thus the total energy is

$$E = \tfrac{1}{2}mv^2 - mgx.$$

The initial condition is $v = 0$ when $x = 0$, so $E = 0$. Hence we have

$$0 = \tfrac{1}{2}v^2 - gx, \quad \text{or} \quad v^2 = 2gx,$$

throughout the motion since energy is conserved. So when $x = 77$, we find that

$$|v| = \sqrt{2 \times 9.81 \times 77} \simeq 38.9 \quad \text{to 3 s.f.}$$

Hence the speed of the marble just before it hits the water is $38.9\,\mathrm{m\,s^{-1}}$.

(The same answer was obtained by direct application of Newton's second law for constant acceleration in Unit 3, Example 4.)

When $v = 20$, we have

$$x = 20^2/(2 \times 9.81) \simeq 20.4 \quad \text{to 3 s.f.}$$

So the marble has fallen $20.4\,\mathrm{m}$ when its speed reaches $20\,\mathrm{m\,s^{-1}}$.

(b) The force of air resistance is a function of velocity. Hence the total force acting on the marble is not dependent on position alone, and the law of conservation of mechanical energy is not applicable.

Solution to Exercise 28

(a) First, we draw a picture. ◀ Draw picture ▶

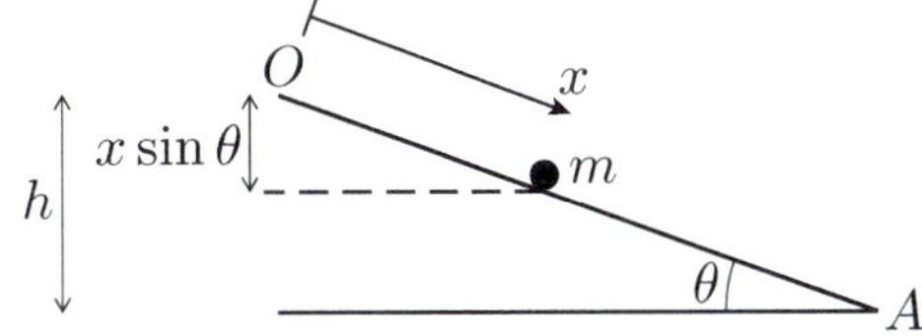

We assume that the particle travels in a straight line down a straight frictionless slope, subject only to the gravitational force and the normal reaction. Air resistance is ignored. ◀ State assumptions ▶

The point A is chosen as the datum, at $x = h/\sin\theta$. ◀ Choose datum ▶

The component of the force down the slope is $mg\sin\theta$, so the potential energy function is ◀ Find potential energy ▶

$$\begin{aligned} U(x) &= -\int_{h/\sin\theta}^{x} F(x)\,dx \\ &= -\int_{h/\sin\theta}^{x} mg\sin\theta\,dx \\ &= mgh - mgx\sin\theta. \end{aligned}$$

So the total mechanical energy of the particle is

$$E = \tfrac{1}{2}mv^2 + mg(h - x\sin\theta).$$

(b) At O we have $x = 0$ and $v = 0$, so $E = mgh$ and ◀ Find E ▶

$$\tfrac{1}{2}mv^2 + mg(h - x\sin\theta) = mgh$$

throughout the motion.

At A we have $x\sin\theta = h$, so $\frac{1}{2}mv^2 = mgh$. Since $v > 0$, this gives ◀ Find x or v ▶

$$v = \sqrt{2gh}.$$

The speed of the particle at A is therefore $\sqrt{2gh}$. ◀ Interpret solution ▶

This result depends on the height h, but not on the angle θ at which the slope is inclined. For a less steep slope, the acceleration of the particle will be less, but then the particle has further to travel down the slope in order to descend a given vertical distance h. The result depends on the assumption of no friction between the particle and the surface, but does not depend on the mass of the particle.

(c) For the given data, we obtain a value at A of

$$v = \sqrt{2gh} = \sqrt{2 \times 9.81 \times 2 \sin \tfrac{\pi}{6}} \simeq 4.4,$$

so the crate of empty bottles reaches the bottom of the slope with speed $4.4\,\mathrm{m\,s^{-1}}$. (This agrees with the answer obtained in Example 5 of Unit 3.)

Solution to Exercise 29

◀ Draw picture ▶

First, we draw a picture.

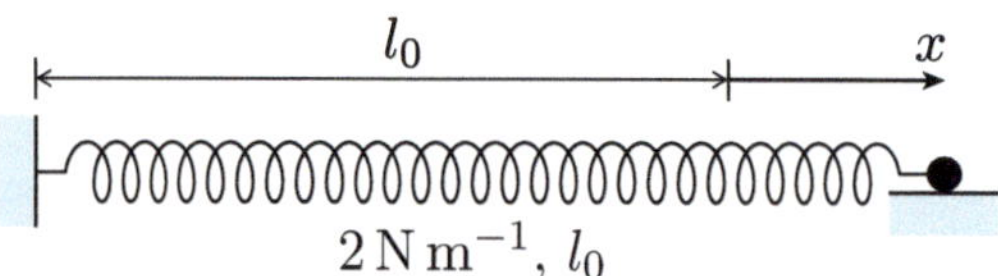

◀ State assumptions ▶

We assume that the track is straight, level and frictionless, that the spring is a model spring, and that air resistance forces are ignored.

◀ Choose datum ▶

Take the datum at $x = 0$ (the point of zero deformation).

◀ Find potential energy ▶

For the force exerted by a spring with deformation x, the potential energy function is $U(x) = \frac{1}{2}kx^2$ (from equation (27)), so the law of conservation of mechanical energy gives

$$\tfrac{1}{2}mv^2 + \tfrac{1}{2}kx^2 = E.$$

◀ Find E ▶

Since $m = 0.5$ and $k = 2$, this becomes

$$\tfrac{1}{4}v^2 + x^2 = E.$$

Initially, $v = -3$ when $x = 2$, so

$$E = \tfrac{9}{4} + 4 = \tfrac{25}{4}.$$

◀ Find x or v ▶

Hence we have

$$v^2 + 4x^2 = 25.$$

From this equation, $|v|$ will take its maximum value when $4x^2$ takes its minimum value, which is zero. So the maximum speed is $|v| = \sqrt{25} = 5$.

The particle is at rest when $v = 0$, that is, at $x = \pm\frac{5}{2}$.

◀ Interpret solution ▶

Summarising, the maximum speed of the particle is $5\,\mathrm{m\,s^{-1}}$, and it is momentarily at rest at $\pm\frac{5}{2}\,\mathrm{m}$ from the equilibrium position.

Solution to Exercise 30

We assume that the spring is a model spring and that the particle only moves up and down, subject only to the spring force and the gravitational force. Air resistance is ignored. ◀State assumptions▶

We choose the fixed top of the spring as the datum for gravitational potential energy. We choose the point of zero deformation as the datum for the spring potential energy. ◀Choose datum▶

Let x be the length of the spring, measured downwards from the datum. The potential energy function is the sum of two contributions, one due to gravity and one due to the spring. The height above the datum is $-x$, so the gravitational potential energy is $-mgx$. The extension of the spring is $x - l_0$, so the spring contributes a term $\frac{1}{2}k(x-l_0)^2$ to the potential energy. So the potential energy of the particle is ◀Find potential energy▶

$$U(x) = \tfrac{1}{2}k(x-l_0)^2 - mgx.$$

Hence by the law of conservation of energy,

$$E = \tfrac{1}{2}mv^2 + \tfrac{1}{2}\frac{10mg}{l_0}(x-l_0)^2 - mgx,$$

where we have substituted in the given value of k.

Initially, the particle is at rest at a depth of $\frac{5}{4}l_0$, so ◀Find E▶

$$\begin{aligned} E &= \tfrac{1}{2}m0^2 + \tfrac{1}{2}\frac{10mg}{l_0}(\tfrac{5}{4}l_0 - l_0)^2 - mg\tfrac{5}{4}l_0 \\ &= 0 + \tfrac{5}{16}mgl_0 - \tfrac{5}{4}mgl_0 \\ &= -\tfrac{15}{16}mgl_0. \end{aligned}$$

We wish to find $|v|$ when the spring has its natural length. Substituting $x = l_0$ into the equation for E gives ◀Find x or v▶

$$-\tfrac{15}{16}mgl_0 = \tfrac{1}{2}mv^2 + \tfrac{1}{2}\frac{10mg}{l_0}(l_0 - l_0)^2 - mgl_0.$$

This leads to $|v| = \sqrt{gl_0/8}$.

The speed of the particle when the spring is of length l_0 is $\sqrt{gl_0/8}$. ◀Interpret solution▶

Solution to Exercise 31

(a) We choose the origin as the datum (you may have made a different choice). The potential energy function is then

$$U(x) = -\int F(x)\,dx = -\int (2-2x)\,dx = -2x + x^2.$$

(b) The total mechanical energy of the particle of mass $m = 2$ is

$$E = \tfrac{1}{2}mv^2 + U(x) = v^2 - 2x + x^2.$$

Now $v = 0$ when $x = -1$, so

$$E = 0 + 2 + 1 = 3.$$

(If you used a different datum for the potential energy, so that $U(x) = -2x + x^2 + C$, for C a constant, then your answer for the total mechanical energy will be $E = 3 + C$.)

(c) By the law of conservation of mechanical energy,

$$3 = v^2 - 2x + x^2,$$

so

$$v = \pm\sqrt{3 + 2x - x^2}.$$

(d) The region of motion is given by

$$3 + 2x - x^2 \geq 0,$$

which factorises to give

$$(1 + x)(3 - x) \geq 0,$$

that is,

$$-1 \leq x \leq 3.$$

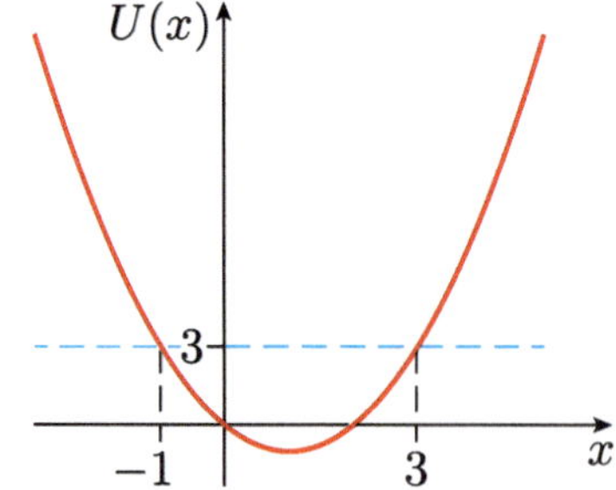

At the midpoint (corresponding to the minimum value for $U(x)$ – see the sketch in the margin) $x = 1$, the particle's speed is

$$|v| = \sqrt{3 + 2 - 1} = 2.$$

Solution to Exercise 32

Differentiating gives $U'(x) = 3x^2 - 3$, so $U'(x) = 0$ when $3x^2 - 3 = 0$, which has solutions $x = \pm 1$. These are the positions of equilibrium.

Solution to Exercise 33

(a) We choose $x = \infty$ as the datum. The potential energy function is then

$$U(x) = -\int F(x)\,dx = -\int \frac{2}{x^2}\,dx = \frac{2}{x}.$$

(If you chose a different datum, then your answer should be $U(x) = 2x^{-1} + C$, where C is a constant.)

(b) The law of conservation of mechanical energy is

$$\tfrac{1}{2}mv^2 + U(x) = E,$$

which gives

$$\tfrac{3}{2}v^2 + \frac{2}{x} = E.$$

Initially, $v = -1$ when $x = 10$, which leads to

$$E = \tfrac{17}{10} \quad \text{(in joules)}.$$

(If you used a different datum for the potential energy function, as above, then you should have found $E = \frac{17}{10} + C$.)

(c) At the point of closest approach, the particle must be stationary, so $v = 0$. Then from the law of conservation of mechanical energy, the corresponding value of x is given by

$$\tfrac{3}{2} \times 0^2 + \frac{2}{x} = \tfrac{17}{10}.$$

Hence the point of closest approach is $x = \frac{20}{17}$, that is, about 1.18 m.

(d) The region of motion is $x \geq \frac{20}{17}$, and a sketch of $U(x)$ is shown in the margin.

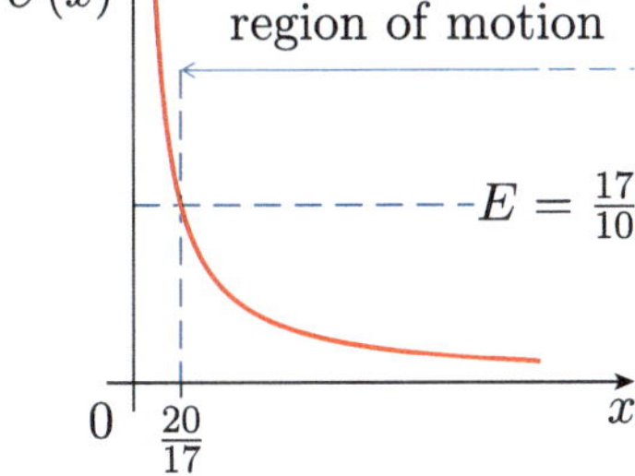

The particle is initially at $x = 10$, moving in the direction of decreasing x. It comes instantaneously to rest at $x = \frac{20}{17}$, and changes its direction of motion. It then continues to move in the direction of increasing x. For large values of x, its velocity will approach $v = \sqrt{17/15}$ (again using the law of conservation of mechanical energy), that is, about $1.06\,\mathrm{m\,s^{-1}}$.

(e) The graph shows no extrema, so there are no equilibrium points for this potential energy function.

Solution to Exercise 34

When the buffer spring is at maximum compression, the spring potential energy (using the natural length as the datum) is

$$\tfrac{1}{2} \times \text{stiffness} \times (\text{deformation})^2 = \tfrac{1}{2} \times 10^5 \times (0.1)^2 = 500,$$

and the kinetic energy of the truck is zero. So the total mechanical energy of the system is 500 J.

When the truck first comes into contact with the buffers, the spring potential energy is zero and the kinetic energy of the truck is $\frac{1}{2} \times 2000v^2$, where v is the speed of the truck. Using conservation of energy, we have

$$1000v^2 = 500,$$

so $|v| = 1/\sqrt{2}$, that is, the speed of the truck when it first hits the buffers is about $0.71\,\mathrm{m\,s^{-1}}$ to two significant figures.

Solution to Exercise 35

(a) Take Q as a datum for gravitational potential energy; this choice is the easiest for subsequent working.

The position of the particle above the datum is $-x$, so the gravitational potential energy is

$$U_1 = -mgx.$$

(b) The kinetic energy is given by

$$T_1 = \tfrac{1}{2}m\dot{x}^2.$$

(c) The potential energy stored in a stretched (or compressed) spring is given by $\frac{1}{2}kd^2$, where d is the deformation (extension or compression) relative to the natural length of the spring, and k is the stiffness. A tabular approach similar to that used for Exercise 18, and repeating many of the entries, is useful. We use the natural length of each spring as the datum point for the energy of the spring.

Spring	Deformation	Stiffness	Potential energy
QP	$x - l_0$	$2k$	$U_2 = \frac{1}{2}(2k)(x - l_0)^2$
RP	$x - l_0$	$5k$	$U_3 = \frac{1}{2}(5k)(x - l_0)^2$
SP	$\frac{5}{2}l_0 - x$	$2k$	$U_4 = \frac{1}{2}(2k)\left(\frac{5}{2}l_0 - x\right)^2$

So the potential energy stored in each spring is

$$U_2 = k(x - l_0)^2,$$
$$U_3 = \tfrac{5}{2}k(x - l_0)^2,$$
$$U_4 = k\left(\tfrac{5}{2}l_0 - x\right)^2.$$

(d) The total mechanical energy in the system at a general point is the sum of the various energies listed above, that is,

$$\begin{aligned} E &= T_1 + U_1 + U_2 + U_3 + U_4 \\ &= \tfrac{1}{2}m\dot{x}^2 - mgx + k(x - l_0)^2 + \tfrac{5}{2}k(x - l_0)^2 + k\left(\tfrac{5}{2}l_0 - x\right)^2. \end{aligned}$$

This is a constant since this system conserves mechanical energy.

Writing

$$U(x) = -mgx + k(x - l_0)^2 + \tfrac{5}{2}k(x - l_0)^2 + k\left(\tfrac{5}{2}l_0 - x\right)^2$$

gives

$$\begin{aligned} U'(x) &= -mg + 2k(x - l_0) + 5k(x - l_0) - 2k\left(\tfrac{5}{2}l_0 - x\right) \\ &= -mg + 9kx - 12kl_0, \end{aligned}$$

so from equation (31) we have

$$0 = m\ddot{x} - mg + 9kx - 12kl_0,$$

which can be rearranged as

$$m\ddot{x} + 9kx = mg + 12kl_0,$$

in agreement with the solution to Exercise 18(c).

Solution to Exercise 36

(a) The force is given by $F(x)\,\mathbf{i}$ where

$$F(x) = -\frac{dU}{dx} = -x^3 + x^2 + 2x.$$

(b) In each case the system is constrained by the fact that $E \geq U$. The graph below shows the potential energy function and the four values for E.

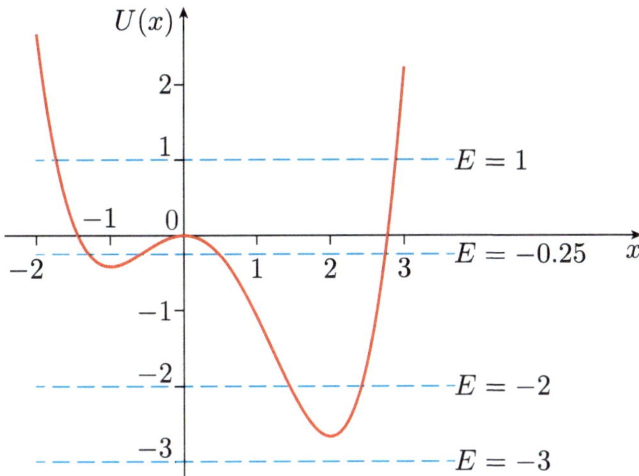

- From the graph it is apparent that the inequality $U(x) \leq -3$ has no solution (i.e. the minimum of $U(x)$ is greater than -3).
- For $E = -2$ the graph gives an approximate range

  $$1.4 < x < 2.4$$

  as a range for possible motion.
- It is clear from the graph that there are two possible ranges of motion when $E = -0.25$. The graph shows two disjoint ranges of x where $E \geq U$, namely

  $$-1.3 < x < -0.6 \quad \text{and} \quad 0.5 < x < 2.7.$$

- The range of possible values for $E = 1$ is

  $$-1.7 < x < 2.9.$$

(c) We have

$$U'(x) = x^3 - x^2 - 2x = x(x-2)(x+1),$$

so there are three equilibrium points, -1, 0 and 2. It is clear from the graph that -1 and 2 are minima, thus oscillations will take place around these two equilibrium points.

Acknowledgements

Grateful acknowledgement is made to the following sources:

Figure 1: Rita Greer (2004) 'Portrait of Robert Hooke'. This file is licensed under the Copyleft Free Art License 1.3 http://artlibre.org/licence/lal/en.

Figure 3: Wellcome Library, London. This file is licensed under the Creative Commons Attribution Licence http://creativecommons.org/licenses/by/4.0.

Every effort has been made to contact copyright holders. If any have been inadvertently overlooked, the publishers will be pleased to make the necessary arrangements at the first opportunity.

Unit 10

Forcing, damping and resonance

Introduction

In Unit 9 we studied models of vibrating systems based on the concept of a model spring. These models predicted simple harmonic motion. In particular, we looked at a simple experiment with a home-made oscillating system, using steel nuts hanging from a metal ruler (Figure 1). In Unit 9 certain qualitative features of the motion of the ruler were modelled well by assuming that the ruler could be modelled as a model spring. There was one feature of the motion that was not satisfactorily modelled in Unit 9 – the fact that the oscillations die away until the ruler is at its equilibrium position. This feature is known as *damping* and is the first topic studied in this unit.

Figure 1 The vibrating ruler experiment

Rather than continuing to model this experiment, you will meet some new oscillating system experiments in Section 1, which we will attempt to model in this unit. Section 1 introduces a linear model for damping, then Section 2 introduces the *model damper*, which provides a convenient diagrammatic and algebraic representation of linear damping, just as the model spring provides a convenient representation of a linear restoring force. After application to some everyday examples, the various types of motion brought about by the damping of an oscillatory system are summarised mathematically.

Section 3 looks at models of systems that undergo sinusoidal *forcing*, that is, an additional sinusoidal force applied to the system. Forcing is needed to keep a system vibrating in spite of any damping. For example, a pendulum clock may have a wound spring to force its oscillations, and a wave machine in a swimming pool is designed to keep the water vibrating. A vehicle driving over a bumpy surface, formed perhaps by cobbles or a cattle grid, is forced to vibrate by the bumps that its wheels encounter.

Resonance refers to the fact that the amplitude of the vibrations with which a system responds to periodic forcing may be much greater at some frequencies of the forcing than at others. This effect may be desired, as in the tuning of a radio. However, resonant oscillations can be unwanted and even destructive in mechanical systems, bridges and tall buildings. For this reason, platoons of soldiers are often ordered not to march in step over bridges. Designers try to choose spring stiffnesses and damping devices within system components so as to eliminate unwanted resonant behaviour over the range of frequencies that are likely to be encountered. The amplitude of the forced oscillations of a vehicle going over a set of bumps depends very much on the speed or frequency with which it encounters the bumps. Rumble strips on roads are therefore designed to take this into account, so that resonant vibrations are likely to occur at higher speeds and drivers are encouraged to slow down. Section 4 shows that resonance is predicted under certain circumstances by the model developed earlier for forced oscillations, and that the experimental particle–spring system also exhibits resonance.

1 Damping and vibrations

The model of a vibrating system that was developed in Unit 9 led to the prediction of *simple harmonic motion*. For a particle with position $x\mathbf{i}$ at time t, the corresponding graph of $x(t)$ is shown in Figure 2(a). The motion, about an equilibrium position $x = x_{\text{eq}}$, has amplitude A and period τ. However, in the real world no system when left to its own devices maintains oscillations of constant amplitude as predicted by this model. In reality the amplitude *decreases* with time, as illustrated in Figure 2(b), and the motion eventually dies away. This section therefore introduces a new ingredient whose inclusion in a revised model leads to predictions that match better the observed behaviour of *damped vibration*. This ingredient is a resistive force that depends on the velocity of the particle.

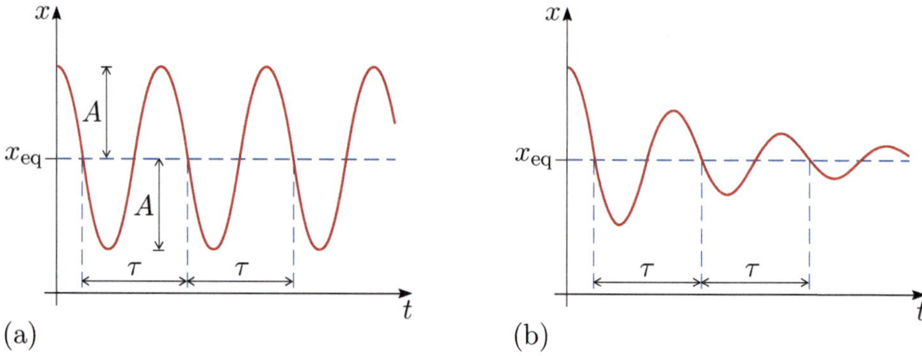

Figure 2 Graphs of (a) an undamped vibration, (b) a damped vibration

Subsection 1.1 introduces a linear model for damping, which is applied in Subsection 1.2 to an experiment that involves a magnet falling down tubes with differing resistance properties. Subsection 1.3 then models a damped particle–spring system. The model is tested experimentally in Subsection 1.4.

1.1 A linear damping model

The reason for the decreasing amplitude of vibrations shown in Figure 2(b) is the effect of some form of *resistance* due either to the air in which the system is vibrating or to some other form of friction. In the context of vibrations, such frictional effects are commonly referred to as **damping** and the corresponding motion is called a **damped vibration**.

Damping can be due to a number of physical causes and can be incorporated in a variety of mathematical models. As you saw in Unit 3, air resistance is commonly modelled as a force opposed to motion, of magnitude proportional to some power of the speed of the body – for example, on an object falling from the Clifton Suspension Bridge.

The emphasis here is not to model one particular physical cause of resistance, but to set up a general mathematical model that could apply in many situations.

In Unit 3 linear and quadratic models for air resistance were investigated: the magnitude of the resistance force was assumed to be proportional either to the magnitude of the velocity or to its square, where in each case the resistance force opposes the motion. In mathematical modelling it is a good principle to start with the *simplest* available mathematical model that appears reasonable, and not to complicate matters unless this initial model turns out to be inadequate. Consequently, we will begin here by assuming that a linear model applies, so we model the resistance to motion as being proportional to the velocity, but in the opposite direction.

If the velocity of the particle is $\dot{x}\mathbf{i}$, then the resistance force is assumed to be $\mathbf{R} = r(-\dot{x}\mathbf{i}) = -r\dot{x}\mathbf{i}$, where r is a positive constant, and the vector $-\dot{x}\mathbf{i}$ shows that the direction of $\mathbf{R}$ opposes the motion. This model is called *linear damping*, and the constant r is called the *damping constant*. The model has the virtue of simplicity in that it is mathematically easy to handle, which is such an advantage that the model is sometimes used even in situations where it may not be strictly appropriate. In such circumstances you would not expect the model to give results that are accurate in detail, but it could still show the effects of resistance and friction in general terms.

In Unit 3 when air resistance was considered, r was taken to be Dc_1, where D is the effective diameter of the object and c_1 is a constant.

The SI units of the damping constant r are $\mathrm{N\,s\,m^{-1}}$ (force divided by speed), or $\mathrm{kg\,s^{-1}}$.

The **linear damping** model assumes that the resistance force acting on a particle is proportional to its velocity, but in the opposite direction. If the velocity is $\dot{x}\mathbf{i}$, then the resistance force is modelled by

$$\mathbf{R} = -r\dot{x}\mathbf{i}, \qquad (1)$$

where r is a positive constant, called the **damping constant**.

The following example should help you to recall from Unit 3 how the linear damping model applies to motion with air resistance. It also provides some revision of the methods for solving second-order differential equations from Unit 1. The result obtained will be of use in Subsection 1.2.

Example 1

An object of mass m falls vertically under gravity. The object is to be modelled as a particle that experiences linear damping, with damping constant r, due to air resistance. Suppose that the particle has fallen a distance x after time t.

(a) Derive the equation of motion of the particle.

(b) Find the particular solution of the equation of motion for which the particle is dropped from the origin at time $t = 0$.

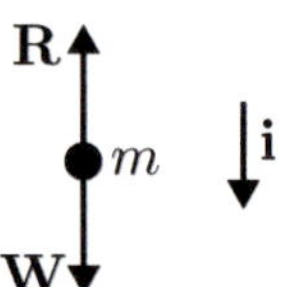

Figure 3 Force diagram

Solution

(a) The choice of direction for the x-axis (downwards) is determined by the fact that x is the distance fallen. The force diagram is shown in Figure 3. The weight of the particle is $\mathbf{W} = mg\mathbf{i}$ (acting vertically downwards). According to the linear damping model, the air resistance force is given by $\mathbf{R} = r(-\dot{x}\mathbf{i}) = -r\dot{x}\mathbf{i}$, which opposes the motion as required.

By Newton's second law,

$$m\ddot{x}\mathbf{i} = \mathbf{W} + \mathbf{R} = mg\mathbf{i} - r\dot{x}\mathbf{i}.$$

Resolving in the **i**-direction gives

$$m\ddot{x} + r\dot{x} = mg.$$

(b) This differential equation may be solved in several different ways. Since it is linear, of second order and with constant coefficients, the methods of Unit 1 may be applied.

The differential equation is inhomogeneous, so the solution will be the sum of the complementary function (the general solution of the associated homogeneous differential equation) and a particular integral.

The associated homogeneous differential equation is

$$m\ddot{x} + r\dot{x} = 0,$$

whose auxiliary equation is

$$m\lambda^2 + r\lambda = 0.$$

This has solutions $\lambda = -r/m$ and $\lambda = 0$. The complementary function is therefore

$$x_{\text{c}} = Ae^{-(r/m)t} + Be^{0t} = Ae^{-rt/m} + B,$$

where A and B are arbitrary constants.

Exceptional cases are discussed in Unit 1, Subsection 3.3.

The right-hand side of the original differential equation is mg. Since this is a constant, we might think of trying a constant for the particular integral. However, there is already a constant term, B, in the complementary function, so we need here to use a trial solution of the form $x_{\text{p}} = pt$, where p is a constant to be determined. Substituting this into the differential equation gives $0 + rp = mg$, so $p = mg/r$. Thus $x_{\text{p}} = mgt/r$ is a particular integral. On adding the complementary function to this, we obtain the general solution of the equation of motion as

$$x = x_{\text{c}} + x_{\text{p}} = Ae^{-rt/m} + B + \frac{mgt}{r}.$$

'Dropped' here means that the particle is initially at rest.

In order to find the appropriate particular solution, we need to refer to the initial conditions. Since the object is dropped from the origin at time $t = 0$, these conditions are $x(0) = 0$ and $\dot{x}(0) = 0$. Substituting the first of these into the general solution gives $0 = A + B + 0$, so $B = -A$.

The derivative of the general solution is

$$\dot{x} = -\frac{r}{m}Ae^{-rt/m} + \frac{mg}{r},$$

so the second initial condition gives $0 = -(r/m)A + mg/r$. Hence we obtain $A = m^2g/r^2$ and $B = -m^2g/r^2$. We conclude that

$$x = \frac{mgt}{r} - \frac{m^2g}{r^2}\left(1 - e^{-rt/m}\right)$$

is the required particular solution.

Another approach was adopted in Example 7(a) of Unit 3, giving a solution that agrees with the one obtained here.

Look at the form of the particular solution obtained in this example, and think about its interpretation. What does the model predict will happen?

The second term in the solution contains a negative exponential. As t increases, this exponential becomes negligibly small, so the second term in the solution tends to a constant. The first term is linear in t. In the absence of any other significant term involving t, this represents motion at the constant speed mg/r. So the model predicts that for large values of t, the motion will be at the constant velocity $(mg/r)\mathbf{i}$, where mg/r is the terminal speed of the object.

Terminal speed was defined in Unit 3.

1.2 An experiment with damping

When a magnet moves inside a copper tube, an electromagnetic effect causes its motion to be damped. The level of damping depends on the thickness and diameter of the tube. This subsection examines the results of a number of experiments carried out by the module team, starting with the motion of a magnet falling down three different copper tubes under gravity, and down a glass tube where only air resistance counteracts the descent (Figure 4). In each case, the resistive force may be modelled by linear damping as in Subsection 1.1.

Phenomena like electromagnetic damping are beyond the scope of this module. For current purposes, you do not need to understand how this effect comes about.

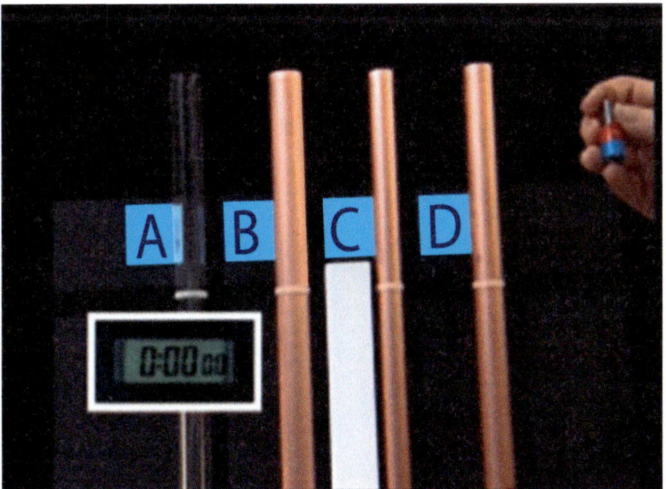

Figure 4 A magnet is dropped down each of four tubes

The equation of motion that arises is that of Example 1(a), and the appropriate particular solution is given by the equation found in Example 1(b). Using this equation and other measurements made, it is possible to determine an estimate for the damping constant r for each tube.

The magnet is then attached to a particle–spring system, like those that you met in Unit 9, and the system is set in motion. The model of motion of an undamped system is modified in order to take the damping into account. A more detailed account of this modelling is provided in Subsection 1.3.

Let $\mathbf{x}$ denote the position vector for the magnet, and let $\mathbf{i}$ be a unit vector pointing down the tube. Since the motion is one-dimensional, we have $\mathbf{x} = x\mathbf{i}$, $\dot{\mathbf{x}} = \dot{x}\mathbf{i}$ and $\ddot{\mathbf{x}} = \ddot{x}\mathbf{i}$. Thus the model for linear damping is

$$\mathbf{R} = -r\dot{x}\mathbf{i}.$$

The motion of the magnet down a tube is modelled with the origin at the top of the tube and the x-direction down the tube. This leads to the equation of motion

$$m\ddot{x} + r\dot{x} = mg,$$

which you met in Example 1(a). Since the magnet is dropped from the origin, the initial conditions are the same as in Example 1. Hence the particular solution is, once more,

$$x = \frac{mgt}{r} - \frac{m^2 g}{r^2}\left(1 - e^{-rt/m}\right). \tag{2}$$

The time t for the magnet to fall the length of each tube was measured. The length of each tube was 1 m. The magnet's mass was found to be 0.038 kg, and we take the magnitude of the acceleration due to gravity to be $g = 9.81\,\mathrm{m\,s^{-2}}$. Using these values, equation (2) can be solved numerically, to find the value of r for each tube. The results are in Table 1.

Table 1

Tube	Material	Time of descent (s)	Damping constant r ($\mathrm{N\,s\,m^{-1}}$)
A	Glass	0.64	0.15
B	Copper	2.52	0.92
C	Copper	3.60	1.33
D	Copper	22.60	8.42

These values for the damping constant r will be used to predict the motion of the magnet in each tube when it is attached to a particle–spring system.

1.3 Damping the spring motion

The next set of experiments with the magnet and tubes involves the magnet being attached to a particle–spring system (Figure 5).

Figure 5 A magnet attached to a spring, and the four tubes

The model of Unit 9 is modified by the inclusion of a linear damping force $\mathbf{R}$ as well as the model spring force $\mathbf{H}$ and the weight $\mathbf{W}$ of the particle, to give the equation

$$m\ddot{x}\mathbf{i} = \mathbf{W} + \mathbf{H} + \mathbf{R}. \tag{3}$$

In this subsection, we will show that with an appropriate choice of origin for x, the equation of motion can be written in the homogeneous form

$$m\ddot{x} + r\dot{x} + kx = 0,$$

where m is the mass of the particle, r is the damping constant, and k is the spring stiffness. To do this, we look again at the modelling involved.

The experimental apparatus and the corresponding force diagram are shown in Figure 6. This force diagram corresponds to the situation in which the spring is extended and the particle is descending, at which time both the spring force $\mathbf{H}$ and the damping force $\mathbf{R}$ will be directed upwards.

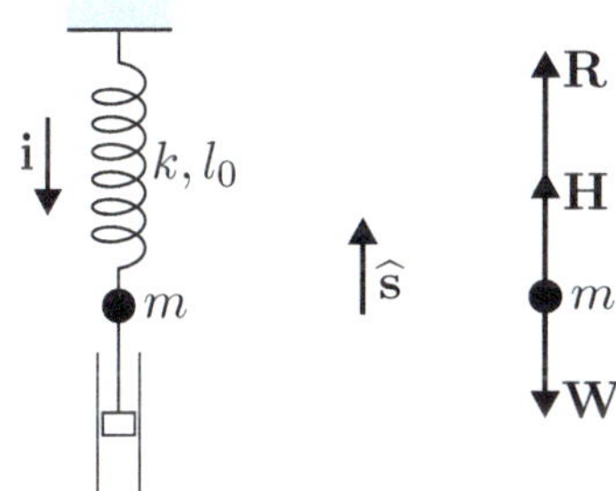

Figure 6 The magnet in the tube and the force diagram

As before, we take the x-axis to be directed downwards, but for the moment we leave open the choice of origin. Regardless of this choice, the acceleration of the particle is $\ddot{x}\mathbf{i}$ and its velocity is $\dot{x}\mathbf{i}$, so the linear damping force is $\mathbf{R} = -r\dot{x}\mathbf{i}$. The weight of the particle, $\mathbf{W} = mg\mathbf{i}$, is independent of where the origin is chosen. In fact, only the expression for the spring force $\mathbf{H}$ in terms of x will depend on the choice of origin.

Hooke's law states that the force exerted by a model spring is given by

Hooke's law was introduced in Unit 9.

$$\mathbf{H} = k(l - l_0)\,\widehat{\mathbf{s}},$$

where k, l and l_0 are respectively the stiffness, length and natural length of the spring, and $\widehat{\mathbf{s}}$ is a unit vector in the direction from the end where the particle is attached towards the centre of the spring. Here we have $\widehat{\mathbf{s}} = -\mathbf{i}$, so $\mathbf{H} = k(l - l_0)(-\mathbf{i})$.

Newton's second law (equation (3)) therefore gives

$$m\ddot{x}\mathbf{i} = \mathbf{W} + \mathbf{H} + \mathbf{R} = mg\mathbf{i} - k(l - l_0)\mathbf{i} - r\dot{x}\mathbf{i},$$

which after resolution in the **i**-direction leads to

$$m\ddot{x} + r\dot{x} + kl = mg + kl_0. \tag{4}$$

When the system is in equilibrium, we have $\dot{x} = 0$ and $\ddot{x} = 0$, so the equilibrium length l_{eq} of the spring is given by

$$0 + 0 + kl_{\text{eq}} = mg + kl_0,$$

that is,

$$l_{\text{eq}} = l_0 + \frac{mg}{k}.$$

This is the expression for the equilibrium spring length that was derived in Unit 9, Example 3. It is the sum of the natural length of the spring, l_0, and its equilibrium extension mg/k.

Hence equation (4) can be written as

$$m\ddot{x} + r\dot{x} + kl = kl_{\text{eq}}.$$

It follows that if we take $x = l - l_{\text{eq}}$, then the equation of motion takes the homogeneous form

$$m\ddot{x} + r\dot{x} + kx = 0.$$

This choice for x corresponds to taking $x = 0$ at the point where $l = l_{\text{eq}}$, that is, choosing the origin to be *at the equilibrium position* of the particle.

Exercise 1

In the next set of experiments, the total mass suspended from a spring of stiffness $23\,\text{N}\,\text{m}^{-1}$ is $0.711\,\text{kg}$. Find the equilibrium extension of the spring, taking $g = 9.81\,\text{m}\,\text{s}^{-2}$.

Exercise 2

Use equation (4) to write down the equation of motion for the case in which the origin for x is chosen to be at the fixed upper end of the spring.

Exercise 2 shows that the equation of motion takes the form

$$m\ddot{x} + r\dot{x} + kx = kx_{\text{eq}}, \tag{5}$$

where x_{eq} is the equilibrium displacement of the particle relative to the chosen origin. We will normally use the simplest form,

This equation (using different notation) was investigated in Unit 1, Subsection 2.2.

$$m\ddot{x} + r\dot{x} + kx = 0, \tag{6}$$

for which $x_{\text{eq}} = 0$, where the origin is chosen to be the equilibrium point of the system.

Notice that all the terms on the left-hand side of equation (6) (or equation (5)) have the same sign, and that if damping were to be removed (by putting r equal to zero), then the resulting equation would revert to that of simple harmonic motion, as derived in Unit 9.

In order to use our model to predict the motion of the damped particle–spring system, we need first to solve the equation of motion (6). This is a linear constant-coefficient second-order differential equation, so once again the methods of Unit 1 may be applied to find the form of its solution. The auxiliary equation is

The process of solving equation (6) will be revisited in Section 2, so do not dwell on the mathematical details here. The main point is that both graphical (Figure 7 below) and numerical (Table 2 below) predictions of the behaviour of the damped particle–spring system can be obtained from the model.

$$m\lambda^2 + r\lambda + k = 0,$$

whose roots are

$$\lambda_1 = \frac{-r + \sqrt{r^2 - 4mk}}{2m} \quad \text{and} \quad \lambda_2 = \frac{-r - \sqrt{r^2 - 4mk}}{2m}.$$

Provided that $\lambda_1 \neq \lambda_2$, the solution is

$$x(t) = Be^{\lambda_1 t} + Ce^{\lambda_2 t},$$

where B and C are arbitrary constants that depend on the initial conditions. If λ_1 and λ_2 are real and distinct, which they will be if $r^2 - 4mk$ is positive, then the solution is the sum of two real exponential terms. As you may recall from Unit 1, if the roots of the auxiliary equation are complex (in this case, when $r^2 - 4mk$ is negative), then the solution can be written in real form as an exponential times a sinusoid, that is, as

$$x(t) = e^{-\rho t}(B\cos\nu t + C\sin\nu t),$$

where $\rho = r/(2m)$, $\nu = \sqrt{4mk - r^2}/(2m)$, and B and C are arbitrary constants that depend on the initial conditions. As in Unit 9, the sine and cosine terms can be combined to write this solution in the form

$$x(t) = Ae^{-\rho t}\cos(\nu t + \phi), \qquad (7)$$

Algebraic manipulation reveals that $B = A\cos\phi$ and $C = -A\sin\phi$ (see Unit 9).

where A and ϕ are arbitrary constants.

It can be seen that under certain circumstances, the model predicts oscillations (due to the sinusoidal factor) of decreasing amplitude (since the exponential term has a negative exponent).

The measured values $m = 0.711$ and $k = 23$ are the same in each case, and the damping constant r varies with each tube. Where $r^2 < 4mk = 65.41$, so $r < 8.09$, the model suggests that the magnet will oscillate with decreasing amplitude. The calculated values for r for tubes A to D are 0.15, 0.92, 1.33 and 8.42, so the model predicts that the magnet will oscillate for tubes A to C but not for tube D.

See Table 1.

In the cases of tubes A–C, the model can be used to predict the period of each oscillation and the rate at which the oscillations decay. For these three tubes the period of oscillations (in seconds) is

$$\tau = \frac{2\pi}{\nu} = \frac{4\pi m}{\sqrt{4mk - r^2}},$$

and that over each complete oscillation, the amplitude decays by a factor $e^{-\rho\tau}$, where $\rho = r/(2m)$. The predicted values are given in the last two columns of Table 2.

Table 2

Tube	Damping constant r ($\mathrm{N\,s\,m^{-1}}$)	Period τ (s)	Amplitude decay factor per cycle $e^{-\rho\tau}$
A	0.15	1.10	0.89
B	0.92	1.11	0.49
C	1.33	1.12	0.35

With these values we can sketch the graphs of the behaviour of the magnet in each of the four tubes (Figure 7).

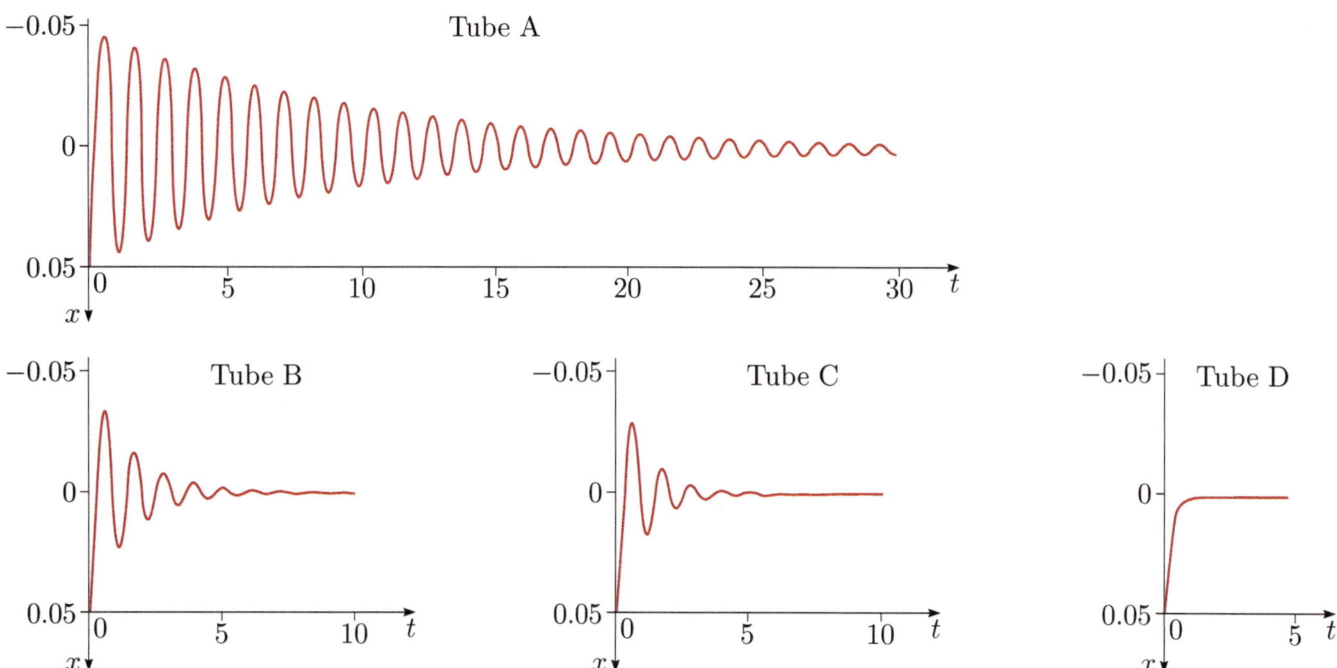

Figure 7 Predictions from the model for each tube

What are the most significant features in each case?

Tube A: The graph of the solution clearly shows the amplitude of the oscillations decreasing exponentially. The amplitude corresponds to the exponential factor $Ae^{-\rho t}$ identified earlier, while the sinusoidal factor gives oscillations of constant frequency. The model predicts that the oscillations die away relatively slowly.

See equation (7).

Tube B: The model again predicts decaying oscillations, though these die down more quickly than for tube A.

Tube C: The model predicts motion very similar to that for tube B.

Tube D: For this tube, with $r = 8.42$, the expression $4mk - r^2$ is negative, that is, $r^2 - 4mk$ is positive. In this case the auxiliary equation has two negative real roots and the solution is purely exponential. Hence there are no oscillations in this case, as confirmed by the corresponding graph in Figure 7. The model predicts no oscillations at all and a rapid return directly to the equilibrium position.

1.4 Comparing predictions with reality

Figure 7 provides predictions for the behaviour of the damped particle–spring system for each of the tubes. Do these predictions bear any relationship to reality?

The match between the model's graphical predictions in Figure 7 and the traces from the experiment in Figure 8 is striking.

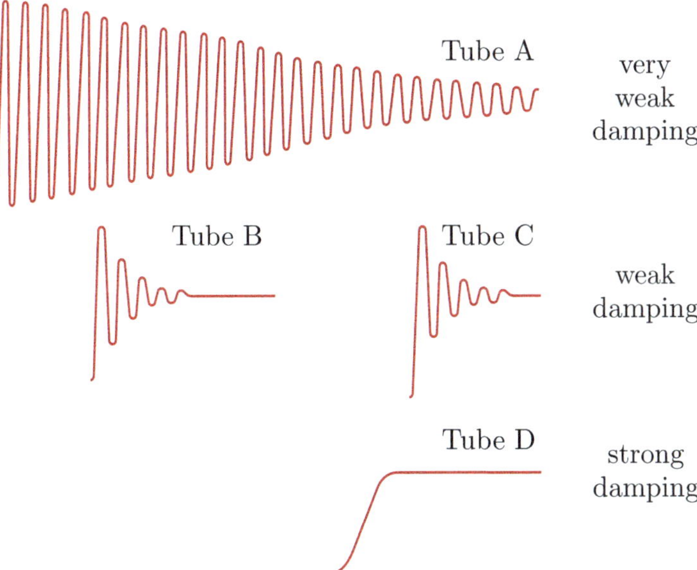

Figure 8 Traces of experimental results for each tube

These traces do not give an indication of the timescale involved. Timings provide an estimate of 1.1 seconds for the period of oscillations for tube A. The corresponding periods for tubes B and C are very similar, but harder to measure due to the few oscillations that occur. This period estimate is in agreement with those given in Table 2.

The tabulated predictions for the amplitude decay factor also appear to be borne out by the experiments, to the accuracy with which measurements can be made from the traces. For example, the prediction for tube A is that the amplitude should halve over about six cycles (since $0.89^6 \simeq 0.5$), whereas for tube B the amplitude is predicted to halve approximately every cycle.

We conclude that both qualitatively and quantitatively, the model is validated quite well by the experimental results. (There are nevertheless some discrepancies. For example, the actual decay of the amplitude in tube A does not look exponential throughout the range shown.)

When the magnet is in tube A, the damping is very light. With the magnet in either tube B or tube C, the damping level is higher but there are still some oscillations. The even higher level of damping in tube D prevents oscillations altogether. Damping like this, for which there are no oscillations, is called **strong damping**. Damping where there are decaying oscillations, as for the other three tubes, is called **weak damping**.

For a system in which the level of damping can be varied continuously, there will be a crossover point between weak and strong damping, and the corresponding level of damping is called **critical damping**. In the case of the experiments, critical damping lies between the damping levels of tubes C and D. Critical damping gives the most rapid return to the equilibrium position, given an initial displacement, whereas very strong damping can cause a significant delay in the return to equilibrium.

In terms of the mass m, spring stiffness k and damping constant r, we have weak damping if $r^2 - 4mk < 0$, strong damping if $r^2 - 4mk > 0$, and critical damping if $r^2 - 4mk = 0$. More is said about the mathematical aspects of damping in Section 2. To conclude this section, we ask you to consider briefly what type of damping might be required in various real-world systems.

For the particle–spring system discussed in Subsection 1.3, critical damping occurs when $r = 8.09$.

Different levels of damping are appropriate in different situations. A baby bouncer (which we will consider in Example 3) is more fun for the child the longer its oscillations continue, following an initial displacement, so here very weak damping is desirable.

For kitchen scales to be useful, any oscillations should die down quickly, so that readings may be taken. In this case the damping should be close to critical damping. For given kitchen scales, the values of r and k are fixed. There is then only one value m of the mass placed on the scales for which critical damping can be achieved exactly, since in this case $r^2 = 4mk$. Scales are usually designed to give a speedy return to the equilibrium position for a specified range of values of the mass, and since a few small oscillations initially are acceptable, damping that is weak but close to critical is preferred. Strong damping would result in the scales taking a longer time to return to the equilibrium position, and hence hold up the taking of a reading.

Vehicle suspension systems are designed to smooth out the ride, so here strong damping is better. However, the damping should not be *too* strong. The spring needs to be returned close to its equilibrium position in order to be able to absorb the next jolt from the road. The time between jolts will obviously depend on the road surface and the speed at which the vehicle is travelling. When the vehicle is carrying a heavy load, m will be larger. Designers must aim to include a level of damping that is appropriate for both the heaviest and the lightest loads envisaged, as well as for the different terrains and speeds that are likely to be encountered.

Exercise 3

What level of damping would be appropriate for each of the following mechanisms?

(a) Buffers at the end of a railway line, which are intended to halt a train that comes into the station too fast

(b) A device that prevents a door from slamming shut

(c) A mechanism linking the fuel gauge of a vehicle and a float in the fuel tank, which is designed to damp fluctuations in the gauge reading caused by travel over an uneven surface

(d) A tow-bar mechanism, which is designed to minimise the transfer of jolts from the towing vehicle to the towed vehicle, or vice versa

(e) Bathroom scales

2 Spring–damper models of motion

This section is designed to give you practice in modelling systems where both model spring forces and linear damping forces act. As in earlier units in this module, this involves drawing appropriate diagrams and deriving the equation of motion. The emphasis is on setting up the model, including appropriate initial conditions, and interpreting solutions in terms of the physical system concerned.

In order to represent linear damping diagrammatically, we introduce in Subsection 2.1 the concept of a *model damper*. This is applied to various modelling examples described in Subsection 2.2. A mathematical summary of the various types of motion that can be caused by model springs and dampers is given in Subsection 2.3.

2.1 The model damper

In Unit 9 we introduced the concept of a *model spring*. This is a convenient means of indicating diagrammatically and describing algebraically the presence of a force that depends linearly on the length l of the spring. The force exerted on a particle connected to an end of the model spring is given by Hooke's law,

See Unit 9, Subsection 1.1.

$$\mathbf{H} = k(l - l_0)\,\widehat{\mathbf{s}}, \qquad (8)$$

where k is the stiffness and l_0 is the natural length of the spring. The vector $\widehat{\mathbf{s}}$ is a unit vector in the direction from the particle towards the centre of the model spring (see Figure 9).

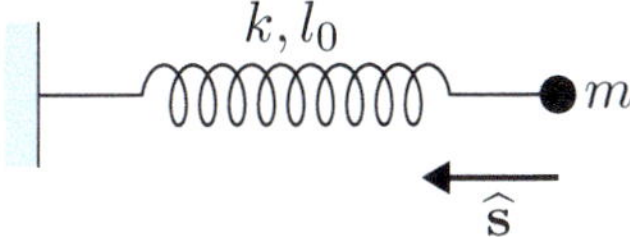

Figure 9 A model spring

In Unit 9 you saw this force specification applied to model springs that had one end attached to a particle and the other kept fixed. However, the same expression for the force on the particle applies even when the other end of the model spring is in motion. You have seen applications of this type in Unit 9, and will see further applications in Section 3 of this unit and in Unit 18.

We now define a *model damper*. Like the model spring, this is a hypothetical one-dimensional system component that can be included in diagrams and with which a certain vector expression for force is associated.

When one end of the damper is fixed, it embodies the linear damping model defined in Subsection 1.1.

A **model damper**, with a particle attached to one of its ends, provides a force on the particle that opposes its motion relative to the other end. The magnitude of this resistance force is proportional to the rate of change of length of the model damper. The force provided by the model damper (when compressing or extending) is therefore

$$\mathbf{R} = r\dot{l}\,\widehat{\mathbf{s}}, \tag{9}$$

where $\dot{l}$ is the rate of change of length of the model damper, r is a positive constant (called the **damping constant**), and $\widehat{\mathbf{s}}$ is a unit vector in the direction from the particle towards the centre of the model damper.

A model damper has zero mass. It is represented diagrammatically as shown in Figure 10.

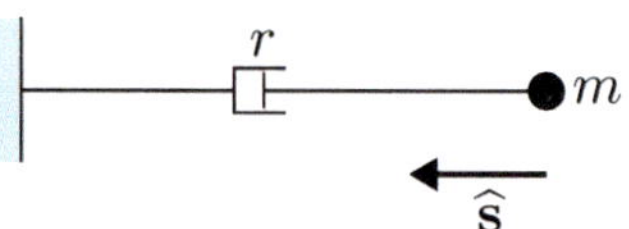

Figure 10 A model damper

The diagrammatic representation of a model damper, as shown in Figure 10, is a cross-sectional picture of an actual physical device known as a **dashpot**, which involves the motion of a piston within a circular cylinder. However, you could also think of it as a picture of the magnet within a copper tube from Section 1, where the resistance was electromagnetic.

One form of dashpot is used as the shock-absorber on a car, where the cylinder is full of oil and the relative motion of the piston causes oil to flow through the small annular gap between piston and cylinder. This provides resistance to the relative motion.

When used in a diagram, the model damper simply indicates the presence of a force of the type described by equation (9), just as a model spring in a diagram stands for a force of the type given by equation (8). Thus we could represent the modelling of the damped particle–spring system as shown in Figure 11. This is a more abstract form of the system in Figure 6, but it indicates clearly the assumed presence of linear damping, which the earlier diagram does not do.

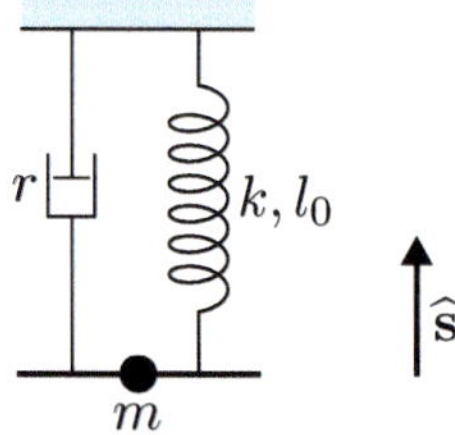

Figure 11 Modelling a damped particle–spring system

Although we represent a model spring and a model damper as independent elements, they may correspond to a single physical entity (a real spring, say) that exhibits to some extent both types of force behaviour. For example, the metal ruler from which nuts were hung in Unit 9 shows spring-like behaviour but also seems to exhibit considerable 'internal friction', which may be a more significant factor than air resistance in reducing the amplitude of oscillations. If we were to model this internal friction as being linear, then Figure 11 as it stands would suffice to represent our model for the nuts hanging from the ruler.

Exercise 4

(a) Suppose that a model damper is attached to a particle at one end and to a fixed point at the other, where the damper lies along the direction of motion of the particle (described by an x-axis). Show that if $\mathbf{i}$ is a unit vector in the positive x-direction, then equation (9) leads in this case to the expression

$$\mathbf{R} = -r\dot{x}\mathbf{i}$$

(as in the definition of linear damping in Subsection 1.1) for either of the possible choices of x-direction.

(b) Suppose now that the end of the model damper not attached to the particle is made to move in such a way that its position on the x-axis is given at time t by $y(t)\,\mathbf{i}$. Show that equation (9) now leads to

$$\mathbf{R} = -r(\dot{x} - \dot{y})\mathbf{i}.$$

Note that the model damper is very much a first model of damping effects, and may not describe accurately what occurs in real systems except over small ranges of the relative velocity. However, its simplicity makes it convenient to use, and it is capable (as you saw in Section 1) of providing reasonable representations of certain damped systems.

2.2 Applying model dampers

The examples and exercises in this subsection present a number of situations where there is some damping. In each case you are invited to consider carefully the setting up of the model, by drawing diagrams and then using Newton's second law to obtain the equation of motion. Do not focus too much here on how to solve the differential equations that arise. Concentrate rather on the modelling, including the specification of initial conditions and the way in which they enable values for the arbitrary constants to be found. Think also about the interpretation of the solutions.

The methods required to solve these differential equations are from Unit 1.

Example 2

A toy train of mass 2 kg, travelling on a straight horizontal track, freewheels into buffers at a speed of $0.25\,\mathrm{m\,s^{-1}}$.

(a) Model the buffers as a model spring with stiffness $25\,\mathrm{N\,m^{-1}}$ together with a model damper with damping constant $15\,\mathrm{N\,s\,m^{-1}}$. Derive the equation of motion, and determine its general solution.

(b) Use this model to predict the subsequent motion. In particular, by how much will the buffers be compressed, and what happens to the train thereafter?

Solution

(a) The model of the buffers is shown in Figure 12. The train is represented by a particle of mass m, and the model spring of stiffness k and natural length l_0, and model damper of damping constant r, are assumed to be joined at their non-fixed ends. The origin is taken to be where the free end of the spring is situated when the length of the spring is equal to its natural length. This is the point at which the front of the train will first come into contact with the buffers. The train's initial direction of travel is taken to be the positive x-direction.

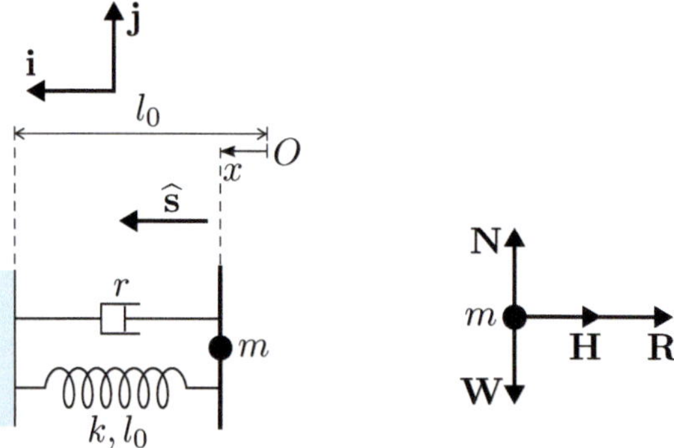

Figure 12 The buffers and the force diagram

The corresponding force diagram is also shown in Figure 12. Since the weight $\mathbf{W}$ of the train and normal reaction $\mathbf{N}$ on it from the track are vertical forces, they balance each other, as there is no motion in this direction. As with model springs, there is in general no single choice of direction for the model damper force that is correct at all times, but for the purposes of the force diagram, *either of the possible choices will do.* The directions for $\mathbf{H}$ and $\mathbf{R}$ in Figure 12 correspond to the model spring being compressed and the model damper shortening. The same unit vector $\widehat{\mathbf{s}} = \mathbf{i}$ can be used for the spring and damper, as the directions from the particle to the centre of the spring and centre of the damper are the same. The length of the spring is $l = l_0 - x$, so $\dot{l} = -\dot{x}$.

From Hooke's law, the model spring force is

$$\mathbf{H} = k(l - l_0)\,\widehat{\mathbf{s}} = k(l_0 - x - l_0)\mathbf{i} = -kx\mathbf{i},$$

and from equation (9) and Exercise 4(a), the resistance force is

$$\mathbf{R} = r\dot{l}\,\widehat{\mathbf{s}} = r(-\dot{x})\mathbf{i} = -r\dot{x}\mathbf{i}.$$

The equation of motion is therefore

$$\begin{aligned} m\ddot{x}\mathbf{i} &= \mathbf{H} + \mathbf{R} + \mathbf{W} + \mathbf{N} \\ &= -kx\mathbf{i} - r\dot{x}\mathbf{i} - mg\mathbf{j} + |\mathbf{N}|\,\mathbf{j}, \end{aligned}$$

which after resolution in the $\mathbf{i}$-direction gives

$$m\ddot{x} + r\dot{x} + kx = 0.$$

This is equation (6) once more.

Substituting in the given values $m = 2$, $k = 25$ and $r = 15$, we have

$$2\ddot{x} + 15\dot{x} + 25x = 0.$$

Note that in this case $r^2 - 4mk = 25$, which is positive. According to the criteria obtained near the end of Section 1, this confirms that the buffers will provide strong damping, with real exponential terms in the solution, rather than decaying oscillations.

The corresponding auxiliary equation is $2\lambda^2 + 15\lambda + 25 = 0$, which has roots $\lambda = -5$ and $\lambda = -2.5$. Hence the general solution of the differential equation is

$$x = Ae^{-5t} + Be^{-2.5t},$$

where A and B are arbitrary constants that depend on the initial conditions. In order to find the appropriate particular solution, we need to formulate these initial conditions.

(b) Choose the origin of time as the instant at which the particle representing the train makes contact with the buffers. The initial conditions are then $x(0) = 0$ and $\dot{x}(0) = 0.25$. Substituting the first of these into the general solution gives $0 = A + B$, so we have $B = -A$. To apply the second initial condition, we first need to differentiate the general solution, obtaining

$$\dot{x} = -5Ae^{-5t} - 2.5Be^{-2.5t}.$$

Substituting $\dot{x}(0) = 0.25$ and $B = -A$ here gives $0.25 = -5A + 2.5A$, so $A = -0.1$ and $B = 0.1$. Putting these values for A and B into the general solution gives the required particular solution as

$$x = 0.1(e^{-2.5t} - e^{-5t}).$$

The graph of this position function is shown in Figure 13.

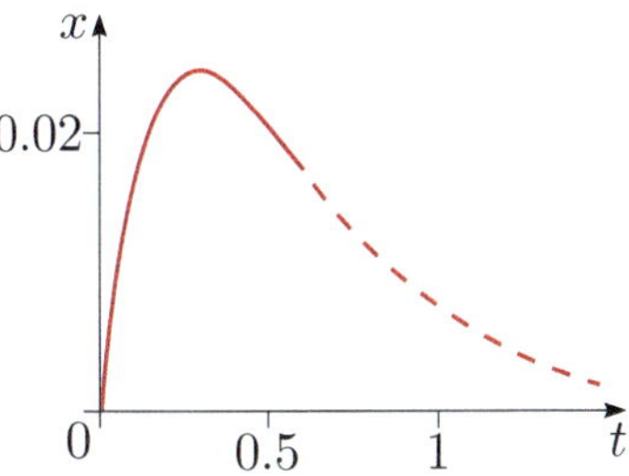

Figure 13 Graph of the position function x

We now need to interpret this solution in order to predict the motion of the train after it meets the buffers. Clearly, the buffers will be compressed. The maximum compression is achieved when x is a maximum, for which $\dot{x} = 0$ (the train's velocity is zero). This occurs when

$$\dot{x} = -0.25e^{-2.5t} + 0.5e^{-5t} = 0, \quad \text{that is,} \quad 1 = 2e^{-2.5t}.$$

Here we have divided through by $e^{-2.5t}$ and noted that $e^{-5t} = (e^{-2.5t})^2$.

The corresponding time is $t = 0.4\ln 2 \simeq 0.28$, and (putting this time into the particular solution) the maximum compression is $x = 0.1\left(\frac{1}{2} - \frac{1}{4}\right) = 0.025$. So the model predicts that the buffers will be compressed by 2.5 cm.

For this to be reasonable, the spring must have a natural length longer than this! Otherwise, the model predicts that the train will compress the spring to zero length without being brought to a halt, which might result in the train breaking through the buffers or becoming derailed, or being brought to a standstill very quickly.

It remains to consider what happens to the train after it has been instantaneously brought to rest at the point of maximum compression for the model spring. As expected, and as the graph in Figure 13 indicates, it then starts to move back in the direction from which it arrived. However, the spring–damper is not attached to the train thus can only 'push' it, not 'pull' it. If at some time the model predicts a total force on the particle that is in the positive x-direction, then the model has become invalid. This will correspond to a moment at which the train has lost contact with the buffers.

According to Newton's second law, the total force on the particle is equal to $m\ddot{x}\mathbf{i}$, so the train leaves the buffers where $\ddot{x}$ becomes positive. This is where the graph in Figure 13 changes to a broken line. Differentiation of the expression above for $\dot{x}$ gives

Notice that the curvature of the graph reverses where the graph changes to a broken line; there is a point of inflection, where $\ddot{x}$ changes sign.

$$\ddot{x} = 0.625e^{-2.5t} - 2.5e^{-5t} = 0.625e^{-2.5t}(1 - 4e^{-2.5t}).$$

This expression becomes positive when $4e^{-2.5t}$ is less than 1, which first happens when $t = 0.4 \ln 4 \simeq 0.55$. Therefore the corresponding position of the particle is $x = 0.1\left(\frac{1}{4} - \frac{1}{16}\right) \simeq 0.019$, which is 0.6 cm back from the point of maximum compression. The velocity at this time is $\dot{x}\mathbf{i}$, where $\dot{x} = -0.25 \times \frac{1}{4} + 0.5 \times \frac{1}{16} \simeq -0.031$.

We conclude that provided that the buffers are long enough to sustain the maximum compression, they are predicted to turn an incoming speed of $0.25\,\mathrm{m\,s^{-1}}$ into an outgoing speed of $0.031\,\mathrm{m\,s^{-1}}$. This is the sort of effect that buffers are intended to have!

The speed decrease is also evidence of a very significant decrease in the train's kinetic energy. There is no space in this unit to focus further on the topic of energy, beyond pointing out that the presence of damping in a system will always entail loss of energy, and that heavier damping means a greater rate of loss of energy.

In the example above, we chose the x-axis to be directed from the track towards the buffers, with origin at the point where the train first comes into contact with the buffers. If the opposite direction were to be chosen for the x-axis, with the same origin, then the equation of motion would not be altered. The initial condition $x(0) = 0$ is unchanged, while that for $\dot{x}$ has the opposite sign, that is, $\dot{x}(0) = -0.25$. The solution for x, correspondingly, has its sign reversed.

If, on the other hand, we selected a different point for the origin (with either direction for the x-axis), the equation of motion would become

This is equation (5) once more.

$$m\ddot{x} + r\dot{x} + kx = kx_{\mathrm{eq}},$$

where x_{eq} is the value of x where the train meets the buffers. The initial conditions are now $x(0) = x_{\mathrm{eq}}$, $\dot{x}(0) = \pm 0.25$, where the sign for $\dot{x}(0)$ depends on the choice of x-direction, as discussed above. The solution for x is altered by the addition of x_{eq} to the expression obtained previously (when the origin is at the point where the train meets the buffers).

As you would expect, the eventual answers to the problem in Example 2 do not depend on the choices of origin or direction for the x-axis. The interpretation is the same in each case.

Exercise 5

A miniature train of mass 40 kg, travelling on a straight horizontal track, freewheels into buffers at a speed of $1\,\mathrm{m\,s^{-1}}$. The buffers are to be modelled by a model spring with stiffness $140\,\mathrm{N\,m^{-1}}$, together with a model damper with damping constant $180\,\mathrm{N\,s\,m^{-1}}$. The x-axis is chosen directed away from the buffers down the track (in the direction opposite to the incoming train), with origin at the fixed end of the model spring. The natural length of the model spring is 0.5 m.

(a) Show that the equation of motion can be written as

$$4\ddot{x} + 18\dot{x} + 14x = 7.$$

(b) Write down a pair of initial conditions for the motion of the train while it is in contact with the buffers.

(c) The solution of the equation of motion that satisfies the initial conditions of part (b) is

$$x = 0.5 - 0.4e^{-t} + 0.4e^{-3.5t}.$$

What is the maximum compression of the buffers? At what point, and with what speed, does the train leave the buffers?

Example 3

A baby bouncer's suspension (see Figure 14) is modelled by a model spring with stiffness $200\,\mathrm{N\,m^{-1}}$. (This is equivalent to having two equal springs of stiffness $100\,\mathrm{N\,m^{-1}}$; see Unit 9, Exercise 4.) There is some internal damping in the suspension and there is some air resistance, which together can be modelled by a model damper with damping constant $0.2\,\mathrm{N\,s\,m^{-1}}$. The bouncer is designed for a baby whose mass is about 10 kg.

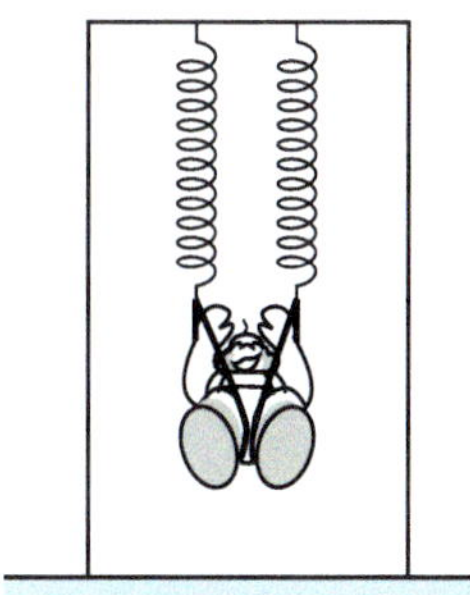

Figure 14 A baby bouncer

Set up the equation of motion for the model, and find the particular solution for the case in which the baby is released from rest when 0.3 m above its equilibrium position. Use the model to predict how long it takes before the amplitude of the oscillations drops below 0.05 m, if the baby is not pushed in any way.

Solution

The system and force diagram are shown in Figure 15. We model the baby as a particle of mass m. Take the origin at the top of the spring, with the x-axis pointing downwards and the unit vector $\mathbf{i}$ pointing in the positive x-direction, so $l = x$ and $\dot{l} = \dot{x}$.

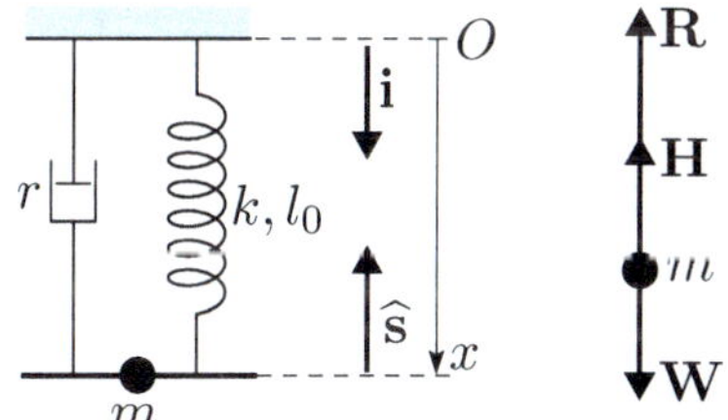

Figure 15 Spring–damper system and force diagram

The forces acting on the baby are the weight $\mathbf{W} = mg\mathbf{i}$, the model spring force $\mathbf{H} = k(x - l_0)(-\mathbf{i})$, and the resistance force $\mathbf{R} = r\dot{x}(-\mathbf{i})$. Newton's second law gives

$$m\ddot{x}\mathbf{i} = \mathbf{W} + \mathbf{H} + \mathbf{R} = mg\mathbf{i} - k(x - l_0)\mathbf{i} - r\dot{x}\mathbf{i},$$

which leads, after resolution in the $\mathbf{i}$-direction, to the equation of motion

$$m\ddot{x} + r\dot{x} + kx = mg + kl_0.$$

The right-hand side can be written as kx_{eq}, where $x_{\mathrm{eq}} = mg/k + l_0$ is the equilibrium position of the baby. Hence $x_{\mathrm{p}} = x_{\mathrm{eq}}$ is a particular integral of the differential equation.

After substituting the given values for m, k and r, the associated homogeneous differential equation has auxiliary equation

$$10\lambda^2 + 0.2\lambda + 200 = 0,$$

which has solutions $\lambda = -0.01 \pm 4.5i$ (to two significant figures). This leads to the complementary function

$$x_c = Ae^{-0.01t}\cos(4.5t + \phi),$$

where A and ϕ are arbitrary constants, so the general solution is

$$x = x_c + x_p = Ae^{-0.01t}\cos(4.5t + \phi) + x_{eq}.$$

This represents decaying oscillations about the equilibrium position.

In order to find the particular solution, we need the initial conditions. Since the baby is released from rest at 0.3 m above the equilibrium position, the initial conditions are $x(0) = x_{eq} - 0.3$ and $\dot{x}(0) = 0$. These give

$$-0.3 = A\cos\phi \quad \text{and} \quad 0 = A(-0.01\cos\phi - 4.5\sin\phi).$$

Solving these equations leads to $\phi \simeq \pi - 0.0022$ and $A \simeq 0.30$, so the amplitude of the decaying oscillations is about $0.30e^{-0.01t}$ m. This reduces to 0.05 m when $0.30e^{-0.01t} = 0.05$, or $t = 100\ln 6 \simeq 180$ (about 3 minutes).

Exercise 6

A sit-ski is a seat attached above a ski, with a spring–damper suspension (see Figure 16). The spring is chosen according to the weight of the skier.

Figure 16 A sit-ski on a ski slope

The damper can be adjusted according to the skier's weight and the terrain. Suppose that a particular sit-ski suspension is modelled as a model spring–damper system, with a spring of natural length 0.2 m and stiffness 30 000 N m^{-1}, and damping constant 6300 N s m^{-1}. The skier (plus seat) is modelled as a single particle of mass 60 kg. Take the magnitude of the acceleration due to gravity as $g = 9.81$ m s^{-2}. Figure 17 represents the sit-ski with skier.

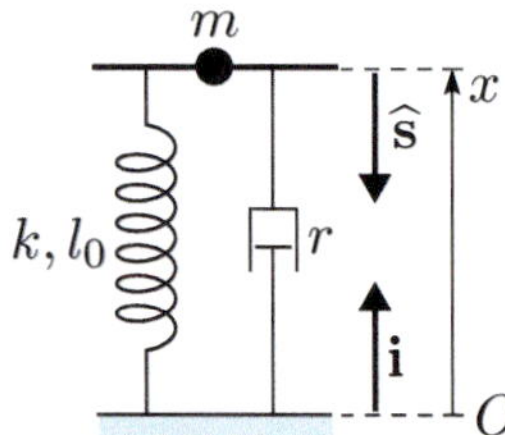

Figure 17 Spring–damper system for the sit-ski and skier

(a) Draw the diagram for the forces acting on the skier. Take the x-axis to be directed upwards, with origin at ground level, and obtain the corresponding equation of motion.

(b) Suppose that the skier is lowered onto the seat and then released from rest when the spring has its natural length. Write down the corresponding initial conditions for the skier's subsequent motion.

(c) Given that the particular solution of the equation of motion is $x = x_{\text{eq}} + 0.021e^{-5t} - 0.001e^{-100t}$ (in metres), find approximately how long the model predicts that it will take before the skier's displacement is within 0.001 m (1 mm) of the equilibrium position.

We look next at a different type of situation, although it leads to an equation of motion that is very similar to those seen already.

The fuel gauge for a vehicle is connected to a mechanism that monitors the level of fuel in the tank. This mechanism is designed to damp oscillations in the gauge reading after disturbances such as those caused by travel over bumps on the road. The mechanism includes a float on the surface of the liquid in the fuel tank, and the buoyancy force of the liquid on this float acts in an analogous way to a spring.

You first met buoyancy in Unit 9, Exercise 19 and the text above it, in the context of simple harmonic motion.

A buoyancy force, sometimes called an *upthrust*, is experienced by any object that is wholly or partly immersed in a liquid. If the object floats in equilibrium on the surface of the liquid, then the upthrust from the liquid balances the weight of the object (see Figure 18). If you push the object down further into the liquid, then there is a greater upthrust pushing it back up. When you lift the object up a little from its equilibrium position, the upthrust is less and the weight of the object pulls it down again.

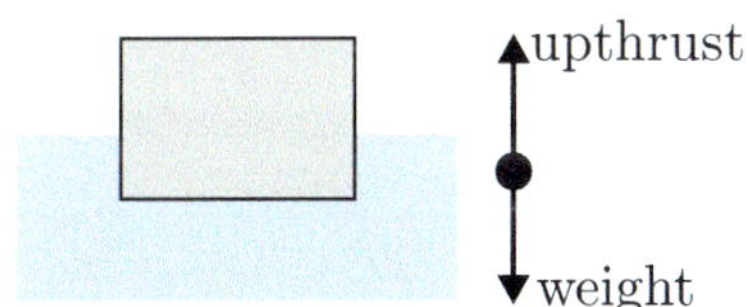

Figure 18 A float on the surface of a liquid

According to Archimedes' principle (see Unit 9), the upthrust is directed vertically upwards and is equal in magnitude to the weight of liquid displaced by the object. It follows that if the object has a constant horizontal cross-section, then the magnitude of the upthrust on it is proportional to the depth of its base below the surface of the liquid. In fact, if the displacement of the base (measured downwards from the surface of the liquid) is $x\mathbf{i}$, then the upthrust is $-kx\mathbf{i}$, where k is a constant that depends on the density of the liquid and the cross-sectional area of the floating object. (For this model to be valid, the object must be at least partly immersed in the liquid, but not wholly submerged.)

In fact, $k = \rho A g$, where ρ is the liquid density, A is the cross-sectional area, and g is the magnitude of the acceleration due to gravity.

Hence the float system in a vehicle's fuel tank can be modelled by a model spring–damper system, provided that the length of the spring is regarded as being equal to its natural length when its non-fixed end is at the level specified by the liquid surface. However, it may be more straightforward to write down the upthrust force directly, as in Example 4 below. The return to equilibrium should be rapid, for ease of reading the fuel gauge, hence the system requires near-critical damping.

Example 4

This formula for the upthrust is specific to this example. It corresponds to the equilibrium position of the float being half in and half out of the fuel.

The mechanism in a particular fuel tank is to be modelled by a model spring–damper system. The upthrust from the liquid fuel on the float is equal in magnitude to twice the weight of the float times the proportion of the float below the surface. The mass of the float is $m = 0.1$, its vertical height is $d = 0.01$, and the damping constant is $r = 21$. Take $g = 9.81\,\mathrm{m\,s^{-2}}$.

Find the equation of motion, and solve it for the case in which the motion begins with the base of the float 0.01 m below the surface, with zero velocity. Hence predict when the float will be less than 0.001 m (1 mm) from its equilibrium position, assuming that there is no further disturbance.

Solution

Take x as the downward displacement of the bottom of the float from the surface of the liquid (see Figure 19). Then the unit vector $\mathbf{i}$ points downwards, and the unit vector $\widehat{\mathbf{s}}$, for both the upthrust (spring) and the damper, points upwards.

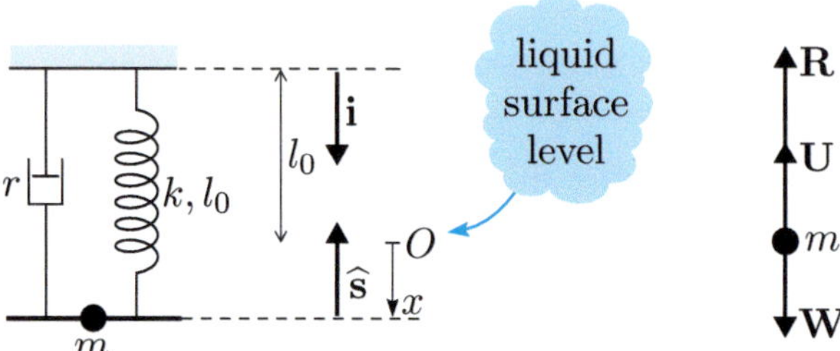

Figure 19 Spring–damper system and force diagram

From the form of $\mathbf{U}$, the stiffness of the model spring equivalent to the upthrust is $k = 2mg/d$.

The upthrust from the liquid is $\mathbf{U} = 2mg(x/d)\,\widehat{\mathbf{s}} = 2mg(x/d)(-\mathbf{i})$, the weight is $\mathbf{W} = mg\mathbf{i}$, and the damping resistance force is $\mathbf{R} = r\dot{x}\,\widehat{\mathbf{s}} = r\dot{x}(-\mathbf{i})$. Hence, using Newton's second law, we obtain

$$m\ddot{x}\mathbf{i} = \mathbf{W} + \mathbf{U} + \mathbf{R} = mg\mathbf{i} - (2mg/d)x\mathbf{i} - r\dot{x}\mathbf{i},$$

which leads to the equation of motion

$$m\ddot{x} + r\dot{x} + (2mg/d)x = mg.$$

Once the parameter values have been substituted, the general solution is found to be

$$x = Ae^{-200t} + Be^{-9.80t} + 0.005,$$

where A and B are constants.

The initial conditions are $x(0) = 0.01$ and $\dot{x}(0) = 0$, which give the particular solution

$$x = 0.005\,26\,e^{-9.80t} - 0.000\,257\,e^{-200t} + 0.005.$$

There are three terms in the solution. The last represents the equilibrium position of 0.005 m. The second term involves e^{-200t}, which dies away very quickly, and the other term is the dominant term in the variable part of the solution, namely $0.005\,26\,e^{-9.80t}$. This will reduce to 0.001 m when $e^{-9.80t} = 0.190$, that is, when $t = -(\ln 0.190)/9.80 \simeq 0.17$.

Hence the model predicts that the displacement of the float will be within 1 mm of its equilibrium position in less than a fifth of a second.

Exercise 7

Modify the model from Example 4 for a mechanism where the upthrust from the liquid fuel is again equal in magnitude to twice the weight of the float times the proportion of the float below the surface level, and the mass of the float is again $m = 0.1$, but now the vertical height of the float is $d = 0.02$ and the damping constant is $r = 11$. Take $g = 9.81\,\mathrm{m\,s^{-2}}$.

(a) Derive the equation of motion, and write down the initial conditions for the case in which the motion begins with the base of the float at the liquid surface with zero velocity.

(b) Given that the particular solution for the initial conditions described in part (a) is $x = 0.001\,08\,e^{-100.2t} - 0.0111\,e^{-9.79t} + 0.01$, predict when the float will be less than 0.001 m (1 mm) from its equilibrium position, assuming that there is no further disturbance.

Exercise 8

Bathroom scales can be modelled by a model spring–damper system, as shown in Figure 20. If the stiffness of the spring is $k = 50\,000$ and the damping constant is $r = 5000$, what mass of person standing on the scales will give critical damping? What would you expect to happen if a slightly heavier or lighter person stood on the scales?

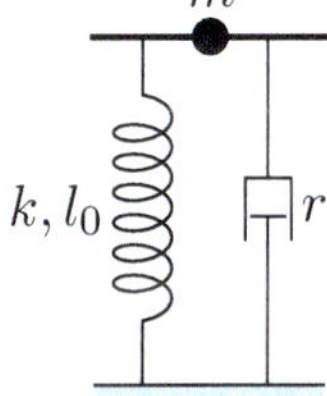

Figure 20 Spring–damper system for bathroom scales

All the models encountered in this subsection have led to equations of motion of the form

$$m\ddot{x} + r\dot{x} + kx = kx_{\mathrm{eq}}, \tag{10}$$

where x_{eq} is the equilibrium position of the particle. The following points should now be apparent.

- In equation (10), only the expression for x_{eq} depends on the choices of origin and direction for the x-axis.
- If the origin is chosen to be the equilibrium position of the particle, then equation (10) reduces to its homogeneous form

  $$m\ddot{x} + r\dot{x} + kx = 0. \tag{11}$$

- Given any solution of equation (11), there is a corresponding solution of equation (10) obtainable by adding the constant x_{eq}.

In fact, $x_{\mathrm{p}} = x_{\mathrm{eq}}$ is a particular integral for equation (10).

Hence equation (11) always describes the motion of the system *relative to the equilibrium position.* In the next subsection, we summarise the mathematical possibilities that arise when solving this differential equation.

We have concentrated on systems with a single model spring and a single model damper, but it is easy to extend the results to situations where more than one model spring or damper is present, acting as before along an x-axis. All that needs to be done is to add an appropriate force term to the right-hand side of Newton's second law for each component present. For example, if there are two model springs, with stiffnesses k_1 and k_2, then for motion relative to the equilibrium position we again obtain equation (11), with $k = k_1 + k_2$. In other words, the combined effect of the two model springs is equivalent to that of a single model spring with stiffness $k_1 + k_2$. A similar result holds for model dampers: the combined effect of two model dampers, with damping constants r_1 and r_2, is equivalent to that of a single model damper with damping constant $r_1 + r_2$.

2.3 Weak, critical and strong damping

A mechanical system whose equation of motion is of the form (10) or (11) is called a **damped linear harmonic oscillator**, or **damped harmonic oscillator** for short. In the absence of damping, the equation reduces to that of a simple harmonic oscillator, whose motion you studied in Unit 9.

During this subsection, you may like to refer to Procedure 5 in Unit 1.

In this subsection we return to the types of behaviour that can occur for a damped harmonic oscillator that we saw in Section 1. We concentrate on the equation of motion (11), for which the motion of the particle is described relative to its equilibrium position. The auxiliary equation for equation (11) is

$$m\lambda^2 + r\lambda + k = 0,$$

whose roots are

$$\lambda_1 = \frac{-r + \sqrt{r^2 - 4mk}}{2m} \quad \text{and} \quad \lambda_2 = \frac{-r - \sqrt{r^2 - 4mk}}{2m}. \tag{12}$$

The form of solution falls into one of three types, depending on whether the expression $r^2 - 4mk$ is positive, negative or zero. These three cases correspond respectively to strong, weak and critical damping.

Before examining each case in turn, we introduce the **damping ratio**

Note that by definition, $\alpha > 0$.

$$\alpha = \sqrt{\frac{r^2}{4mk}} = \frac{r}{2\sqrt{mk}}, \tag{13}$$

which makes some of the mathematical descriptions more transparent. For example, since $r^2 - 4mk = 4mk(\alpha^2 - 1)$, the conditions for strong, weak and critical damping can be expressed in terms of the damping ratio as $\alpha > 1$, $\alpha < 1$ and $\alpha = 1$, respectively. Whereas the damping *constant* r provides an absolute value for the damping force per unit speed exerted on the particle, the damping *ratio* α gives a measure of how important damping is *relative* to the mass m and spring stiffness k of the system.

Note that α is a dimensionless quantity, since the dimensions of r are the same as those of $\sqrt{mk}$.

Exercise 9

Increasing the damping constant r (while keeping the mass m and spring stiffness k fixed) will increase the damping ratio α. What other changes in parameters will increase α?

Exercise 10

For the spring–damper system in Exercise 1, the mass was $m = 0.711$ and the spring stiffness was $k = 23$. The damping constants r for tubes A–D were, respectively, 0.15, 0.92, 1.33, 8.42. Find the corresponding damping ratios, and verify that tubes A–C provide weak damping while tube D provides strong damping.

We will now look in turn at each of the three cases of damping identified above, but first, here is a reminder of the situation with no damping that you saw in Unit 9.

No damping

If there is no damping, then we have $r = 0$ and $\alpha = 0$. From equations (12), the roots of the auxiliary equation are

$$\lambda_1 = \frac{\sqrt{-4mk}}{2m} = i\sqrt{\frac{k}{m}} \quad \text{and} \quad \lambda_2 = -\frac{\sqrt{-4mk}}{2m} = -i\sqrt{\frac{k}{m}}.$$

The motion is simple harmonic, as described by the solution

$$x(t) = A\cos(\omega t + \phi),$$

where A and ϕ are arbitrary constants, and $\omega = \sqrt{k/m}$ is the *natural* (undamped) *angular frequency*. In practice, we restrict A to be positive, and call it the *amplitude* of the motion. The *period* of the motion is $\tau = 2\pi/\omega$, and ϕ (restricted to the range $-\pi < \phi \leq \pi$) is called the *phase angle*. A graph of $x(t)$, with $\phi = 0$, is shown in Figure 21.

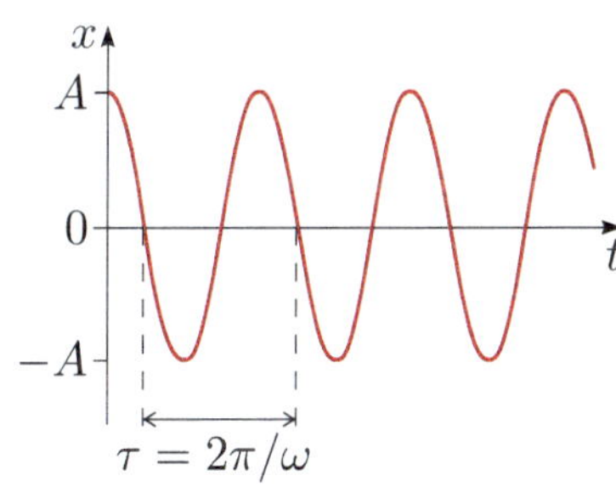

Figure 21 No damping

Weak damping

For weak damping, $r^2 - 4mk < 0$ and $\alpha < 1$. The solution of equation (11) is the product of a decaying exponential and a sinusoidal function, namely

$$x(t) = Ae^{-\rho t}\cos(\nu t + \phi), \tag{14}$$

where $\rho = r/(2m)$, $\nu = \sqrt{4mk - r^2}/(2m)$, and A and ϕ are arbitrary constants. (As before, we restrict A to be positive, and the phase angle ϕ to be within the range $-\pi < \phi \leq \pi$.) A graph of this motion is shown in Figure 22.

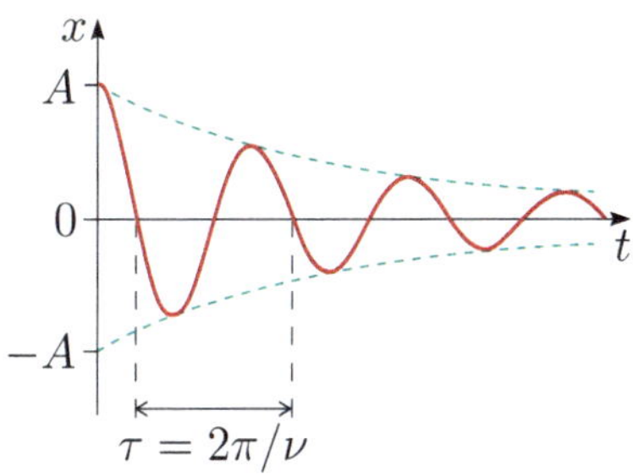

Figure 22 Weak damping

For simple harmonic motion, the period is the time to complete one cycle. Here the motion does not repeat itself, so we generalise the definition. The **period** τ of the motion is the time between successive zeros of x where the particle is moving in the same direction. In equation (14), the angular frequency is ν and the period is $\tau = 2\pi/\nu$.

To investigate equation (14) further, we write the parameters ρ and ν in terms of the damping ratio $\alpha = r/(2\sqrt{mk})$ and the natural angular frequency $\omega = \sqrt{k/m}$. To do this, we first note that

$$\omega\alpha = \sqrt{\frac{k}{m}} \times \frac{r}{2\sqrt{mk}} = \frac{r}{2m} = \rho.$$

The angular frequency ν can also be written in terms of ω and α:

$$\begin{aligned}\nu = \frac{\sqrt{4mk - r^2}}{2m} = \frac{2\sqrt{mk}\sqrt{1 - r^2/(4mk)}}{2m} &= \sqrt{\frac{k}{m}}\sqrt{1-\alpha^2} \\ &= \omega\sqrt{1-\alpha^2}.\end{aligned}$$

The negative exponent $-\omega\alpha t$ ensures that $Ae^{-\omega\alpha t}$ is a decreasing function, since $A > 0$.

This effect was apparent in Table 2, where the periods for tubes A–C were very close to the undamped value of 1.10 seconds.

The broken curves in Figure 22, corresponding to the two graphs $x = \pm Ae^{-\omega\alpha t}$, indicate how quickly the oscillations decay (larger α gives more rapid decay). The angular frequency $\nu = \omega\sqrt{1-\alpha^2}$ is less than the natural angular frequency ω, so the period $\tau = 2\pi/\nu$ is greater than the period of undamped oscillations (becoming larger as α increases). If α is close to zero, then the period τ is very close to its undamped value $2\pi/\omega$ (because τ depends on the square of α, namely $\tau = 2\pi/(\omega\sqrt{1-\alpha^2})$).

For simple harmonic motion, the amplitude is the constant maximum displacement from the mean position. In this motion, the maximum displacement from the mean position is not constant, so we define the **amplitude** to be the positive and continually changing quantity $Ae^{-\omega\alpha t}$. Over one cycle, of period τ, the amplitude of the motion decreases from $Ae^{-\omega\alpha t}$ to $Ae^{-\omega\alpha(t+\tau)}$, which is equivalent to $Ae^{-\omega\alpha t}e^{-\omega\alpha\tau}$, so $Ae^{-\omega\alpha t}$ is multiplied by the factor

$$e^{-\omega\alpha\tau} = \exp\left(-\frac{2\pi\alpha}{\sqrt{1-\alpha^2}}\right).$$

This is the amplitude decay factor per cycle, which was used to make predictions for the magnet motion within tubes A–C in Table 2.

Exercise 11

Suppose that the period of a weakly damped harmonic oscillator is greater by 10% than the corresponding undamped period $2\pi/\omega$. Show that the amplitude of the motion decays by a factor of about 0.056 per cycle.

A conclusion from the result of Exercise 11 is that if the period is very different from the undamped value, then few oscillations will be visible before the motion dies away.

Strong damping

Now consider the solution of equation (11) when $r^2 - 4mk > 0$, that is, when $\alpha > 1$. Here the auxiliary equation has real roots given by equations (12), namely

$$\lambda_1 = \frac{-r + \sqrt{r^2 - 4mk}}{2m} \quad \text{and} \quad \lambda_2 = \frac{-r - \sqrt{r^2 - 4mk}}{2m},$$

and the corresponding solution is

$$x(t) = Be^{\lambda_1 t} + Ce^{\lambda_2 t}, \tag{15}$$

where B and C are arbitrary constants. In terms of the natural angular frequency ω and the damping ratio α, the roots can be written as

$$\lambda_1 = \omega(-\alpha + \sqrt{\alpha^2 - 1}\,) \quad \text{and} \quad \lambda_2 = \omega(-\alpha - \sqrt{\alpha^2 - 1}\,).$$

Both λ_1 and λ_2 are negative since $\alpha > \sqrt{\alpha^2 - 1}$. Hence solution (15) is a sum of two *decaying* exponentials. Solutions of this type predict a return to the equilibrium position without oscillation (see Figure 23), although the graph of x against t may cross the t-axis once.

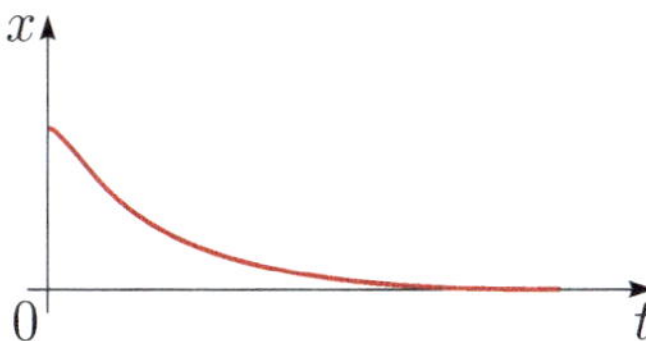

Figure 23 Strong damping

Exercise 12

As α increases from 1, what happens to the values of λ_1 and λ_2? Which of the exponential terms in equation (15) will be the dominant term for very strongly damped systems?

Critical damping

Finally, we consider the solution of equation (11) when $r^2 - 4mk = 0$, that is, when $\alpha = 1$. Here we have equal roots of the auxiliary equation, namely

$$\lambda_1 = \lambda_2 = -\frac{r}{2m} = -\omega\alpha = -\omega.$$

From Procedure 5 of Unit 1, the corresponding solution is

$$x(t) = (Bt + C)e^{-\omega t},$$

where B and C are arbitrary constants. Solutions of this form do not represent oscillations, although the graph of x against t may cross the t-axis once if the initial conditions are such that $t = -C/B > 0$. The graph (see Figure 24) resembles that for strong damping, but the system returns more quickly to close to the equilibrium position.

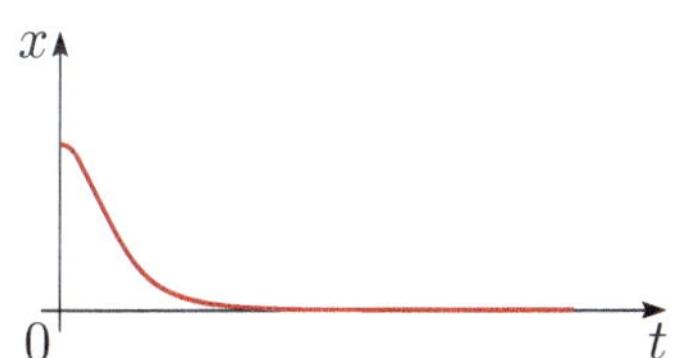

Figure 24 Critical damping

In conclusion, note that there is a continuum of behaviour from weak damping, with marked decaying oscillations, through near-critical damping, with a rapid return towards the equilibrium position, to strong damping, with a slower return towards the equilibrium position. Figures 21 to 24 show snapshots of the possible behaviour along this continuum.

3 Forcing the oscillations

All the damped solutions obtained in Section 2 have the property that they tend over time towards the equilibrium position of the particle, with the motion dying away. Sometimes, in addition to the effect of the model spring and damper, we need to model a further force that acts to keep the system moving. For example, the sit-ski may go over bumpy terrain or an adult may push the child in the baby bouncer. The motion of a model spring–damper system that is subjected to an additional time-dependent force is said to be **forced**.

There are several distinct ways of providing the forcing to a damped particle–spring system. In Subsection 3.1 we look at the case in which a periodic force is applied to the particle itself. This could, for example, model the regular forcing of the baby bouncer motion by pushing the child. In Subsection 3.2 we turn to alternative possibilities in which the force arises due to the prescribed displacement of some point of the system other than the particle itself. The end of the model spring or damper not attached to the particle would be such a point. For example, the baby bouncer motion could be forced by the action of a motor that moves the top end of the spring up and down. For the sit-ski suspension, it is the base of the model spring–damper that is displaced, due to contact with an uneven surface beneath. In Subsection 3.3 we summarise mathematically how the amended equation of motion may be solved.

3.1 Direct forcing

As a first model, we assume that the forcing is not just periodic, but sinusoidal. In Unit 13 you will see that any periodic force can be expressed as a sum of sinusoidal terms of different frequencies. It follows from the principle of superposition that if we can find a particular integral of the equation of motion for a 'typical' sinusoidal input, then by taking an appropriate sum of such solutions, we obtain the particular integral for any periodic input. The assumption of sinusoidal forcing is not therefore as restrictive as it might seem initially.

See Unit 1, Theorem 2.

Example 5

Consider the baby bouncer described in Example 3, with spring stiffness $k = 200$ and damping constant $r = 0.2$. The baby plus parts of the apparatus suspended from the spring have mass 10 kg. Suppose that by alternately pushing downwards and pulling upwards on the baby, an adult exerts a direct sinusoidal force of amplitude 10 newtons and frequency 1 hertz (1 cycle per second).

(a) Modelling the baby as a particle, formulate the equation of motion for the baby.

(b) Find the general solution of the equation of motion, and interpret this to predict what motion the baby will undergo in the long term. Is this affected by the initial conditions?

Solution

(a) The model assumes that the baby does not touch the ground and that the motion is completely vertical. The model spring–damper diagram for the system is shown in Figure 25, along with the corresponding force diagram. The origin is taken, as in Example 3, to be at the top of the spring, with the x-axis and unit vector $\mathbf{i}$ pointing downwards. The unit vector $\widehat{\mathbf{s}}$ for both the spring and the damper points upwards.

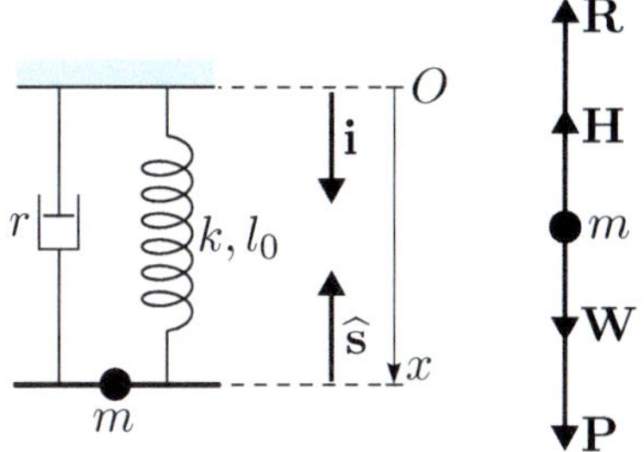

Figure 25 Spring–damper system and force diagram for the baby bouncer

There are four forces to consider. Three of these are the same as in Example 3, namely the weight $\mathbf{W} = mg\mathbf{i}$, the model spring force $\mathbf{H} = k(x - l_0)\widehat{\mathbf{s}} = k(x - l_0)(-\mathbf{i})$, and the resistance force $\mathbf{R} = r\dot{x}\widehat{\mathbf{s}} = r\dot{x}(-\mathbf{i})$. In addition, we have the sinusoidal force provided by the adult. With a suitable choice of time origin, this can be represented as $\mathbf{P} = P\cos(\Omega t)\,\mathbf{i}$, where the amplitude is $P = 10$ and the angular frequency is $\Omega = 2\pi$ (corresponding to 1 hertz). Newton's second law gives

$$m\ddot{x}\mathbf{i} = \mathbf{W} + \mathbf{H} + \mathbf{R} + \mathbf{P} = mg\mathbf{i} - k(x - l_0)\mathbf{i} - r\dot{x}\mathbf{i} + P\cos(\Omega t)\,\mathbf{i},$$

which leads, after resolution in the $\mathbf{i}$-direction, to the equation of motion

$$m\ddot{x} + r\dot{x} + kx = mg + kl_0 + P\cos(\Omega t).$$

The first two terms on the right-hand side can again be written as kx_{eq}, where $x_{\text{eq}} = mg/k + l_0$ is the equilibrium position of the baby in the absence of forcing. Substituting the given values for m, k, r, P and Ω in the equation of motion, we obtain

$$10\ddot{x} + 0.2\dot{x} + 200x = 200x_{\text{eq}} + 10\cos(2\pi t). \tag{16}$$

If the equilibrium position were chosen as the origin, then there would be no constant term on the right-hand side of the equation of motion.

(b) The complementary function is the same as that found in Example 3, which was

$$x_{\text{c}} = Ae^{-0.01t}\cos(4.5t + \phi),$$

where A and ϕ are arbitrary constants that are determined by the initial conditions. The particular integral will be the sum of two terms, the first of which (again as in Example 3) is the constant x_{eq}. According to the principle of superposition, we need to add to this a particular integral corresponding to the sinusoidal term on the right-hand side of equation (16). With a trial function of the form

See Unit 1, Subsection 3.2.

$$x_{\text{p}} = B\cos(2\pi t) + C\sin(2\pi t),$$

where B and C are constants, we find $B = -5.1337 \times 10^{-2}$ and $C = 3.3120 \times 10^{-4}$, so

$$x_{\text{p}} = -5.1337 \times 10^{-2}\cos(2\pi t) + 3.3120 \times 10^{-4}\sin(2\pi t).$$

This can also be written in the alternative sinusoidal form as

$$x_{\text{p}} = 5.1338 \times 10^{-2}\cos(2\pi t - 3.1351).$$

When the numerical values are rounded to two significant figures, the general solution of the equation of motion becomes

$$x = Ae^{-0.01t}\cos(4.5t + \phi) + x_{\text{eq}} + 5.1 \times 10^{-2}\cos(2\pi t - 3.1).$$

Whatever the initial conditions are, the magnitude of the complementary function will decay gradually towards zero, as observed in Section 2. The initial conditions therefore have no influence on the long-term behaviour. After a long time, the motion will be given simply by the remainder of the general solution,

$$x = x_{\text{eq}} + 5.1 \times 10^{-2} \cos(2\pi t - 3.1).$$

The predicted motion settles down to steady oscillations about the mean position x_{eq}, with amplitude 5.1×10^{-2} m (about 5 cm). These oscillations have the same angular frequency 2π as the input sinusoidal force $10\cos(2\pi t)\,\mathbf{i}$, but the output is out of phase with the input by almost π. This means that when the displacement is at its maximum (at the lowest point for the baby), the **i**-component of the force exerted by the adult is at its minimum, and vice versa. The baby reaches the highest point as the adult pushes down hardest, reaches the lowest point as the adult pulls up hardest, and passes through the equilibrium position as the adult momentarily exerts no force.

This example shows several features that are typical of such forcing problems. The complementary function dies away with time, regardless of the initial conditions, since in all cases the complementary function corresponds to one of the damped but unforced systems seen in Section 2. For this reason, the complementary function is referred to in this context as the **transient** part of the solution, while the remainder (corresponding to the particular integral, which does not die away) is called the **steady-state** solution of the equation of motion. The particular integral does not depend on the initial conditions, so neither does the steady-state behaviour. Another common feature is that the frequency of the steady-state solution is the same as that of the sinusoidal input force, but the output is out of phase with the input.

Exercise 13

(a) Without performing any detailed calculations or algebra, say how the solution to Example 5 would alter if the amplitude of the forcing oscillations were reduced to 2 N, say, by the baby pushing with its feet on the floor rather than being pushed by an adult.

(b) Suppose that a heavier or lighter baby is placed in the baby bouncer. Without going into details, say what aspects of the long-term motion predicted in Example 5 will alter, and what will remain the same.

Exercise 14

Bathroom scales are modelled by a model spring–damper system. The spring stiffness is k, and the damping constant is r. A girl of mass m is on the scales. By alternately pushing down against and pulling up on an adjacent towel rail, she manages to alter her effective weight on the scales by an amount modelled as an input force $\mathbf{P} = mg\cos(\Omega t)\,\mathbf{i}$, where the **i**-direction is upwards.

Draw the force diagram for this situation. Taking the origin at the base of the model spring, obtain the equation of motion. What long-term behaviour is predicted by this model?

You have seen that the effect of a directly applied sinusoidal force on a model spring–damper system is modelled by an equation of motion of the form

$$m\ddot{x} + r\dot{x} + kx = kx_{\text{eq}} + P\cos(\Omega t), \tag{17}$$

where P and Ω are the amplitude and angular frequency of the applied force, and x_{eq} is the equilibrium position of the particle in the absence of forcing.

Initially, the behaviour of the system depends on both the particular integral and the complementary function of equation (17), with the arbitrary constants in the complementary function being determined by the initial conditions of the situation. However, the complementary function dies away with time (is transient), and the particular solution then takes the form (equal to the particular integral)

$$x = x_{\text{eq}} + B\cos(\Omega t) + C\sin(\Omega t) = x_{\text{eq}} + A\cos(\Omega t + \phi).$$

This represents sinusoidal oscillations about the equilibrium position, which are independent of the initial conditions.

The model predicts that the steady-state output oscillations have the same frequency as the input forcing, but a different phase. The values of the output amplitude A and phase angle ϕ depend on the configuration of the system, and on the values of the parameters m, k, r, P and Ω.

3.2 Forcing by displacement

In this subsection we look at examples of spring–damper systems in which the forcing is not applied directly to the particle where the mass of the system is concentrated. Suppose, for instance, that the baby bouncer from Example 3 is adapted by the addition of a motor at the top, which moves the top of the spring up and down sinusoidally with time. This situation is represented in Figure 26. The force is not applied directly to the baby, but is applied to the top end of the spring in such a way that the top end of the spring is forced to oscillate.

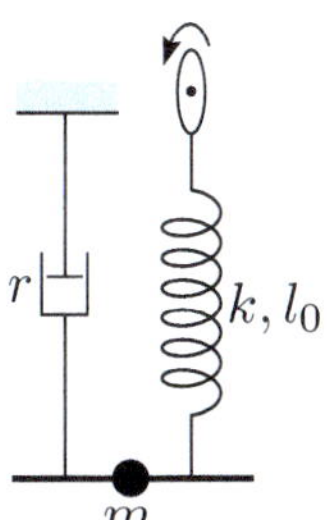

Figure 26 The top of the spring is forced to vibrate

The precise form of the model spring force exerted on the particle that represents the baby and bouncer must differ from that seen before, because now both ends of the spring are in motion. However, Hooke's law continues to apply, so the force on the particle due to the model spring is

$$\mathbf{H} = k(l - l_0)\,\widehat{\mathbf{s}},$$

where the spring has stiffness k and natural length l_0, and $\widehat{\mathbf{s}}$ is a unit vector in the direction from the particle towards the centre of the spring. The novel aspect introduced by this new situation is how the spring length l is changing. The following example illustrates how this is used.

Example 6

Consider the baby bouncer described in Examples 3 and 5, with spring stiffness $k = 200$ and damping constant $r = 0.2$. As before, the baby plus parts of the apparatus suspended from the spring has mass 10 kg. Suppose that a small motor causes the top of the spring to oscillate sinusoidally, with amplitude 0.04 m and period 1 s. Suppose also that the damping is regarded as due to air resistance alone, so that the top of the model damper remains fixed.

(a) Derive the equation of motion for the particle representing the baby and bouncer.

(b) Find the general solution of the equation of motion, and interpret this to predict what motion the baby will undergo in the long term.

(c) Say in general terms what would happen if the period of the oscillations at the top of the spring were to be changed to 2 s.

Solution

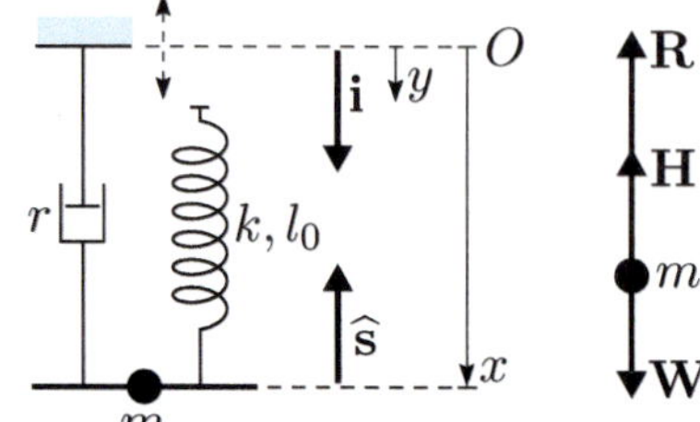

Figure 27 Forcing the top of the baby bouncer to vibrate

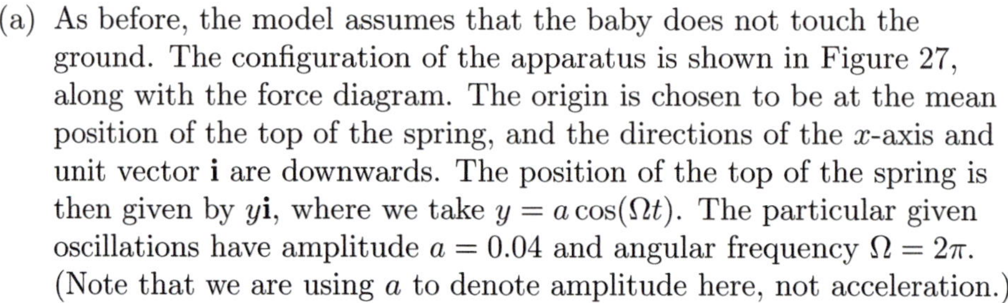

(a) As before, the model assumes that the baby does not touch the ground. The configuration of the apparatus is shown in Figure 27, along with the force diagram. The origin is chosen to be at the mean position of the top of the spring, and the directions of the x-axis and unit vector $\mathbf{i}$ are downwards. The position of the top of the spring is then given by $y\mathbf{i}$, where we take $y = a\cos(\Omega t)$. The particular given oscillations have amplitude $a = 0.04$ and angular frequency $\Omega = 2\pi$. (Note that we are using a to denote amplitude here, not acceleration.)

There are three forces to consider. Two of these are as in the previous analyses of the baby bouncer, namely, the weight $\mathbf{W} = mg\mathbf{i}$ and the resistance force $\mathbf{R} = r\dot{x}(-\mathbf{i})$. Since the model spring has length $l = x - y$, the force that it exerts on the particle is

$$\mathbf{H} = k(l - l_0)\widehat{\mathbf{s}} = k(x - y - l_0)(-\mathbf{i}).$$

Newton's second law gives

$$m\ddot{x}\mathbf{i} = \mathbf{W} + \mathbf{H} + \mathbf{R} = mg\mathbf{i} - k(x - y - l_0)\mathbf{i} - r\dot{x}\mathbf{i},$$

which leads to the equation of motion

$$m\ddot{x} + r\dot{x} + kx = mg + kl_0 + ky.$$

On putting $x_{\text{eq}} = mg/k + l_0$ and $y = a\cos(\Omega t)$, this becomes

$$m\ddot{x} + r\dot{x} + kx = kx_{\text{eq}} + ak\cos(\Omega t). \tag{18}$$

On substituting in the given values for m, k, r, a and Ω, we obtain

$$10\ddot{x} + 0.2\dot{x} + 200x = 200x_{\text{eq}} + 8\cos(2\pi t).$$

(b) The equation of motion is almost identical to that obtained in Example 5 (see equation (16)). The only difference is that the amplitude of the sinusoidal term on the right-hand side is 8, rather than 10. Thus the solution here is obtained by taking that in Example 5(b) but scaling the particular integral x_p for the sinusoidal term by $\frac{8}{10}$. This gives the steady-state solution

A similar approach was adopted in Exercise 13(a).

$$x = x_{eq} + 4.1 \times 10^{-2} \cos(2\pi t - 3.1).$$

The baby undergoes oscillations as before, but now of amplitude about 4 cm.

(c) If the period changes to 2 s, then the forcing angular frequency becomes $\Omega = \pi$. The corresponding steady-state solution will also have this angular frequency, but its amplitude and phase angle depend on Ω as well as on m, k, r and a, so it is not possible to deduce what these are from the previous solution. We would need to use a fresh trial function, of the form $B\cos(\pi t) + C\sin(\pi t)$.

Note that in comparison with Example 5, the equation of motion in Example 6 does not involve an additional force. The effect of the motor is modelled entirely by the inclusion of the term y in the expression for the spring force, and the fact that y is a function of time. Although the mechanism is different, Example 6(b) demonstrates that the effect of the motor displacing the top of the model spring is the same as that obtainable by direct sinusoidal forcing with a suitable amplitude and the same angular frequency. Leaving aside the particular values of the parameters, the equation of motion for direct forcing with amplitude P is equation (17), while the equation of motion for prescribed displacement oscillations of amplitude a is equation (18). The two match, provided that we put $P = ak$, and the mathematical solution and interpretation are then essentially the same in either case.

The model of the baby bouncer in Example 6 assumed that the top of the model spring moved, but that the top of the model damper was fixed. Suppose instead that the chief cause of damping is not air resistance but the internal damping in the spring. Then it is appropriate to model the situation by assuming that the top end of the model damper performs the same motion as the top end of the model spring (see Figure 28).

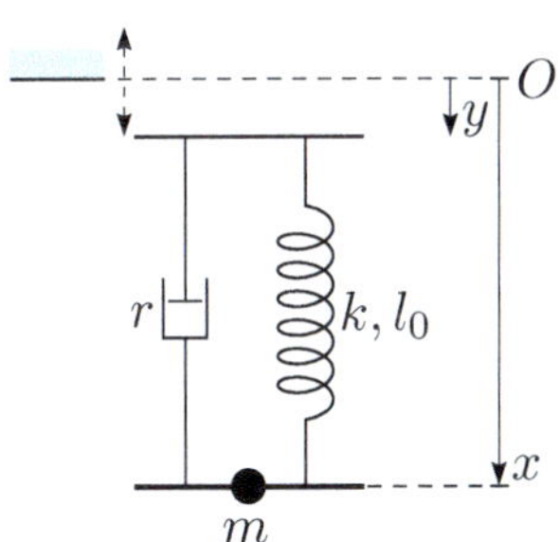

Figure 28 Forcing the spring–damper system to vibrate

This leads to an amended model in which a new expression is required for the damping resistance force. You showed in Exercise 4(b) that if the model damper extends from $x\mathbf{i}$ (where the particle is) to $y\mathbf{i}$, so that the length is $l = x - y$, then the corresponding resistance force on the particle is given by

$$\mathbf{R} = r\dot{l}\,\widehat{\mathbf{s}} = r(\dot{x} - \dot{y})(-\mathbf{i}).$$

Exercise 15

(a) Modify the model of the baby bouncer with motor, from Example 6, to represent the top of the model damper being attached to the top of the model spring and hence experiencing the same forced sinusoidal displacement. How does the equation of motion compare with that for direct forcing?

(b) Describe how you would find the steady-state solution, and hence the type of motion predicted by the model (but do not do the detailed calculations).

Exercise 16

The model sit-ski that you considered in Exercise 6 has a spring of natural length 0.2 m and stiffness 30 000 N m^{-1}, and damping constant 6300 N s m^{-1}. During testing, the sit-ski carries a skier plus seat of mass 60 kg. To simulate the effect of travelling over uneven ground, the base of the sit-ski is subjected to regular sinusoidal oscillations, with amplitude 0.1 m and angular frequency π rad s^{-1}, as shown in Figure 29.

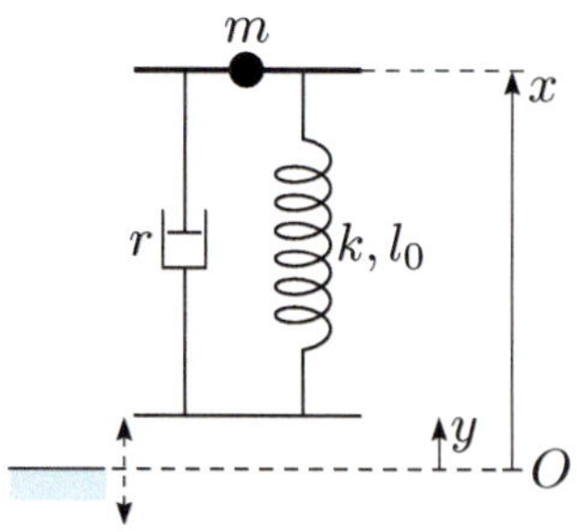

Derive the equation of motion, taking the origin to be at the mean level of the sit-ski base. Hence describe in general terms the long-term motion predicted by the model.

Figure 29 The sit-ski on a rough surface

You have now seen a case where the model spring alone was subjected to a forcing displacement at the end not attached to the particle, and other cases where the forcing point was located on both the model spring and damper together. A third possibility is where the model damper alone is displaced by a forcing displacement (and you are asked to consider this case at the end of the section). Here again, a very similar equation of motion results, with a sinusoid on the right-hand side that is similar to the corresponding term for direct forcing.

See Exercise 19.

Under certain circumstances, the position x of the particle relative to some fixed point may be of less interest than its position z relative to the forcing point. Since $z = x - y$, it is a simple matter to obtain the solution for z from that for x, since the forcing input term y is a known function. Alternatively, we can set up a differential equation directly for z rather than for x.

For example, the equation

$$m\ddot{x} + r\dot{x} + kx = kx_{\text{eq}} + r\dot{y} + ky$$

arose in Exercises 15 and 16. On replacing x by $y + z$, this becomes

$$m\ddot{z} + r\dot{z} + kz = kx_{\text{eq}} - m\ddot{y},$$

which with $y = a\cos(\Omega t)$ gives

$$m\ddot{z} + r\dot{z} + kz = kx_{\text{eq}} + ma\Omega^2\cos(\Omega t).$$

This is yet another occurrence of a differential equation of the form (17), so similar comments apply as to the method of solution and output behaviour.

Whether the forcing is applied via a direct sinusoidal force or by sinusoidal displacement to one end of the spring and/or damper, and whether the output is measured relative to a fixed point (x) or to the forcing point (z), the solution of the equation of motion is the sum of two parts: the complementary function (transient), which dies away and becomes negligible, and the particular integral, representing a steady-state sinusoidal oscillation about the equilibrium position, with the same frequency as the input forcing.

The process of finding a particular integral that corresponds to a sinusoidal forcing function is time-consuming, as described so far, and we have gone over this step in full for only one example. A computer can be used to undertake this task, but in the next subsection we indicate how it is also possible to calculate the output amplitude and phase angle more rapidly from suitable formulas.

Exercise 17

Consider the baby bouncer problem of Example 6. The equation of motion is

$$m\ddot{x} + r\dot{x} + kx = kx_{\text{eq}} + ky,$$

where x is measured from the fixed point (see Figure 27), and the baby bouncer is forced to vibrate using a motor attached to the top of the spring that causes the top of the spring to vibrate sinusoidally with $y = a\cos(\Omega t)$.

Let $z = x - y$ denote the length of the spring. Determine the equation of motion for the baby in terms of z, and find the form of its general solution.

3.3 The steady-state solution

It is tedious to look from scratch for sinusoidal particular integrals for inhomogeneous differential equations. In this subsection we will solve this problem once for a large number of possible cases. While this requires a fair amount of algebraic manipulation, the result obtained saves much further work. It is also a starting point for the discussion in Section 4.

We start from the differential equation

$$m\ddot{x} + r\dot{x} + kx = kx_{\text{eq}} + P\cos(\Omega t), \tag{19}$$

where m, r, k, x_{eq}, P and Ω are constants. This equation can be regarded as representing all the equations of motion that arose earlier in this section, provided that all of the constants that appear in it (except possibly x_{eq}) are positive. Where there is forcing applied to a model damper, the sinusoidal term on the right-hand side includes a phase shift and hence appears initially as $P\cos(\Omega t + \psi)$. However, such a phase shift can readily be 'transformed away' by moving the time origin from $t = 0$ to $t = -\psi/\Omega$. In a similar way, the kx_{eq} term on the right-hand side of equation (19) can be 'transformed away' by choosing a new origin for x at the equilibrium position, $x = x_{\text{eq}}$.

See Exercises 15, 16 and 19.

This permits us to concentrate on the slightly simpler equation

$$m\ddot{x} + r\dot{x} + kx = P\cos(\Omega t). \tag{20}$$

Once a solution of this equation has been obtained, we may, if it is desired, express it in terms of the original x- and t-coordinates, by adding x_{eq} to it and by reversing the phase shift (adding ψ to the phase angle obtained).

The general solution of equation (20) will be the sum of a complementary function and a particular integral. The complementary function was found in Subsection 2.3 and, as shown there, it dies away in all cases (is transient). We concentrate here on finding general expressions that determine the particular integral of equation (20). These provide a complete description of the steady-state behaviour of any system represented by equation (20).

The right-hand side of this differential equation is a sinusoid with angular frequency Ω. This means that the particular integral will also be sinusoidal, with the same frequency, so we start with a trial solution

$$x_{\text{p}} = B\cos(\Omega t) + C\sin(\Omega t),$$

where B and C are constants to be found in terms of m, r, k, P and Ω. The first two derivatives of the trial function are

$$\dot{x}_{\text{p}} = -\Omega B\sin(\Omega t) + \Omega C\cos(\Omega t),$$
$$\ddot{x}_{\text{p}} = -\Omega^2 B\cos(\Omega t) - \Omega^2 C\sin(\Omega t).$$

Substituting into equation (20) gives

$$m\big(-\Omega^2 B\cos(\Omega t) - \Omega^2 C\sin(\Omega t)\big) + r\big(-\Omega B\sin(\Omega t) + \Omega C\cos(\Omega t)\big) + k\big(B\cos(\Omega t) + C\sin(\Omega t)\big) = P\cos(\Omega t).$$

We can compare coefficients here because the cosine and sine functions are *linearly independent*: if $c_1 \sin c_3 x + c_2 \cos c_3 x = 0$ for all x, then $c_1 = c_2 = 0$.

On equating the coefficients of $\cos(\Omega t)$ and of $\sin(\Omega t)$, we have

$$-m\Omega^2 B + r\Omega C + kB = P, \quad -m\Omega^2 C - r\Omega B + kC = 0.$$

The second of these equations gives $B = (k - m\Omega^2)C/(r\Omega)$, and by substituting this expression for B into the first equation, we obtain

$$-\frac{m\Omega^2(k - m\Omega^2)C}{r\Omega} + r\Omega C + \frac{k(k - m\Omega^2)C}{r\Omega} = P,$$

that is,

$$\frac{\left((k - m\Omega^2)^2 + r^2\Omega^2\right)C}{r\Omega} = P.$$

The required expressions for B and C are therefore

$$C = \frac{Pr\Omega}{(k - m\Omega^2)^2 + r^2\Omega^2} \quad \text{and} \quad B = \frac{P(k - m\Omega^2)}{(k - m\Omega^2)^2 + r^2\Omega^2}. \tag{21}$$

The alternative formulation for the sinusoidal particular integral of equation (20) is $x_{\text{p}} = A\cos(\Omega t + \phi)$, where the amplitude A of the motion is given by $A = \sqrt{B^2 + C^2}$. Noting that

$$(Pr\Omega)^2 + (P(k - m\Omega^2)^2 = P^2((k - m\Omega^2)^2 + r^2\Omega^2),$$

we have

$$A = \frac{P}{\sqrt{(k - m\Omega^2)^2 + r^2\Omega^2}}. \tag{22}$$

The phase angle ϕ satisfies the pair of equations

$$A\cos\phi = B, \quad A\sin\phi = -C.$$

Now the expression for C is always positive, so $\sin\phi < 0$ and ϕ lies in the third or fourth quadrant. Since arccos is defined to give values in the first or second quadrant, it follows that the formula $\phi = -\arccos(B/A)$ will apply in all cases, that is,

The fact that ϕ is negative shows that the output lags behind the input.

$$\phi = -\arccos\left(\frac{k - m\Omega^2}{\sqrt{(k - m\Omega^2)^2 + r^2\Omega^2}}\right). \tag{23}$$

For any values of the constants m, r, k, P and Ω, the steady-state solution

$$x = B\cos(\Omega t) + C\sin(\Omega t) = A\cos(\Omega t + \phi)$$

is completely determined by either equations (21) (first form) or equations (22) and (23) (second form).

Exercise 18

(a) In Example 5 we studied the baby bouncer with direct forcing applied, for which $m = 10$, $k = 200$, $r = 0.2$, $P = 10$ and $\Omega = 2\pi$. Use equations (22) and (23) to check the values quoted in Example 5 for the amplitude and phase angle of the steady-state solution.

(b) Example 6 concerned the baby bouncer with forcing at the top of the model spring alone. The values of m, k and r were as in part (a). Use equation (22), with $P = ak$ and $a = 0.04$, to find whether the forcing period of 2 s referred to in Example 6(c) would give a greater or lesser steady-state amplitude than the 4 cm that was found for a forcing period of 1 s.

(c) Use equation (22) to estimate the steady-state output amplitude for the sit-ski testing scenario described in Exercise 16. Here $m = 60$, $k = 30\,000$, $r = 6300$, $a = 0.1$ and $\Omega = \pi$. Take $P = a\sqrt{k^2 + r^2\Omega^2}$ (as explained in the solution to Exercise 15(a)).

Suppose that we are looking at a forced displacement of the model spring alone, as in Example 6. Then we have $P = ak$ in equation (22), which gives

In other cases P will also depend on the amplitude a of the input, but in a different manner.

$$\frac{A}{a} = \frac{k}{\sqrt{(k - m\Omega^2)^2 + r^2\Omega^2}}. \tag{24}$$

Now A/a is the ratio of the amplitude A of the steady-state output oscillations to the amplitude a of the input forcing displacement. In other words, $M = A/a$ is the amplitude **magnification factor** caused by the forcing process.

It is possible to write the right-hand side of equation (24) in terms of two dimensionless constants:

Note that M itself, as the ratio of two lengths, is also dimensionless.

- the damping ratio $\alpha = r/(2\sqrt{mk})$, which was introduced in Subsection 2.3
- the ratio $\beta = \Omega/\omega$ of the forcing angular frequency Ω and the natural angular frequency of the system $\omega = \sqrt{k/m}$.

After some algebraic manipulation, we find that the magnification factor $M = A/a$ is given by

$$M = \left((1-\beta^2)^2 + 4\alpha^2\beta^2\right)^{-1/2}. \tag{25}$$

From this formula, we can predict the extent to which input oscillations are magnified in amplitude for any values of α and β. Thus we can also examine how M varies with changes to α and β, which corresponds to investigating how the system response depends on the input forcing frequency and on other parameters of the system. This is done in the next section.

In this section you have seen how to model the behaviour of systems such as damped harmonic oscillators maintained in a state of vibration by some external sinusoidal force or displacement. There are many examples of forced spring–damper systems, in addition to those that have been modelled so far, for instance: a vehicle suspension going over rumble strips, cobbles or a cattle grid; an idling engine causing the body of a stationary bus to vibrate; people walking in step over a bridge. The model predicts that a sinusoidal forcing input produces a sinusoidal output vibration of the same frequency.

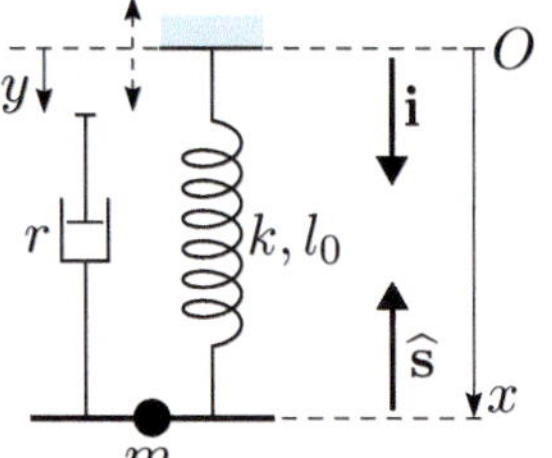

Figure 30 The damper forced to vibrate

Exercise 19

Modify the model of the baby bouncer with motor, from Example 6, to represent the top of the model damper alone being made to undergo a sinusoidal motion, with the top of the model spring held fixed (see Figure 30). Find the form of the equation of motion, without substituting in numerical values. How does this equation of motion compare with the direct forcing equation (19)?

4 Forced vibrations and resonance

At the end of Section 3 we derived alternative formulas (equations (24) and (25)) for the magnification factor by which input oscillations are enlarged in amplitude by a particular type of spring–damper system. As well as providing amplitude output values in specific cases, this enables us to study how the magnification factor depends on the input parameters, and especially on the forcing angular frequency.

It turns out that for some systems, there is a marked peak in amplitude magnification close to a certain forcing angular frequency. This phenomenon is known as *resonance*, and we take a look at this in Subsection 4.1. In Subsection 4.2, the particle–spring system with magnetic damping from Section 1 is modified by a forcing motor at the top of the spring. The model constructed in Section 3 can be applied to predict what magnification factor would occur for certain input frequencies, and the estimates obtained may be compared with experimental outcomes.

4.1 Resonance

Consider once more a spring–damper system in which the forcing point is attached to the model spring only, while the model damper has one end fixed. As pointed out at the end of Section 3, the magnification factor $M = A/a$, from the input forced displacement amplitude a to the steady-state output amplitude A, is given for such a system by

The baby bouncer of Example 6 is such a system.

$$M = \frac{k}{\sqrt{(k - m\Omega^2)^2 + r^2\Omega^2}} = \left((1-\beta^2)^2 + 4\alpha^2\beta^2\right)^{-1/2}, \qquad (26)$$

See equations (24) and (25).

where $\alpha = r/(2\sqrt{mk})$ and $\beta = \Omega/\omega$. Here the system has mass m, spring stiffness k, damping constant r and natural angular frequency $\omega = \sqrt{k/m}$. The input forcing angular frequency is Ω.

For a given damping ratio α (and hence for given values of the system parameters m, k and r), the magnification factor M is a function of β. From the graph of this function, you can read off the rough magnification factor for the system for any input forcing angular frequency. Several graphs of M against β, for different fixed values of α, are shown in Figure 31.

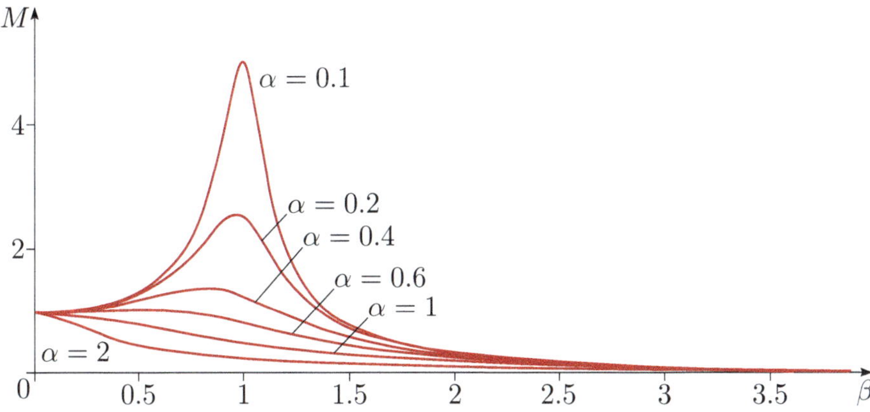

Figure 31 Graphs of the magnification factor M against the ratio $\beta = \Omega/\omega$ for different value of the damping ratio α

The graphs predict that for some (but not all) damping ratios, there is a maximum magnification factor at a certain positive forcing frequency, and this phenomenon is called **resonance**.

If the damping is strong ($\alpha > 1$) or critical ($\alpha = 1$), then no resonance occurs, and the magnification factor decreases throughout as β increases. Nor does resonance occur with weak damping, unless the value of α is beneath a particular threshold level. For smaller values of α, however, there is resonance, and its effect becomes more and more significant as α decreases towards zero. Note that for small values of α, resonance occurs in the vicinity of $\beta = 1$, that is, when the forcing angular frequency Ω is close to the natural angular frequency ω of the system.

You will see below that this threshold value is
$$\alpha = 1/\sqrt{2} \simeq 0.7.$$

Now that we have seen from Figure 31 that the model predicts the phenomenon of resonance, let us see if we can derive this analytically directly from equation (26). We consider $M = M(\beta)$ to be a function of β, and we wish to find the maximum magnification as β varies. The easiest way to do this is to recognise $M(\beta)$ as a composite function, that is, $M(\beta) = f(g(\beta^2))$, where

Another approach is to differentiate $M(\beta)$ directly.

$$f(u) = u^{-1/2} \quad \text{and} \quad g(x) = (1-x)^2 + 4\alpha^2 x.$$

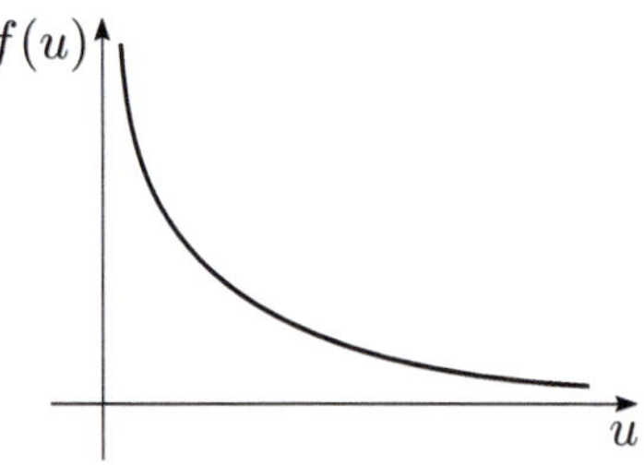

Figure 32 Graph of $f(u) = u^{-1/2}$

Now $f(u)$ is a strictly decreasing function for positive u, with no local maxima or minima (as shown in Figure 32). Also, $g(\beta^2)$ is always positive, since $\alpha > 0$ and $\beta^2 > 0$. So a *minimum* of $g(x)$ corresponds to a *maximum* of $f(g(x))$. Thus we need to find the minima of $g(x)$.

Now consider $g(x)$, which is a quadratic with a positive coefficient of x^2 and so has a single local minimum (which is also the global minimum). To find the location of the minimum, we differentiate $g(x)$:

$$\begin{aligned} g'(x) &= 2(1-x) \times (-1) + 4\alpha^2 \\ &= 2x + 4\alpha^2 - 2. \end{aligned}$$

So the derivative is zero when $x = 1 - 2\alpha^2$. (Alternatively, the minimum can be found without calculus, by completing the square.)

Putting the above results together gives that $f(g(x))$ has a unique global maximum at $x = 1 - 2\alpha^2$. So $M(\beta) = f(g(\beta^2))$ has a single maximum when $\beta^2 = 1 - 2\alpha^2$. Note that if $1 - 2\alpha^2 < 0$, then there are no real values of β satisfying this equation. These results are worth remembering.

Resonance frequency

The frequency at which resonance occurs is given by

$$\beta = \sqrt{1 - 2\alpha^2}. \tag{27}$$

Resonance can occur when (and only when) $1 - 2\alpha^2 > 0$.

By substituting in this value, we can show that the maximum magnification factor is given by
$$M = \frac{1}{2\alpha\sqrt{1-\alpha^2}}.$$

Let us now return to the baby bouncer example.

Exercise 20

The baby bouncer from Example 6, with a baby plus seat of mass 10 kg, has damping ratio

$$\alpha = \frac{r}{2\sqrt{mk}} = \frac{0.2}{2\sqrt{10 \times 200}} \simeq 0.002,$$

and natural angular frequency

$$\omega = \sqrt{\frac{k}{m}} = \sqrt{\frac{200}{10}} \simeq 4.47.$$

What amplitude of oscillations for the baby is predicted for a forcing input (at the top of the spring) with amplitude $a = 0.04$ and angular frequency ω? Comment on your result.

Resonance is found in many physical systems, and in some is desirable. Thus a radio receiver may be tuned to an input signal of a particular carrier frequency, and inputs of other carrier frequencies have much lower magnification factor at that frequency. However, there are other situations where resonance is most undesirable. In the case of a vehicle suspension, a large magnification factor when the vehicle goes over regular bumps would certainly be uncomfortable, and might also be dangerous and destructive.

4.2 Forcing in practice

Consider again the experiments discussed in Section 1. Here we consider the addition of a motor to the particle–spring system with magnetic damping.

Figure 33 The addition of a motor to force the spring–mass system to oscillate

This motor has the effect of forcing the top of the spring to undergo a sinusoidal displacement, while the tubes that cause the damping remain fixed. An appropriate model for this situation is the same as that developed for the baby bouncer in Example 6 (see Figure 27). The corresponding equation of motion is as before,

$$m\ddot{x} + r\dot{x} + kx = kx_{\text{eq}} + ak\cos(\Omega t),$$

if the origin is taken at the mean position of the top of the spring. However, with the alternative choice of origin at the equilibrium position of the particle, this becomes

$$m\ddot{x} + r\dot{x} + kx = ak\cos(\Omega t),$$

which is of the form of equation (20) with $P = ak$. Equation (26) gives the magnification factor, which is the ratio of the amplitude of the sinusoidal particular integral to the amplitude of the input forcing displacement.

The experiments indicated that the results for tubes B and C were similar, so we look at only tube C here.

Examine the forced motion for each of tubes A, C and D, for different sets of initial conditions. The model predicts that the transient part of the solution dies away with time, leaving a sinusoidal steady-state solution. This is illustrated in Figure 34 for each of the tubes, where the input forcing has angular frequency $\Omega = \frac{4}{3}\pi$ (equivalent to $\frac{2}{3}$ hertz or 40 cycles per minute).

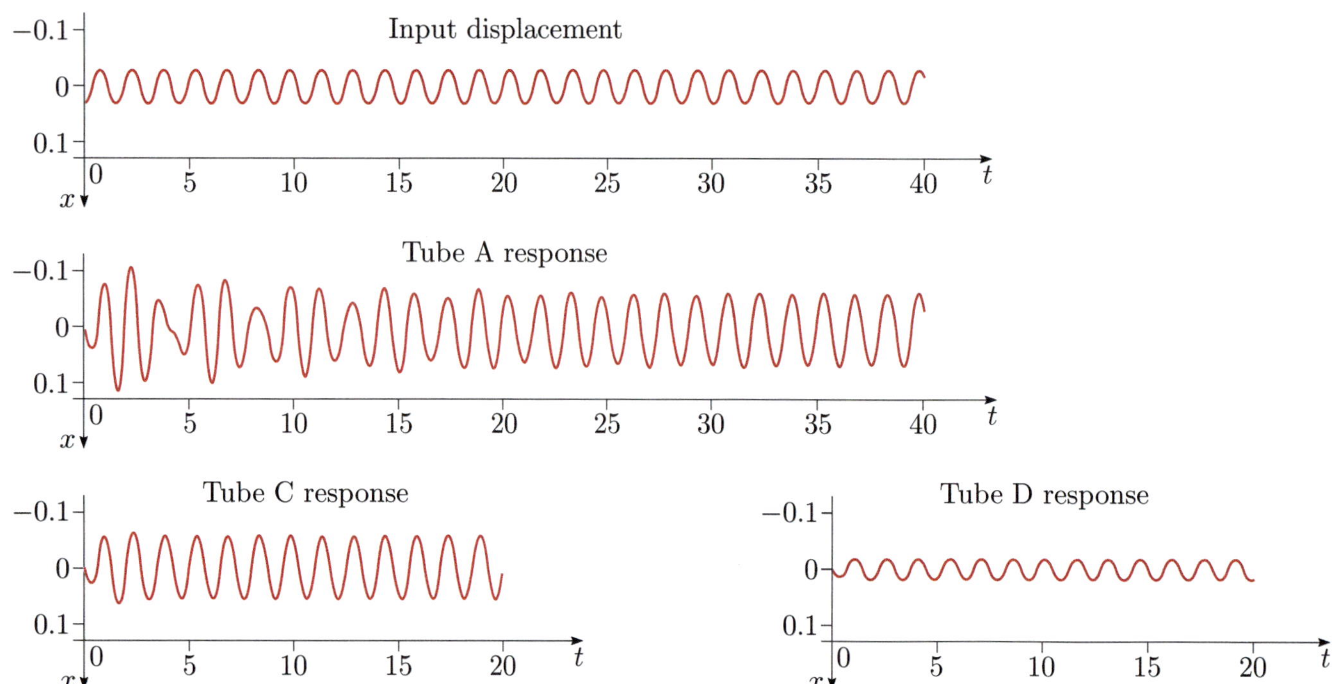

Figure 34 Predictions from the model for tubes A, C and D, with $\Omega = \frac{4}{3}\pi$, $x(0) = 0$ and $\dot{x}(0) = 0$

As before, the experimental apparatus has particle mass $m = 0.711$ and model spring stiffness $k = 23$. The amplitude of the input forcing is $a = 0.03$, hence the value of $P = ak$ is 0.69 N. The damping constants (as found for Table 1) are

$$r = 0.15 \text{ (tube A)}, \quad r = 1.33 \text{ (tube C)}, \quad r = 8.42 \text{ (tube D)}.$$

The corresponding damping ratios were found in Exercise 10. According to equation (26), the magnification factors for the steady-state output amplitude as compared with the input amplitude are

$$M = 2.2 \text{ (tube A)}, \quad M = 1.9 \text{ (tube C)}, \quad M = 0.6 \text{ (tube D)}. \tag{28}$$

These magnifications are visible on the graphs in Figure 34.

The graphs and the values (28) for M are predictions of the model, which may be compared with the outcomes of actual experiments, shown in Figure 35.

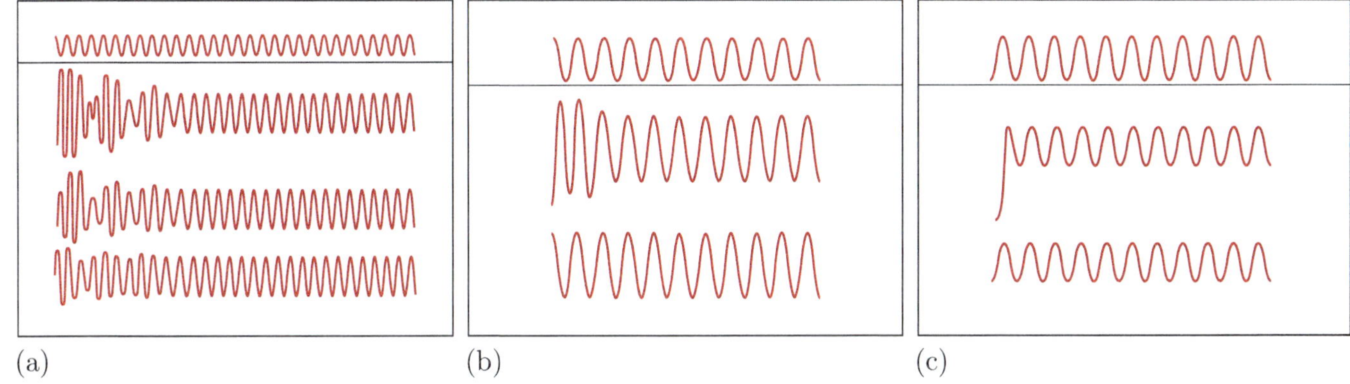

(a) (b) (c)

Figure 35 Input (at the top) and output traces with $\Omega = \frac{4}{3}\pi$ for (a) tube A, (b) tube C, (c) tube D

These experiments confirm the predictions of the model in the following respects.

- For each tube, there is an apparently sinusoidal steady-state solution, independent of the specific initial conditions (see Figure 35), together with a transient part that dies away.
- The transient part of the motion dies away more rapidly when the damping is stronger.
- The angular frequency of the steady-state solution is identical to that of the input forcing.
- The amplitude magnification factors for this forcing frequency (which can be measured from Figure 35) decline as the amount of damping increases, from tube A through tube C to tube D.
- For tubes A and C, the output has larger amplitude than the input, while for tube D there is attenuation (lower output than input amplitude).

It is possible to perform these measurements, and to compare the values obtained with the values (29) given below.

The measured values of the magnification factor are approximately given by

$$M = 2 \text{ (tube A)}, \quad M = 1.8 \text{ (tube C)}, \quad M = 0.8 \text{ (tube D)}, \tag{29}$$

which may be compared with the predicted values (28) of the model.

For each tube, the steady-state oscillations of the suspended mass alter with changes to the input forcing frequency. The experiment is run with the forcing oscillations at 40, 60 and 80 cycles per minute, for which the angular frequency Ω (in rad s^{-1}) has the respective values $\frac{4}{3}\pi$, 2π and $\frac{8}{3}\pi$. The corresponding predictions of the model for the magnification factors are given in Table 3. These values may be obtained from equation (26).

Table 3 Magnification factors predicted by the model

Tube	Damping constant r (N s m^{-1})	Damping ratio α	Magnification factor M $\Omega = \frac{4}{3}\pi$	$\Omega = 2\pi$	$\Omega = \frac{8}{3}\pi$
A	0.15	0.02	2.2	4.5	0.9
C	1.33	0.16	1.9	2.4	0.8
D	8.42	1.04	0.6	0.4	0.3

Note that the values predicted at the intermediate angular frequency, $\Omega = 2\pi$, for tubes A and C are higher than those at the higher or lower frequency. This amounts to a prediction that resonance will occur. Indeed, the model predicts resonance in these cases close to the natural angular frequency of the system, which is

$$\omega = \sqrt{\frac{k}{m}} = \sqrt{\frac{23}{0.711}} \simeq 5.7.$$

This corresponds to forcing at a frequency of about 54 cycles per minute.

The results for the experiments with forcing at angular frequency $\Omega = \frac{4}{3}\pi$ are as shown earlier, in Figure 35, with magnification factors as given in (29). The traces for the other experiments are shown in Figures 36–38.

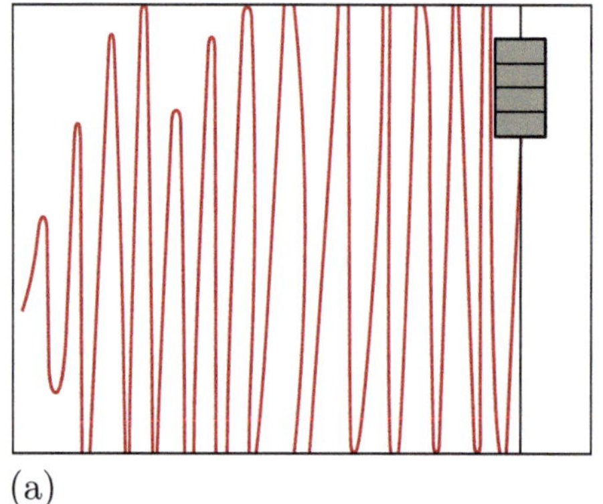

(a)

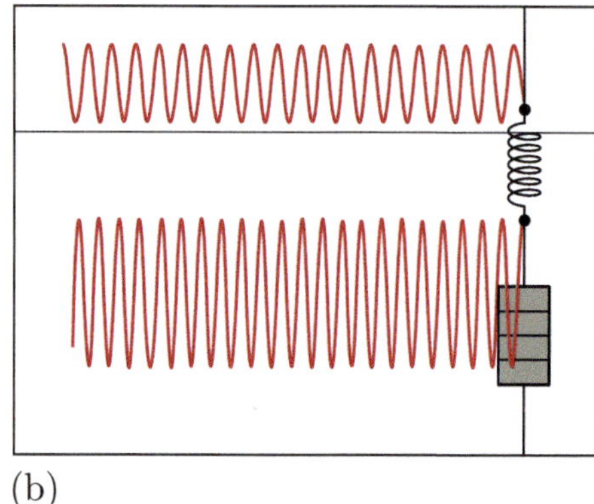

(b)

Figure 36 Output traces for tube A with (a) $\Omega = 2\pi$, (b) $\Omega = \frac{8}{3}\pi$

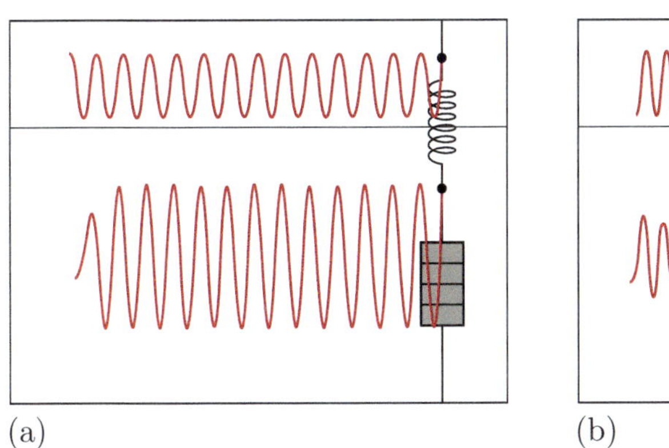

(a)

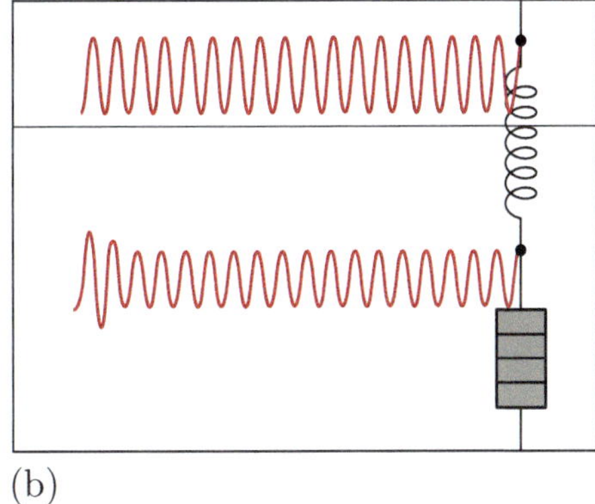

(b)

Figure 37 Output traces for tube C with (a) $\Omega = 2\pi$, (b) $\Omega = \frac{8}{3}\pi$

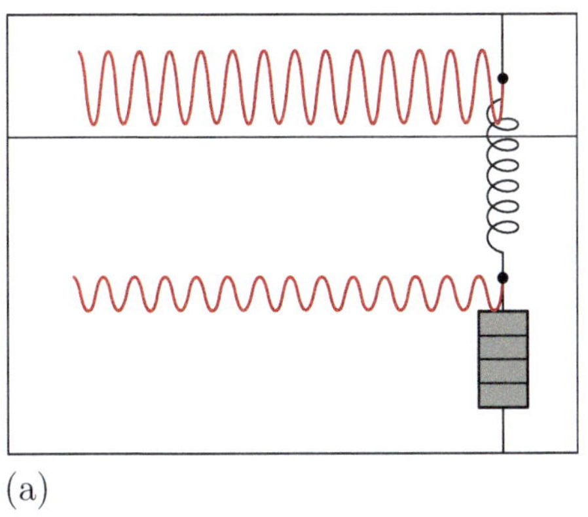

(a)

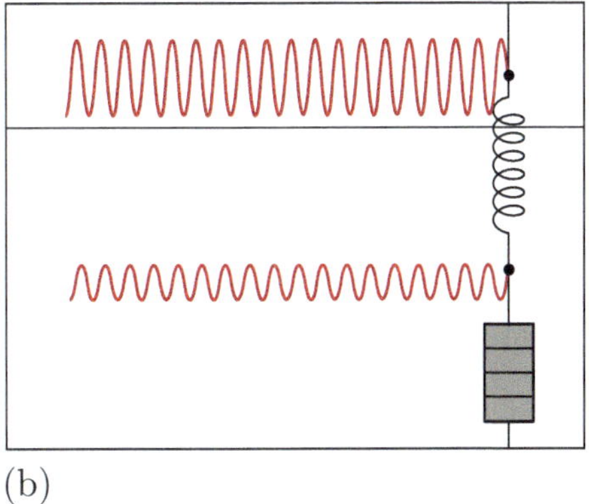

(b)

Figure 38 Output traces for tube D with (a) $\Omega = 2\pi$, (b) $\Omega = \frac{8}{3}\pi$

From these traces, the approximate magnification factors in Table 4 may be derived. (The trace for tube A with $\Omega = 2\pi$ in Figure 36(a) shows that the motion went off the scale, indicating an amplitude magnification greater than 4.)

It is possible to estimate these experimental magnification factors by taking measurements directly from the traces.

Table 4 Magnification factors obtained from experiment

Tube	Magnification factor M		
	$\Omega = \frac{4}{3}\pi$	$\Omega = 2\pi$	$\Omega = \frac{8}{3}\pi$
A	2	>4	2
C	1.8	2.3	0.7
D	0.8	0.5	0.5

Comparing these values with those predicted by the model in Table 3, the qualitative agreement is quite good. The model predicts correctly that the magnification factors for the weakly damped systems of tubes A and C will be greater than 1 at certain frequencies, and that those for the strongly damped motion of tube D will be less than 1 in each case. The predictions of resonance for tubes A and C, and more markedly for A, are borne out by the actual experiments.

The experiments were carried out once more for tubes A and C, at a forcing frequency of 50 cycles per minute (for which $\Omega = \frac{5}{3}\pi$). The observed magnification factor for tube C was about 2.7, while the behaviour for tube A was again beyond the limits of the apparatus. The corresponding predictions from the model are, respectively, $M = 2.95$ and $M = 6.4$.

The model makes a number of simplifying assumptions: the model spring behaviour is one assumption, linear damping is another, and a pure sinusoidal input forcing displacement is a third. Hence it is not surprising that the numerical predictions are somewhat at odds with the experimental values obtained.

Despite this, there is a significant degree of qualitative agreement between the behaviour of the model and that of the real system. In particular, the phenomenon of resonance was observed as predicted.

Understanding resonant behaviour in oscillating systems and how it can affect performance, and how it can be avoided, is very important in many practical situations. The experiments carried out for this unit have clearly demonstrated that resonance can and does occur at particular forcing frequencies, when large vibrations can be observed.

Learning outcomes

After studying this unit, you should be able to:

- understand the meanings of damping, forcing and resonance
- explain and apply the linear damping model
- distinguish between weak, critical and strong damping, and be aware of the distinctive features of each case
- appreciate the role of the damping ratio
- apply the model spring and model damper force specifications to situations in which either spring or damper, or both, may undergo a forced displacement at the end away from the particle
- model direct forcing to the particle where appropriate
- derive an equation of motion, based on Newton's second law, for models that feature model springs and model dampers, with or without forcing, and formulate the initial conditions for a particular motion
- formulate and solve the equation of motion with the origin either at the equilibrium position of the particle or at some other fixed point
- interpret the solutions of an equation of motion in terms of the situation from which the model arose
- understand the terms transient and steady-state, as applied to the solutions for forced and damped harmonic oscillators, and explain the essential features of each of these
- find via formulas the amplitude and phase angle of a steady-state solution in terms of the amplitude and angular frequency of the input forcing and other parameters of the system
- find the magnification factor for the steady-state output amplitude as compared with an input forced displacement amplitude for the model spring
- identify whether resonance may occur or will not occur in a system, and where it may occur, say what approximate input angular frequency will cause it for small damping ratios.

Solutions to exercises

Solution to Exercise 1

We have

$$\frac{mg}{k} = \frac{0.711 \times 9.81}{23} \simeq 0.30,$$

so the equilibrium extension is about 0.3 m.

Solution to Exercise 2

Equation (4) is

$$m\ddot{x} + r\dot{x} + kl = mg + kl_0.$$

If the origin for x is chosen at the fixed top end of the spring, then the displacement of the particle from that origin is $x = l$, and the equation of motion takes the inhomogeneous form

$$m\ddot{x} + r\dot{x} + kx = mg + kl_0.$$

This can also be written as

$$m\ddot{x} + r\dot{x} + kx = kx_{\text{eq}},$$

where $x_{\text{eq}} = l_0 + mg/k$ is the equilibrium displacement of the particle. (With this choice of origin, we have $x_{\text{eq}} = l_{\text{eq}}$.)

Solution to Exercise 3

(a) Strong damping is needed, as the train should bounce back as little as possible, with the buffers absorbing most of the energy.

(b) Strong damping is called for, but not so strong as to make the door close too slowly.

(c) Near-critical damping is appropriate, as the fuel gauge should revert quickly to a true reading and not oscillate much.

(d) Strong damping is needed, as oscillations between the vehicles could be dangerous and should be avoided.

(e) Near-critical damping is required, perhaps slightly on the weak side of critical to allow for the envisaged range of weights. This situation is similar to that of kitchen scales, discussed earlier.

Solution to Exercise 4

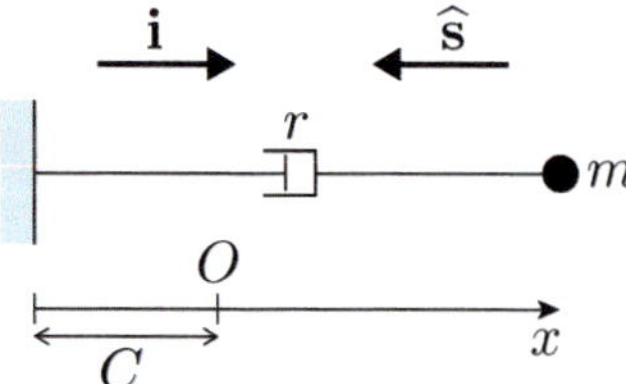

(a) Suppose that the x-axis is in the horizontal direction from left to right, as shown in the figure in the margin. The direction of $\widehat{\mathbf{s}}$ is from the particle towards the centre of the damper, so $\widehat{\mathbf{s}} = -\mathbf{i}$. We have $l = C + x$, where C is a constant that depends on the position of the origin for x, so $\dot{l} = \dot{x}$ regardless of the choice of origin. Hence the resistance force is

$$\mathbf{R} = r\dot{l}\,\widehat{\mathbf{s}} = r\dot{x}(-\mathbf{i}) = -r\dot{x}\mathbf{i}.$$

If, on the other hand, the x-axis is in the opposite direction, from right to left, with the unit vector $\mathbf{i}$ in the positive x-direction, then $\widehat{\mathbf{s}} = \mathbf{i}$. Choosing the origin to be a distance L from the fixed point, we have $x = L - l$, where l is the length of the damper. Thus $\dot{x} = -\dot{l}$, so the same expression for $\mathbf{R}$ results.

(b) The damper has length $l = x - y$, and $\widehat{\mathbf{s}}$ is in the opposite direction to $\mathbf{i}$. Hence we have

$$\mathbf{R} = r\dot{l}\,\widehat{\mathbf{s}} = r(\dot{x} - \dot{y})(-\mathbf{i}) = -r(\dot{x} - \dot{y})\mathbf{i}.$$

(This approach also provides an alternative way of tackling part (a), where y is constant thus $\dot{y} = 0$.)

Solution to Exercise 5

(a) A diagram of the buffer system and a force diagram are given below.

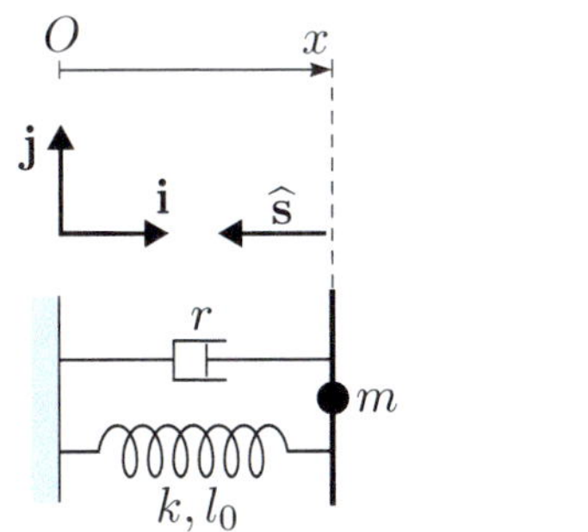

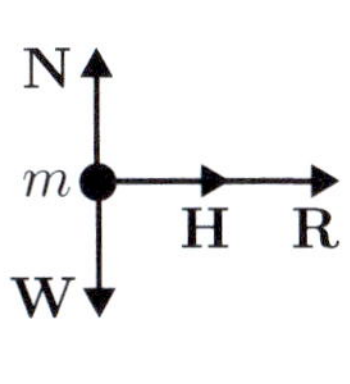

The train of mass m hits the buffers of stiffness k, damping constant r and natural length l_0, at time $t = 0$. The length of the buffer is $l = x$, and the unit vector $\mathbf{i}$ points in the direction of the x-axis.

From Hooke's law, with $\widehat{\mathbf{s}}$ pointing to the left, the model spring force is

$$\mathbf{H} = k(l - l_0)\,\widehat{\mathbf{s}} = k(x - l_0)(-\mathbf{i}).$$

From Exercise 4(a), the resistance force is

$$\mathbf{R} = r\dot{l}\,\widehat{\mathbf{s}} = r\dot{x}(-\mathbf{i}) = -r\dot{x}\mathbf{i}.$$

The equation of motion is therefore

$$\begin{aligned} m\ddot{x}\mathbf{i} &= \mathbf{H} + \mathbf{R} + \mathbf{W} + \mathbf{N} \\ &= -k(x - l_0)\mathbf{i} - r\dot{x}\mathbf{i} - mg\mathbf{j} + |\mathbf{N}|\,\mathbf{j}, \end{aligned}$$

which after resolution in the $\mathbf{i}$-direction gives

$$m\ddot{x} + r\dot{x} + kx = kl_0.$$

Substituting in $k = 140$, $r = 180$, $m = 40$ and $l_0 = 0.5$, we have

$$40\ddot{x} + 180\dot{x} + 140x = 70,$$

that is,

$$4\ddot{x} + 18\dot{x} + 14x = 7.$$

(b) The train meets the buffers first at $x = l_0 = 0.5$, at time $t = 0$. It is then moving in the negative x-direction, with speed $1\,\mathrm{m\,s^{-1}}$. Hence the appropriate initial conditions are

$$x(0) = 0.5, \quad \dot{x}(0) = -1.$$

(c) The solution is given as

$$x = 0.5 - 0.4e^{-t} + 0.4e^{-3.5t}.$$

Maximum compression of the buffers occurs when

$$\dot{x} = 0.4e^{-t} - 1.4e^{-3.5t} = 0,$$

that is, when $e^{-2.5t} = \frac{2}{7}$ or $t = 0.4\ln 3.5 \simeq 0.50$. The corresponding value of x is $0.33\,\mathrm{m}$, so the maximum compression is $0.50 - 0.33 = 0.17\,\mathrm{m}$.

As explained in Example 2, the train leaves the buffers when

$$\ddot{x} = -0.4e^{-t} + 4.9e^{-3.5t} = 0,$$

that is, when $e^{-2.5t} = \frac{4}{49}$ or $t = 0.8\ln 3.5 \simeq 1.00$. The corresponding value of x is $0.37\,\mathrm{m}$, and the corresponding velocity (with which the train leaves the buffers) is $\dot{x}\mathbf{i}$, where $\dot{x} = 0.10$, that is, velocity $0.10\,\mathrm{m\,s^{-1}}$.

Solution to Exercise 6

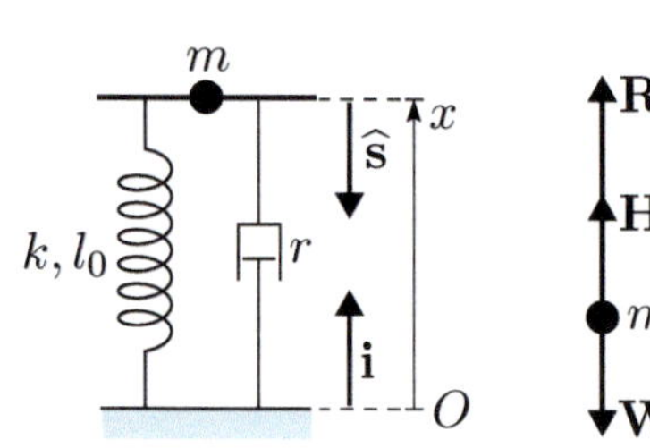

(a) The figure in the margin shows the set-up and the force diagram.

With $\mathbf{i}$ pointing upwards, $\widehat{\mathbf{s}}$ pointing downwards, $l = x$ and $\dot{l} = \dot{x}$, the model spring force is $\mathbf{H} = k(x - l_0)\widehat{\mathbf{s}} = k(x - l_0)(-\mathbf{i})$, the weight is $\mathbf{W} = -mg\mathbf{i}$, and the resistance force is $\mathbf{R} = r\dot{x}\widehat{\mathbf{s}} = r\dot{x}(-\mathbf{i}) = -r\dot{x}\mathbf{i}$. Newton's second law gives

$$m\ddot{x}\mathbf{i} = \mathbf{H} + \mathbf{W} + \mathbf{R} = -k(x - l_0)\mathbf{i} - mg\mathbf{i} - r\dot{x}\mathbf{i},$$

which leads to the equation of motion

$$m\ddot{x} + r\dot{x} + kx = kx_{\mathrm{eq}},$$

where $x_{\mathrm{eq}} = l_0 - mg/k$. On substituting the values given for the parameters, we have

$$x_{\mathrm{eq}} = 0.2 - \frac{60 \times 9.81}{30\,000} \simeq 0.180$$

and

$$60\ddot{x} + 6300\dot{x} + 30\,000x = 0.2 \times 30\,000 - 60 \times 9.81 = 5411.4,$$

that is,

$$\ddot{x} + 105\dot{x} + 500x = 90.19.$$

(b) The required initial conditions are

$$x(0) = 0.2, \quad \dot{x}(0) = 0.$$

(c) As given, but with $x_{\text{eq}} = 0.180$, the particular solution is

$$x = 0.180 + 0.021e^{-5t} - 0.001e^{-100t}.$$

The third term decays very quickly, so we are most interested in the second term. This reduces to 0.001 when

$$e^{-5t} = 0.001/0.021 \simeq 0.048,$$

that is, when

$$t = -0.2\ln 0.048 \simeq 0.61.$$

The model predicts that the sit-ski will take just over half a second for the amplitude to reduce to 0.001 m and hence for the displacement of the skier to be that close to the equilibrium position.

Solution to Exercise 7

(a) Take x as the downward displacement of the bottom of the float from the surface of the liquid, as shown in Figure 19, so that $l = x$ and $\dot{l} = \dot{x}$. The upthrust from the liquid is $\mathbf{U} = -2mg(x/d)\mathbf{i}$, the weight is $\mathbf{W} = mg\mathbf{i}$, and the damping resistance force is $\mathbf{R} = -r\dot{x}\mathbf{i}$.

Hence, using Newton's second law, we obtain

$$\begin{aligned} m\ddot{x}\mathbf{i} &= \mathbf{W} + \mathbf{U} + \mathbf{R} \\ &= mg\mathbf{i} - (2mg/d)x\mathbf{i} - r\dot{x}\mathbf{i}, \end{aligned}$$

which leads to the equation of motion

$$m\ddot{x} + r\dot{x} + (2mg/d)x = mg.$$

Once the parameter values have been substituted, this becomes

$$0.1\ddot{x} + 11\dot{x} + 98.1x = 0.981,$$

that is,

$$\ddot{x} + 110\dot{x} + 981x = 9.81.$$

The initial conditions are $x(0) = 0$ and $\dot{x}(0) = 0$.

(b) The particular solution is given as

$$x = 0.001\,08\,e^{-100.2t} - 0.0111\,e^{-9.79t} + 0.01.$$

The last term represents the equilibrium position. The first term involves $e^{-100.2t}$, which dies away very quickly, and the other term is the dominant term in the variable part of the solution, namely $-0.0111\,e^{-9.79t}$. The magnitude of this will reduce to 0.001 m when $e^{-9.79t} = 0.090$, that is, when $t = -(\ln 0.090)/9.79 \simeq 0.25$.

Hence the model predicts that the displacement of the float will be within 1 mm of its equilibrium position in about a quarter of a second.

Solution to Exercise 8

Critical damping will occur when $r^2 - 4mk$ is zero. The corresponding mass is therefore

$$m = \frac{r^2}{4k} = 125.$$

So a person whose mass is exactly 125 kg will produce critical damping when he or she stands on the scales. However, if a person with a slightly larger mass stands on the scales, then $r^2 - 4mk$ will be negative and there will be weak damping (decaying oscillations). If a person with a slightly smaller mass stands on the scales, then $r^2 - 4mk$ will be positive and there will be strong damping.

(This is a correct answer, but in reality you would not be able to detect any noticeable difference in the way the scales behaved for slight variations around critical damping. In the case that is technically weak damping, any oscillations would die down so quickly that they would be imperceptible. In the case that is technically strong damping, the return towards the equilibrium position would be slightly slower than with critical damping, but imperceptibly so.)

Solution to Exercise 9

The damping ratio $\alpha = r/(2\sqrt{mk})$ will be increased if either m is decreased (with r and k fixed) or k is decreased (with r and m fixed).

Solution to Exercise 10

For tube A, we have the damping ratio

$$\alpha = \frac{r}{2\sqrt{mk}} = \frac{0.15}{2\sqrt{0.711 \times 23}} \simeq 0.02.$$

Similarly, we obtain $\alpha \simeq 0.11$ for tube B, $\alpha \simeq 0.16$ for tube C, and $\alpha \simeq 1.04$ for tube D. The first three values satisfy $\alpha < 1$, so the motion in tubes A–C is weakly damped. For tube D, we see that $\alpha > 1$, confirming strong damping. (However, this is not far from critical damping, for which $\alpha = 1$.)

Solution to Exercise 11

If $\tau = 2\pi/\nu = 1.1 \times 2\pi/\omega$, where $\nu = \omega\sqrt{1-\alpha^2}$, then we have $1.1\sqrt{1-\alpha^2} = 1$, with solution $\alpha \simeq 0.417$. The amplitude decay factor per cycle is therefore

$$\exp\left(-\frac{2\pi\alpha}{\sqrt{1-\alpha^2}}\right) \simeq 0.056.$$

Solution to Exercise 12

As α increases from 1, the value of $\sqrt{\alpha^2-1}$ becomes increasingly close to α, so the magnitude of λ_1 decreases towards zero, while λ_2 tends towards $-2\omega\alpha$, which increases in magnitude with α.

Since both λ_1 and λ_2 are negative, it is the exponential with the exponent of smaller magnitude that dominates for large α, that is, the $e^{\lambda_1 t}$ term.

(In fact, for large α, we have $\lambda_1 \simeq -\omega/(2\alpha)$, since the product of the two roots of the auxiliary equation is $\lambda_1\lambda_2 = k/m = \omega^2$.)

Solution to Exercise 13

(a) The solution would proceed in the same way as in Example 5, only this time P would be 2 N instead of 10 N. Since the differential equation is linear, this change has the effect of scaling the constants B and C by the factor 0.2. Hence the predicted motion about the equilibrium position in the long term would consist of oscillations with the same angular frequency and phase angle as before, but with one-fifth of the amplitude. The baby would then bounce with an amplitude of about 10^{-2} m, or 1 cm.

(b) If the mass m of the baby is changed, then this will change the complementary function, but the latter still dies away with time. The value of x_{eq} will be changed (becoming smaller if the baby plus seat is lighter than 10 kg, or larger if the baby is heavier). The steady-state behaviour will still consist of sinusoidal oscillations about the (new) equilibrium position, with the same angular frequency, but the values of the constants B and C will be different. The amplitude and phase angle of the output oscillations may both be different.

Solution to Exercise 14

The set-up and force diagram are shown in the figure in the margin.

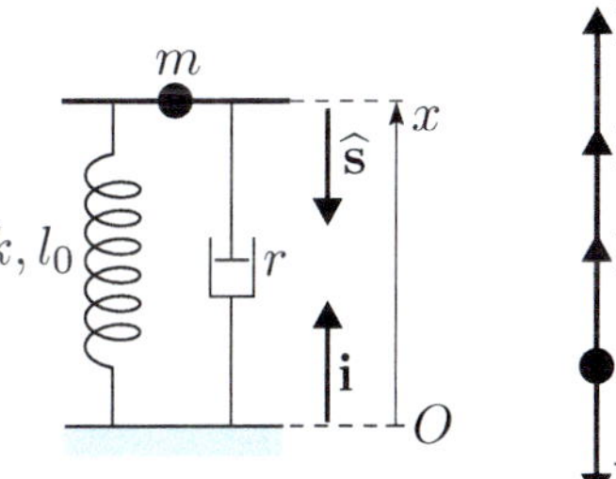

The forces acting are the weight $\mathbf{W} = -mg\mathbf{i}$, the spring force $\mathbf{H} = k(x - l_0)(-\mathbf{i})$, the damping resistance $\mathbf{R} = r\dot{x}(-\mathbf{i})$, and the girl's input force $\mathbf{P} = mg\cos(\Omega t)\,\mathbf{i}$, so Newton's second law gives

$$\begin{aligned} m\ddot{x}\mathbf{i} &= \mathbf{W} + \mathbf{H} + \mathbf{R} + \mathbf{P} \\ &= -mg\mathbf{i} - k(x - l_0)\mathbf{i} - r\dot{x}\mathbf{i} + mg\cos(\Omega t)\,\mathbf{i}. \end{aligned}$$

This leads to the equation of motion

$$\begin{aligned} m\ddot{x} + r\dot{x} + kx &= kl_0 - mg + mg\cos(\Omega t) \\ &= kx_{\text{eq}} + mg\cos(\Omega t), \end{aligned}$$

where $x_{\text{eq}} = l_0 - mg/k$. This differential equation is of the same form as that derived for the baby bouncer in Example 5. The values of the parameters will differ, but the model predicts the same overall long-term behaviour, namely, oscillations of angular frequency Ω about the equilibrium position, with a steady-state displacement function of the form

$$x = x_{\text{eq}} + A\cos(\Omega t + \phi).$$

The values of A and ϕ here may be calculated (using a sinusoidal trial function in the manner of Example 5) from the values for m, k, r and Ω.

Solution to Exercise 15

(a) The weight $\mathbf{W} = mg\mathbf{i}$ and model spring force $\mathbf{H} = k(x - y - l_0)(-\mathbf{i})$ are as in Example 6. The damping resistance force is now $\mathbf{R} = r(\dot{x} - \dot{y})(-\mathbf{i})$. This change leads to the amended equation of motion

$$m\ddot{x} + r\dot{x} + kx = mg + kl_0 + r\dot{y} + ky.$$

On putting $x_{\text{eq}} = mg/k + l_0$ and $y = a\cos(\Omega t)$ (so that also $\dot{y} = -a\Omega\sin(\Omega t)$), this becomes

$$m\ddot{x} + r\dot{x} + kx = kx_{\text{eq}} + ka\cos(\Omega t) - ra\Omega\sin(\Omega t).$$

On substituting in the values for m, k, r, a and Ω given in Example 6, we obtain

$$10\ddot{x} + 0.2\dot{x} + 200x = 200x_{\text{eq}} + 8\cos(2\pi t) - 0.016\pi\sin(2\pi t).$$

Comparing the equation with parameters with the direct forcing equation (17), the form of the sinusoid on the right-hand side is different, in that a sine appears as well as the cosine term. This is to some extent a superficial difference, since we can re-express the sinusoid on the right-hand side using its alternative form, that is,

$$\begin{aligned} ka\cos(\Omega t) - ra\Omega\sin(\Omega t) &= P\cos(\Omega t + \psi) \\ &= P\cos\Omega t\cos\psi - P\sin\Omega t\sin\psi. \end{aligned}$$

Hence the connection between the two forms is given by

$$P\cos\psi = ka, \quad P\sin\psi = ra\Omega,$$

so that

$$P = a\sqrt{k^2 + r^2\Omega^2}, \quad \psi = \arctan(r\Omega/k).$$

The equation of motion then becomes

$$m\ddot{x} + r\dot{x} + kx = kx_{\text{eq}} + P\cos(\Omega t + \psi),$$

and this now differs from the form of the direct forcing equation (17) only by a shift of phase.

(b) The steady-state solution will be the constant x_{eq} plus a sinusoid of the form $B\cos(\Omega t) + C\sin(\Omega t)$, which can also be written in the form $A\cos(\Omega t + \phi)$. In order to find the values for B and C (and hence subsequently A and ϕ), substitute the trial particular integral $B\cos(\Omega t) + C\sin(\Omega t)$ into the differential equation and equate coefficients of $\cos(\Omega t)$ and $\sin(\Omega t)$, to obtain simultaneous equations in B and C.

Then $A = \sqrt{B^2 + C^2}$ gives the amplitude of the vibration (with $A > 0$), and the phase angle ϕ is the solution of the pair of equations $\cos\phi = B/A$, $\sin\phi = -C/A$. Hence the model again predicts steady-state oscillations, of the same frequency as the input forcing displacement.

(In fact, because the given damping constant r is so small compared with the mass m and stiffness k, the solution for the case here is almost indistinguishable from that obtained in Example 6.)

Solution to Exercise 16

The set-up and force diagram are shown in the figure below.

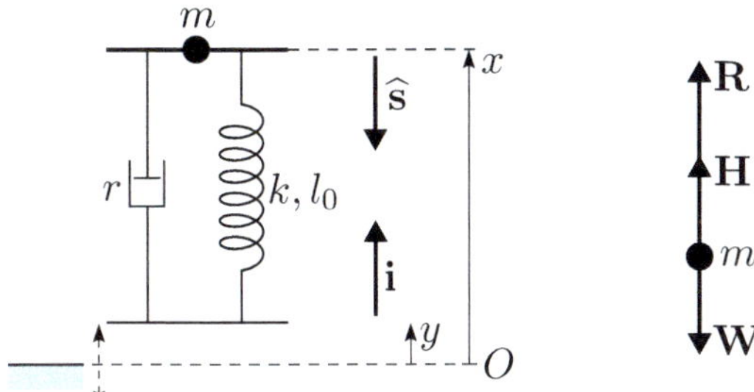

The forces acting are the weight $\mathbf{W} = -mg\mathbf{i}$, the model spring force $\mathbf{H} = k(l - l_0)\,\widehat{\mathbf{s}} = k(x - y - l_0)(-\mathbf{i})$, and the damping resistance $\mathbf{R} = r(\dot{x} - \dot{y})\,\widehat{\mathbf{s}} = r(\dot{x} - \dot{y})(-\mathbf{i})$, where $y = a\cos(\Omega t)$. This leads to the equation of motion

$$\begin{aligned} m\ddot{x} + r\dot{x} + kx &= kl_0 - mg + r\dot{y} + ky \\ &= kx_{\text{eq}} + ka\cos(\Omega t) - ra\Omega\sin(\Omega t), \end{aligned}$$

where $x_{\text{eq}} = l_0 - mg/k$. With the numerical values inserted (and taking $g = 9.81\,\text{m}\,\text{s}^{-2}$), this becomes

$$60\ddot{x} + 6300\dot{x} + 30\,000x = 30\,000x_{\text{eq}} + 3000\cos(\pi t) - 630\pi\sin(\pi t),$$

or

$$\ddot{x} + 105\dot{x} + 500x = 90.2 + 50\cos(\pi t) - 10.5\pi\sin(\pi t).$$

The form of this equation is the same as that considered in Exercise 15, hence the same approach applies to finding a particular integral and general conclusions. The model predicts steady-state oscillations of the same angular frequency $\pi\,\text{rad}\,\text{s}^{-1}$ as the input forcing displacement.

Solution to Exercise 17

We have $\dot{z} = \dot{x} - \dot{y}$ and $\ddot{z} = \ddot{x} - \ddot{y}$. Replacing x with $z + y$ gives

$$m(\ddot{z} + \ddot{y}) + r(\dot{z} + \dot{y}) + k(z + y) = kx_{\text{eq}} + ky,$$

so

$$m\ddot{z} + r\dot{z} + kz = kx_{\text{eq}} - m\ddot{y} - r\dot{y}.$$

The left-hand side of the differential equation for z is the same as that for x, so the complementary function will be the same. The particular integral will be of the form $z_{\text{p}} = x_{\text{eq}} + C\cos(\Omega t + \phi)$, which is the same form as x_{p} with different amplitude and phase angle, but with the same frequency Ω.

Of course, we could just have said that $z = x - y = x_{\text{c}} + x_{\text{p}} - a\cos(\Omega t)$, so $z_{\text{p}} = x_{\text{p}} - a\cos(\Omega t)$.

Solution to Exercise 18

(a) Equation (22) gives $A \simeq 0.051\,338$, and equation (23) gives $\phi \simeq -3.1351$, in agreement with the result quoted in the solution to Example 5.

(b) Here we have $P = ak = 8$. For a period of 2 s, the angular frequency is $\Omega = \pi$. According to equation (22), the corresponding output amplitude in the steady state is $A \simeq 0.079$. This is about 8 cm, which is twice the amplitude that was found for a forcing period of 1 s.

(c) Here we have

$$P = 0.1\sqrt{30\,000^2 + 6300^2\pi^2} \simeq 3594,$$

which leads to $A \simeq 0.101$. This output amplitude is almost the same as that of the input forcing.

Solution to Exercise 19

The solution is very similar to that for Exercise 15. As there, the weight is $\mathbf{W} = mg\mathbf{i}$ and the damping resistance is $\mathbf{R} = r(\dot{x} - \dot{y})(-\mathbf{i})$, but now the model spring force is $\mathbf{H} = k(x - l_0)(-\mathbf{i})$. The resulting equation of motion is

$$m\ddot{x} + r\dot{x} + kx = mg + kl_0 + r\dot{y}.$$

On putting $x_{\text{eq}} = mg/k + l_0$ and $y = a\cos(\Omega t)$ (so that $\dot{y} = -a\Omega\sin(\Omega t)$), this becomes

$$m\ddot{x} + r\dot{x} + kx = kx_{\text{eq}} - ra\Omega\sin(\Omega t).$$

This is of the same form as equation (19) except for a phase shift, since

$$-ra\Omega\sin(\Omega t) = P\cos(\Omega t + \psi)$$

provided that $P = ra\Omega$ and $\psi = \frac{\pi}{2}$.

Solution to Exercise 20

We have $\alpha \simeq 0.002$ and $\beta = \Omega/\omega = 1$. According to equation (26), the corresponding magnification factor is

$$M = \left((1-\beta^2)^2 + 4\alpha^2\beta^2\right)^{-1/2} = (2\alpha)^{-1} \simeq 250.$$

The steady-state output oscillations are therefore predicted to have amplitude

$$A = Ma \simeq 250 \times 0.04 = 10.$$

The system will not in fact be able to sustain an oscillation of this amplitude! For one thing, the natural length of the model spring used to represent a baby bouncer will be nowhere near 10 m. However, the model may be useful to the extent of predicting a potential catastrophe that needs to be avoided.

(Note that the forcing angular frequency of $4.47\,\text{rad}\,\text{s}^{-1}$, which is predicted to cause such a breakdown, lies between $\pi\,\text{rad}\,\text{s}^{-1}$ and $2\pi\,\text{rad}\,\text{s}^{-1}$, for which we found earlier that the predicted output amplitudes were only 8 cm and 4 cm, respectively. The large magnification factors occur for a fairly narrow range of values of the input forcing frequency.)

Unit 11

Normal modes

Introduction

In Unit 9 you studied the simple harmonic motion of oscillating mechanical systems, in particular those involving model springs. A typical example of such a system is shown in Figure 1, where an object modelled as a particle is connected by a model spring (of natural length l_0 and stiffness k) to a fixed wall; the particle is constrained to move in a straight line on a frictionless horizontal surface. In Unit 9 the equation of motion for this system was found to be

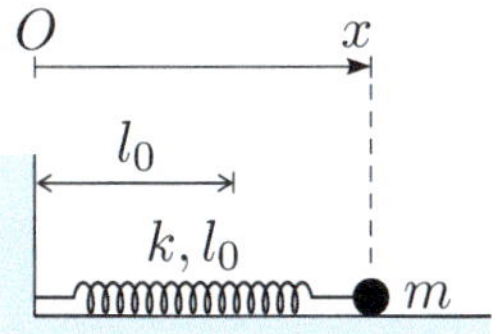

Figure 1 An oscillating system

$$m\ddot{x} + kx = kl_0.$$

The solution of this differential equation can be written in the form

$$x(t) = l_0 + A\cos(\omega t + \phi),$$

where x is the displacement from the wall, and ω $(= \sqrt{k/m})$, A and ϕ $(-\pi < \phi \leq \pi)$ are, respectively, the angular frequency, the amplitude and the phase angle of the oscillations executed by the particle.

Damping and forcing were incorporated in the simple harmonic motion model in Unit 10 to make it more realistic. This unit extends the basic model in a different way, to take account of another aspect of the motion of oscillating mechanical systems – the fact that usually more than one part of the system is free to move.

Damping and forcing are not considered in this unit.

A simple mechanical system that is typical of those considered in this unit, and its schematic representation, are shown in Figure 2. The diagram shows two particles connected by model springs to one another and to two fixed walls. Each particle is free to move in a straight line, but the motion of the system is not as straightforward as the simple harmonic motion considered above – unlike simple harmonic motion, it is, in general, not sinusoidal. But there are particular solutions of the equation of motion for such systems that *do* correspond to each part of the system oscillating backwards and forwards sinusoidally with the same frequency. These particular solutions are called *normal modes*. This unit is concerned with finding the normal modes of simple oscillating mechanical systems and showing how *any* motion of such a system can be built up from normal modes.

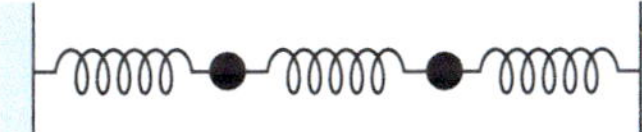

Figure 2 A two-particle system

Section 1 provides an overview of the unit and introduces many of the ideas that underpin it; notably, it contains several important concepts and definitions. Section 2 looks at systems that are confined to one dimension. Section 3 looks at a two-dimensional problem, modelling the behaviour of a guitar string.

1 Oscillations and normal modes

The important concepts of a normal mode and degrees of freedom are introduced in Subsection 1.1. Subsection 1.2 goes on to show how normal modes constitute the building blocks for modelling the motion of an oscillating mechanical system. Subsection 1.3 then explores how certain eigenvectors of a matrix can be used to determine the initial conditions for the normal mode motion of such a system.

1.1 What is a normal mode?

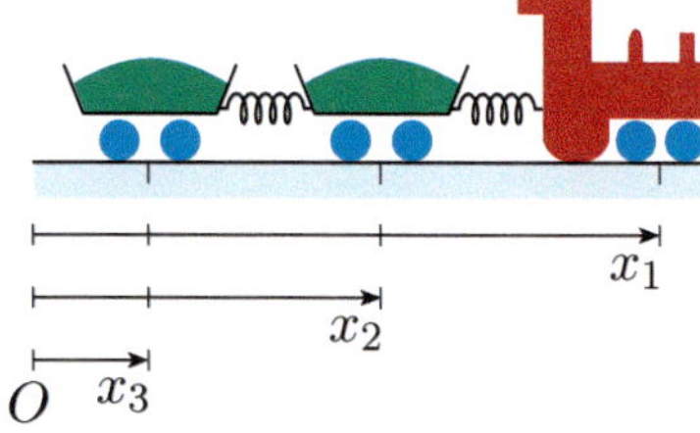

Figure 3 A railway engine and two trucks

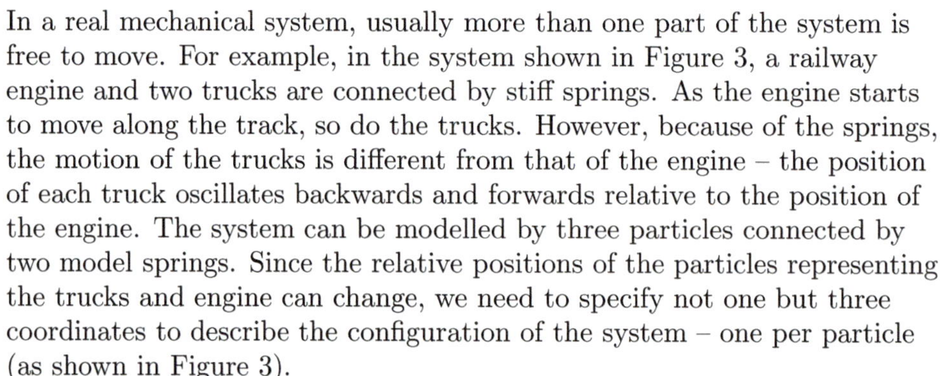

In a real mechanical system, usually more than one part of the system is free to move. For example, in the system shown in Figure 3, a railway engine and two trucks are connected by stiff springs. As the engine starts to move along the track, so do the trucks. However, because of the springs, the motion of the trucks is different from that of the engine – the position of each truck oscillates backwards and forwards relative to the position of the engine. The system can be modelled by three particles connected by two model springs. Since the relative positions of the particles representing the trucks and engine can change, we need to specify not one but three coordinates to describe the configuration of the system – one per particle (as shown in Figure 3).

> The number of **degrees of freedom** of a system is the smallest number of coordinates needed to describe its configuration (i.e. the positions of the constituent parts of the system) at any instant in time.

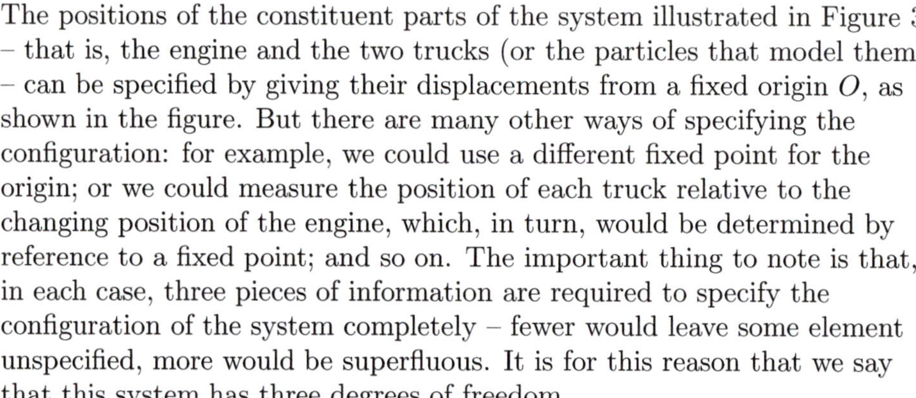

The positions of the constituent parts of the system illustrated in Figure 3 – that is, the engine and the two trucks (or the particles that model them) – can be specified by giving their displacements from a fixed origin O, as shown in the figure. But there are many other ways of specifying the configuration: for example, we could use a different fixed point for the origin; or we could measure the position of each truck relative to the changing position of the engine, which, in turn, would be determined by reference to a fixed point; and so on. The important thing to note is that, in each case, three pieces of information are required to specify the configuration of the system completely – fewer would leave some element unspecified, more would be superfluous. It is for this reason that we say that this system has three degrees of freedom.

In general, any convenient fixed point can be chosen as an origin. Often, for simplicity, an equilibrium position of a system is taken to be the origin.

Exercise 1

State the number of degrees of freedom of each of the mechanical systems shown in Figure 4. Assume that the railway track is straight and flat, that the engine and trucks are modelled as particles, and that the springs are modelled as model springs.

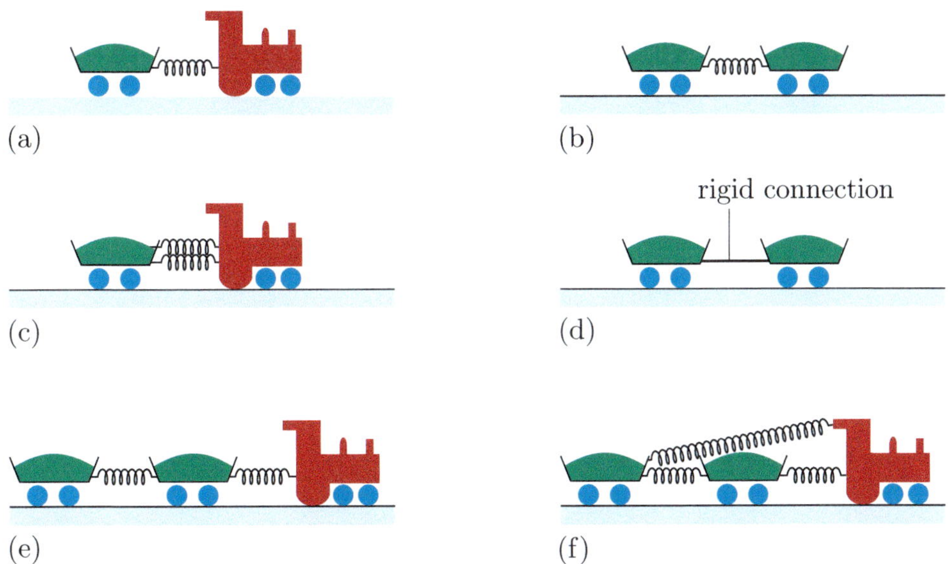

Figure 4 How many degrees of freedom?

The concept of degrees of freedom relates to other situations besides trains travelling on straight railway tracks. It applies to systems ranging from pendulums to the motion of individual molecules. However, determining the number of degrees of freedom of a system is not always as straightforward as in the case of the train considered above – more detailed analysis of the system is often necessary. For example, consider the simple pendulum shown in Figure 5. This is a mechanical system that is constrained to move in two dimensions, that is, in a vertical plane, so we could specify the position of the pendulum bob using a pair of Cartesian coordinates. You might therefore expect that the system would have two degrees of freedom. But the pendulum bob is constrained to move along a circular path in the vertical plane, so only one coordinate, the angle θ, is needed to specify its position. Hence the system has one degree of freedom. To show this, and to help you to make sense of what follows, the equation of motion for small oscillations of a simple pendulum is derived and solved.

Figure 5 A simple pendulum

For the simple pendulum of length l, moving in a vertical plane, let the datum for measuring potential energy be the point O at the top of the light rod. Consider the mechanical energy of the bob of mass m. Its potential energy is $U(\theta) = -mgl\cos\theta$, where θ is measured in radians and the minus sign indicates that the bob is below the datum. The length of the pendulum is constant, and the distance s of the bob along the arc of a circle can be measured as $s = l\theta$, where $s = 0$ when the rod is vertical.

Newton's second law will be used to derive the equations of motion for a *double* pendulum in Subsection 1.2.

Differentiating this equation to get the velocity of the bob along the path gives $\dot{s} = l\dot{\theta}$. Hence the kinetic energy of the bob is given by $T(\theta) = \frac{1}{2}m\dot{s}^2 = \frac{1}{2}ml^2\dot{\theta}^2$. By the conservation of mechanical energy, the total energy of the bob is

$$E(\theta) = T(\theta) + U(\theta) = \tfrac{1}{2}ml^2\dot{\theta}^2 - mgl\cos\theta = \text{constant}.$$

Note that using the chain rule,
$$\frac{d}{dt}(\cos\theta) = \frac{d}{d\theta}(\cos\theta)\,\frac{d\theta}{dt}.$$

Differentiating this equation with respect to t gives

$$\tfrac{1}{2}ml^2\,\frac{d}{dt}(\dot{\theta}^2) - mgl\,\frac{d}{dt}(\cos\theta) = 0,$$

so

$$ml^2\dot{\theta}\ddot{\theta} + mgl\dot{\theta}\sin\theta = 0.$$

Assuming that $\dot{\theta} \neq 0$, this equation can be divided through by $ml^2\dot{\theta}$ to obtain

$$\ddot{\theta} + \frac{g}{l}\sin\theta = 0.$$

For example, when $\theta = \pi/6 = 0.523$, $\sin\theta = 0.5$, and the approximation is within 5% of the true value. For $-\pi/6 < \theta < \pi/6$, the approximation will be within 5%, and it will be much more accurate as $\theta \to 0$.

The small-angle approximation $\sin\theta \simeq \theta$, which is a good approximation when θ is small, gives rise to a simple model for the motion of the pendulum bob:

$$\ddot{\theta} + \frac{g}{l}\theta = 0.$$

This is the equation for simple harmonic motion that you first solved in Unit 1, and studied in Unit 9. Its general solution can be written as $\theta(t) = A\cos\omega t + B\sin\omega t$, where $\omega = \sqrt{g/l}$, and A and B are arbitrary constants. This can also be written as $\theta(t) = C\cos(\omega t + \phi)$, where C and ϕ are arbitrary constants. The small-angle approximation $\sin\theta \simeq \theta$ can be used to obtain an equation for x, the horizontal displacement of the bob, as

$$x = l\sin\theta \simeq l\theta = lC\cos(\omega t + \phi) = D\cos(\omega t + \phi),$$

where $D = lC$ is another arbitrary constant.

Since the motion of the pendulum can be described using one variable, it has one degree of freedom. Other systems with one degree of freedom include the spring systems that you studied in Units 9 and 10. This unit, however, concentrates on systems with more than one degree of freedom.

You could investigate the behaviour of a double pendulum at home, using two lengths of light string and two steel nuts, for example. Use one string to join the two nuts, and use the other string attached at one end to one nut and suspended from a fixed point at the other end. Hold the nuts out sideways, so that the system lies in a vertical plane, then release them simultaneously. Observe the subsequent motion of the nuts.

Degrees of freedom are important in the context of normal modes as they indicate the number of normal modes that a system has. Consider small oscillations for the planar motion of a double pendulum, illustrated in Figure 6(a). Two coordinates are needed to specify the configuration of this mechanical system, which therefore has two degrees of freedom. Two possible coordinates are the horizontal displacements of the two pendulum bobs: x_1 (for the upper pendulum bob) and x_2 (for the lower pendulum bob). The system usually moves in a complicated manner, as shown in Figure 6(b). This behaviour is very different from that of the mechanical systems with one degree of freedom studied in Unit 9, where the motion is always sinusoidal.

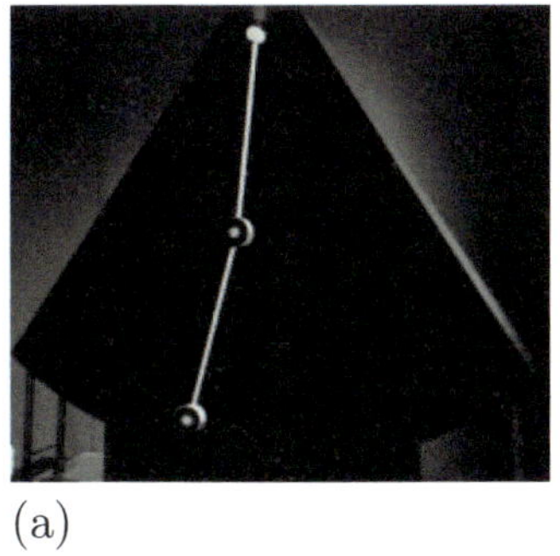

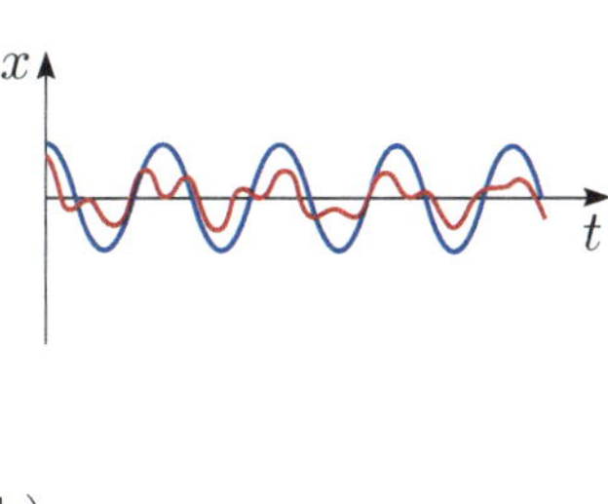

(a) (b)

Figure 6 (a) A double pendulum. (b) Motion of the upper bob (red) and lower bob (blue).

However, the motion is not always so complicated, and there are situations where both particles move sinusoidally with constant frequency. There are two possible simpler motions, the first of which is illustrated in Figure 7.

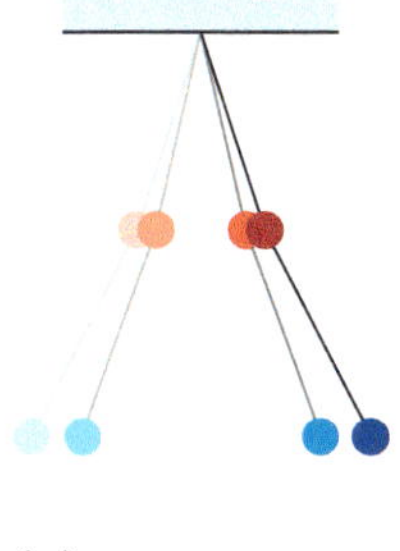

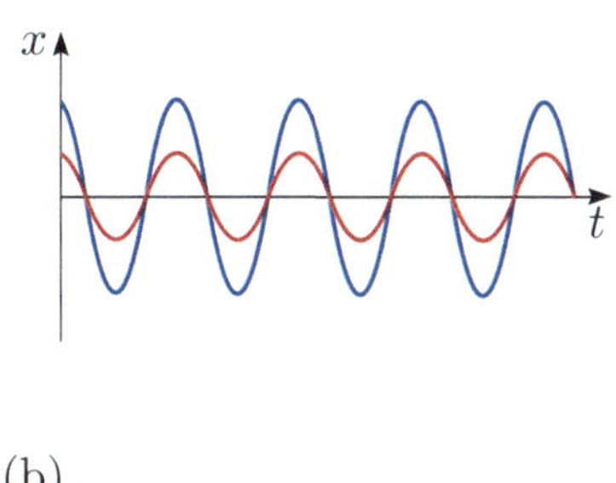

(a) (b)

Figure 7 (a) Snapshots, at intervals of one-sixth of a cycle, of the motion of a double pendulum system moving in a normal mode. (b) Traces of the corresponding sinusoidal motions of the pendulum bobs.

The snapshots in Figure 7(a) include the most extreme points of the motion, when the pendulum bobs are both momentarily at rest.

Comparing Figures 6(b) and 7(b), you can see that the motion is much more regular in the latter: the particles complete each cycle simultaneously, that is, the sinusoidal motion is of the same period (and so of the same frequency) for both. Consequently, the motion of the particles in Figure 7(b) can be modelled by a pair of equations of the form

$$x_1(t) = A_1 \cos(\omega t + \phi), \quad x_2(t) = A_2 \cos(\omega t + \phi), \tag{1}$$

where $|A_1|$ and $|A_2|$ are the amplitudes of the motions of the particles, ω is the common angular frequency, and ϕ is the common phase angle. Such motion is known as a *normal mode* of the system.

Notice that each equation represents simple harmonic motion, as defined in Unit 9.

For a normal mode, the coefficients A_1 and A_2 can be positive or negative.

A **normal mode** is a type of motion of a system of particles in which the coordinates of the particles all vary sinusoidally with the same frequency. The angular frequency of the sinusoidal motion is called the **normal mode angular frequency**.

In this context, *normal* means 'standard', not 'usual'. (In fact, normal modes are unusual in that most modes of motion are not normal modes.)

For each of the systems with *two* degrees of freedom, there are *two* normal modes. More generally, the number of distinct normal modes of a system corresponds to the number of degrees of freedom of the system (though we do not prove this here).

This correspondence occurs in *all* cases if we extend the definition of normal modes to include what is known as *rigid body motion*. Such motion is discussed in Subsection 2.2.

We turn now to another important aspect of motion, which is illustrated in Figure 7. You may have noticed that when one of the pendulum bobs moves from left to right, so does the other, and similarly when one moves from right to left, so does the other. This type of motion, in which two particles move in the same direction, is called **in-phase motion**. Not all normal mode motion is in-phase: it can also be **phase-opposed motion**, which occurs, as you might expect, when two particles move in opposite directions, as illustrated in Figure 8.

If you have set up the home experiment, you may wish to try to reproduce these two normal modes.

The snapshots in Figure 8(a) include the most extreme points of the motion, when the pendulum bobs are both momentarily at rest.

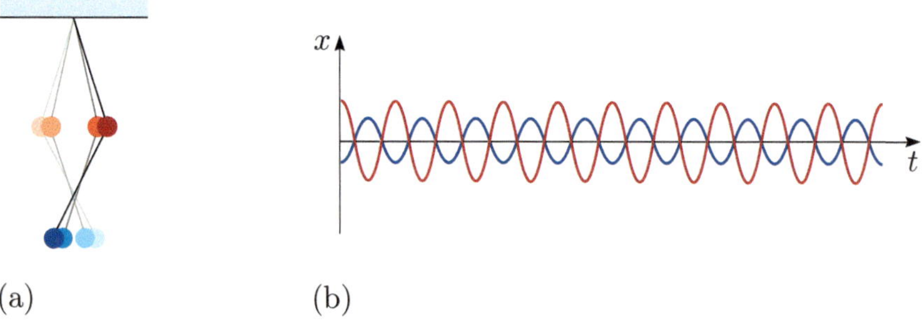

Figure 8 (a) Snapshots, at intervals of one-sixth of a cycle, of the motion of a double pendulum system moving in a phase-opposed normal mode. (b) Traces of the corresponding sinusoidal motions of the pendulum bobs.

If you look at Figure 7(b) you will see that for the in-phase motion, the phase angles of the two sinusoids are the same. In contrast, you can see in Figure 8(b) that for the phase-opposed motion, the phase angles of the two sinusoids differ by π. This is true in general: the sinusoids representing normal mode motion have phase angles that either are the same (in which case the motion is in-phase), or differ by π (in which case the motion is phase-opposed). In both cases, however, equations of the form (1) can be used to represent the normal mode motion.

This is because $\cos(\omega t + \phi + \pi) = -\cos(\omega t + \phi)$, and the negative sign can be absorbed into the coefficient A_i (which can be negative or positive).

If two particles are moving in a normal mode with their motion aligned with a common axis, then there are only three situations that can occur: the particles move in-phase, or they are phase-opposed, or one or both particles are stationary. By extension, in a system with three or more particles, each pair of particles can be considered separately; thus for any particular pair, the particles can be in-phase, or phase-opposed, or stationary, relative to each other.

Exercise 2

For each of the sets of graphs in Figure 9 showing the motion of two or three particles, state whether or not the motion represented is normal mode motion.

If it is, state whether the motion of each pair of particles is in-phase or phase-opposed.

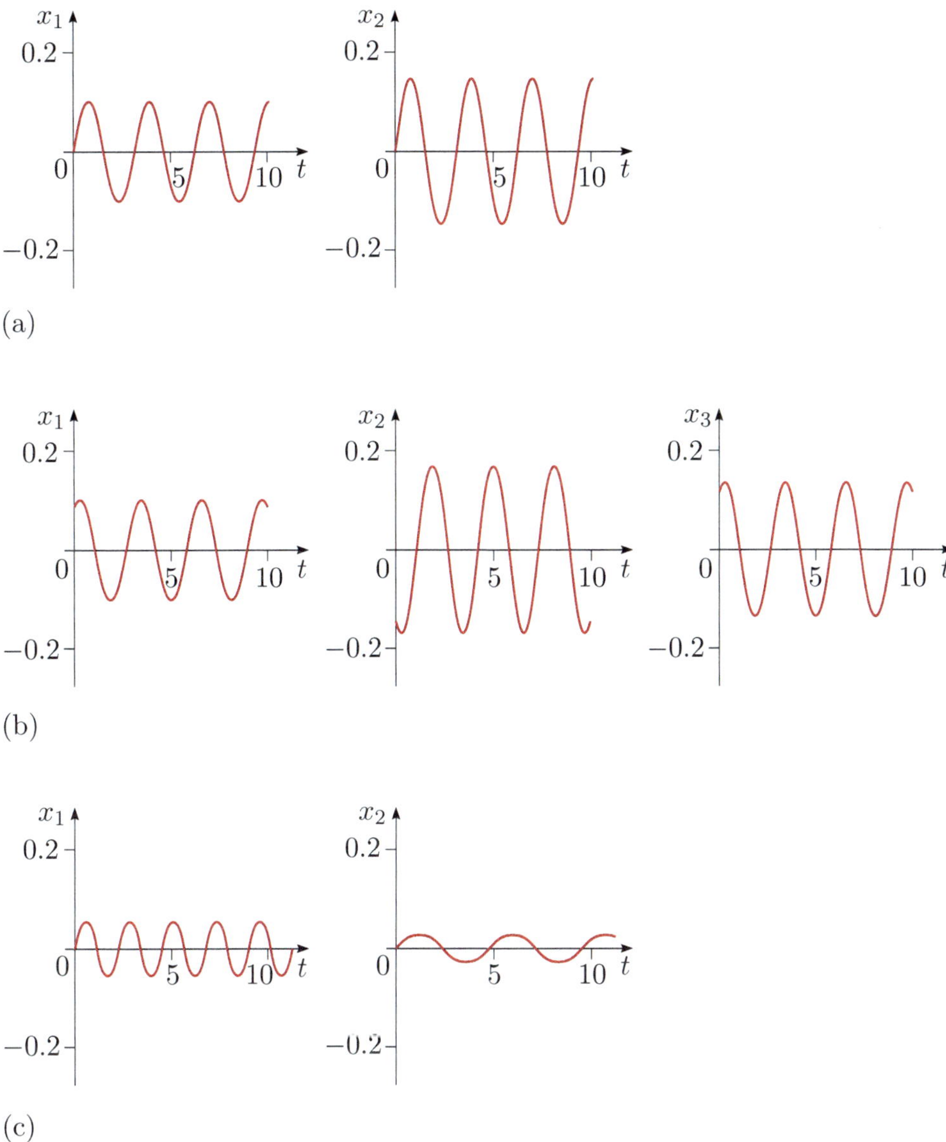

Figure 9 Motion of systems of two or three particles

1.2 Analysing oscillating mechanical systems

In this subsection you will see how to obtain the equations of normal mode motion for an oscillating mechanical system, and how these equations can be combined to model any motion of such a system. We begin by deriving the equations of motion for one of these systems, using techniques familiar from Units 2 and 9.

Example 1

Consider two particles of masses m_1 and m_2, joined to each other and to two fixed walls by three identical model springs of stiffness k and natural length l_0, as shown in Figure 10. The particles are constrained to move across a frictionless horizontal surface in the straight line joining the points of attachment (i.e. the motion is one-dimensional). To simplify the analysis, the distance between the fixed walls is assumed to be exactly three times the natural length of the springs; hence the equilibrium position of the system is that shown in Figure 10(a).

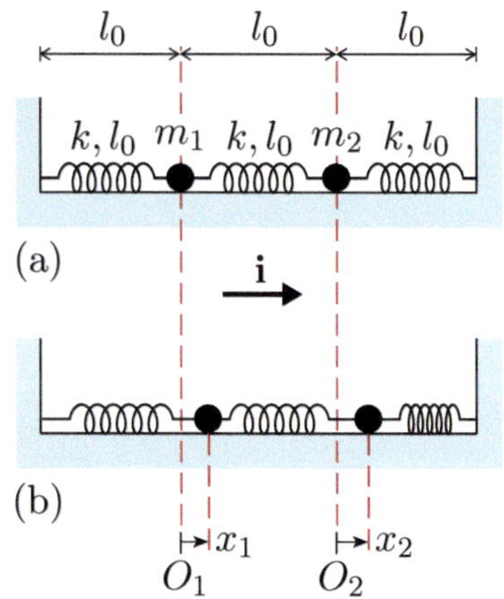

Figure 10 Two masses and three springs

Two coordinates are needed to specify the positions of the particles in the system; therefore the system has two degrees of freedom. The simplest option for the coordinates is to measure the displacement of each particle from its equilibrium position. These displacements are labelled x_1 and x_2 in Figure 10(b), and are measured from the origins O_1 and O_2, respectively.

What are the equations of motion for this oscillating system?

Solution

You have already met the idea (in Unit 2, in relation to particles in equilibrium) that when considering the forces acting on the particles in a system, each particle can be treated separately. This is the first time in the module that this principle has been applied to a system in motion.

To obtain the equations of motion for this system, Newton's second law is applied to each of the particles separately. In order to do this, the first step is to draw two force diagrams – one for each particle (see Figure 11).

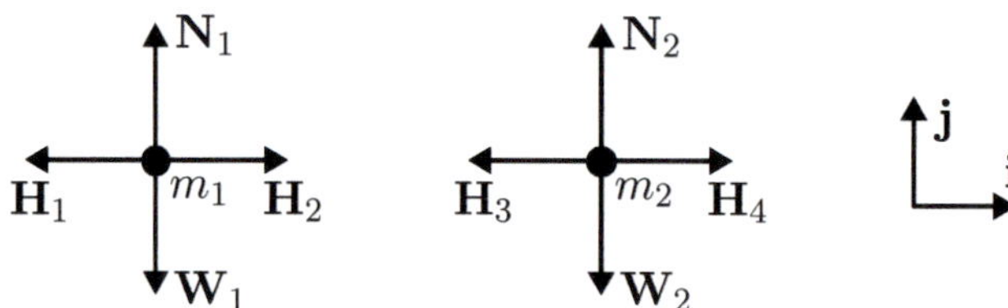

Figure 11 Force diagrams for the two particles

You have seen one-particle systems analysed in this way in Unit 9.

Here the weights of the particles are denoted by $\mathbf{W}_1$ and $\mathbf{W}_2$, the normal reactions of the surface on the particles by $\mathbf{N}_1$ and $\mathbf{N}_2$, and the forces exerted by the springs on the particles by $\mathbf{H}_1$, $\mathbf{H}_2$, $\mathbf{H}_3$ and $\mathbf{H}_4$.

Now we apply Newton's second law to the first particle and obtain

$$m_1\ddot{\mathbf{r}}_1 = \mathbf{H}_1 + \mathbf{H}_2 + \mathbf{W}_1 + \mathbf{N}_1, \tag{2}$$

where $\mathbf{r}_1$ is the position vector of the first particle with respect to its equilibrium position, that is, $\mathbf{r}_1 = x_1\mathbf{i}$, with $\mathbf{i}$ being a unit vector in the direction of positive x_1 and x_2. Similarly, for the second particle,

$$m_2\ddot{\mathbf{r}}_2 = \mathbf{H}_3 + \mathbf{H}_4 + \mathbf{W}_2 + \mathbf{N}_2. \tag{3}$$

The four forces due to the springs can be modelled using Hooke's law. First, recall from Unit 9 that according to Hooke's law, a model spring of natural length l_0 and stiffness k exerts a force on an object attached to one end of the spring, such that the force is given by $\mathbf{H} = k(l - l_0)\,\widehat{\mathbf{s}}$, where l is the length of the spring, and $\widehat{\mathbf{s}}$ is a unit vector directed from the point where the object is attached towards the centre of the spring.

For the force $\mathbf{H}_1$, which is the force exerted by the left-hand spring on the first particle, we have $l = l_0 + x_1$ and $\widehat{\mathbf{s}} = -\mathbf{i}$, so

$$\mathbf{H}_1 = k((l_0 + x_1) - l_0)(-\mathbf{i}) = -kx_1\mathbf{i}.$$

For the force $\mathbf{H}_2$, the distance between the two particles is given by $l = l_0 + x_2 - x_1$, and $\widehat{\mathbf{s}} = \mathbf{i}$, so

$$\mathbf{H}_2 = k((l_0 + x_2 - x_1) - l_0)\mathbf{i} = k(x_2 - x_1)\mathbf{i}.$$

For the force $\mathbf{H}_3$, we can apply Hooke's law again, or we can simply note that all the data are the same as for $\mathbf{H}_2$ (since we are considering the same spring) except that now $\widehat{\mathbf{s}} = -\mathbf{i}$, thus

Hooke's law tells us that the forces exerted at either end of a model spring are always equal in magnitude but opposite in direction. We frequently make use of this property, which is a consequence of Newton's third law. (Compare this property of model springs with that of model strings discussed in Unit 2.)

$$\mathbf{H}_3 = k(x_2 - x_1)(-\mathbf{i}).$$

For the force $\mathbf{H}_4$, we have $l = l_0 - x_2$ and $\widehat{\mathbf{s}} = \mathbf{i}$, so

$$\mathbf{H}_4 = k((l_0 - x_2) - l_0)\mathbf{i} = -kx_2\mathbf{i}.$$

The weights and normal reaction forces are vertical, so can be written as

$$\mathbf{W}_1 = -m_1g\mathbf{j}, \quad \mathbf{W}_2 = -m_2g\mathbf{j}, \quad \mathbf{N}_1 = |\mathbf{N}_1|\,\mathbf{j}, \quad \mathbf{N}_2 = |\mathbf{N}_2|\,\mathbf{j}.$$

Now that all the forces have been modelled, they can be substituted into equations (2) and (3), giving

$$m_1\ddot{\mathbf{r}}_1 = -kx_1\mathbf{i} + k(x_2 - x_1)\mathbf{i} - m_1g\mathbf{j} + |\mathbf{N}_1|\,\mathbf{j}, \qquad (4)$$

$$m_2\ddot{\mathbf{r}}_2 = -k(x_2 - x_1)\mathbf{i} - kx_2\mathbf{i} - m_2g\mathbf{j} + |\mathbf{N}_2|\,\mathbf{j}. \qquad (5)$$

But $\mathbf{r}_1 = x_1\mathbf{i}$ and $\mathbf{r}_2 = x_2\mathbf{i}$, therefore $\ddot{\mathbf{r}}_1 = \ddot{x}_1\mathbf{i}$ and $\ddot{\mathbf{r}}_2 = \ddot{x}_2\mathbf{i}$. Substituting for $\ddot{\mathbf{r}}_1$ and $\ddot{\mathbf{r}}_2$ in equations (4) and (5), resolving in the $\mathbf{i}$-direction and collecting terms gives

$$m_1\ddot{x}_1 = -2kx_1 + kx_2, \qquad (6)$$

$$m_2\ddot{x}_2 = kx_1 - 2kx_2. \qquad (7)$$

These differential equations are the equations of motion for the oscillating mechanical system depicted in Figure 10.

Equations (6) and (7) can be written in matrix form as

$$\ddot{\mathbf{x}} = \begin{pmatrix} \ddot{x}_1 \\ \ddot{x}_2 \end{pmatrix} = \begin{pmatrix} -2k/m_1 & k/m_1 \\ k/m_2 & -2k/m_2 \end{pmatrix} \begin{pmatrix} x_1 \\ x_2 \end{pmatrix} = \mathbf{A}\mathbf{x}. \qquad (8)$$

The equation $\ddot{\mathbf{x}} = \mathbf{A}\mathbf{x}$ is referred to as the *(matrix) equation of motion* for the mechanical system, and the matrix $\mathbf{A}$ is called the **dynamic matrix** of the system.

In Unit 6 you saw that provided that $\mathbf{A}$ has distinct real eigenvalues, the general solution of such an equation can be expressed as a linear combination of exponentials, sinusoids and linear terms. Since equation (8) models the motion of an oscillating mechanical system (without damping or forcing), we would expect only sinusoids in the general solution because sinusoids best represent this type of periodic motion.

See Procedure 6 in Unit 6.

Thus we would anticipate a general solution of the form

$$\mathbf{x}(t) = \mathbf{v}_1(D_1 \cos \omega_1 t + D_2 \sin \omega_1 t) + \mathbf{v}_2(D_3 \cos \omega_2 t + D_4 \sin \omega_2 t),$$

comprising four linearly independent terms, where $\mathbf{v}_1$ and $\mathbf{v}_2$ are eigenvectors corresponding to the eigenvalues λ_1 and λ_2 of $\mathbf{A}$, and $\omega_1 = \sqrt{-\lambda_1}$, $\omega_2 = \sqrt{-\lambda_2}$, while D_1, D_2, D_3 and D_4 are arbitrary constants that depend on the initial conditions. For present purposes, we rewrite this general solution in the form

You saw in Units 9 and 10 that $A \cos \omega t + B \sin \omega t$ can be rewritten in the form $C \cos(\omega t + \phi)$, where $A = C \cos \phi$ and $B = -C \sin \phi$.

$$\mathbf{x}(t) = C_1 \mathbf{v}_1 \cos(\omega_1 t + \phi_1) + C_2 \mathbf{v}_2 \cos(\omega_2 t + \phi_2), \tag{9}$$

where this form of the solution involves the four arbitrary constants C_1, C_2, ϕ_1 and ϕ_2.

One specific solution of equation (9), obtained by putting $C_1 = 1$ and $C_2 = 0$, is $\mathbf{x}(t) = \mathbf{v}_1 \cos(\omega_1 t + \phi_1)$, where $\mathbf{v}_1 = (v_{11} \quad v_{12})^T$. This can be written in the form

In general, any normal mode can be written in vector form as $\mathbf{x}(t) = \mathbf{v} \cos(\omega t + \phi)$, where $\mathbf{v}$ is a constant eigenvector, ω is the normal mode angular frequency, and ϕ is a constant scalar.

$$x_1(t) = v_{11} \cos(\omega_1 t + \phi_1), \quad x_2(t) = v_{12} \cos(\omega_1 t + \phi_1).$$

The interesting thing about this solution is that it represents a normal mode of the system in Example 1. To see this, look back at the general equations for normal mode motion (1), and replace A_1, A_2 and ϕ by v_{11}, v_{12} and ϕ_1, respectively.

Similarly, putting $C_1 = 0$ and $C_2 = 1$ in equation (9) gives the solution $\mathbf{x}(t) = \mathbf{v}_2 \cos(\omega_2 t + \phi_2)$, which represents a second normal mode of the system. Hence equation (9) tells us that the general solution of the equation of motion for the mechanical system in Example 1 can be expressed as a linear combination of normal modes.

This finding concerning the solution of the equation of motion is generally satisfied. This is because the motion of an oscillating mechanical system (without damping or forcing) can be modelled by a system of linear second-order differential equations, such as equation (8), whose coefficient matrix, that is, the dynamic matrix, has real non-positive eigenvalues, all of which usually are negative. If we put aside for the moment the case where some of the eigenvalues are zero, we have a dynamic matrix with negative eigenvalues. From Unit 6, this means that provided that these eigenvalues are distinct, the general solution can be written as a linear combination of sinusoids of the form $\mathbf{v} \cos(\omega t + \phi)$, that is, as a linear combination of normal modes. Furthermore, these normal modes will be linearly independent, since they correspond to linearly independent eigenvectors. Therefore we have the following important result.

We do not prove here that the eigenvalues must be non-positive.

The case where some eigenvalues are zero is considered in Subsection 2.2.

Theorem 1

The general solution of the equation of motion for an oscillating mechanical system (without damping or forcing) whose dynamic matrix has distinct non-zero eigenvalues can be written as a linear combination of linearly independent normal modes of the system.

In this case the number of linearly independent normal modes of a system is the same as the number of degrees of freedom of the system.

Theorem 1 indicates why normal modes are so important: not only are they especially simple oscillations themselves, but *every* motion can be expressed in terms of them.

Having established that the solution of the equation of motion obtained in Example 1 can be expressed as a linear combination of normal modes, we now use the techniques of Units 5 and 6 to find those normal modes, and hence the general solution of the equation of motion.

Solving equations of the form $\ddot{\mathbf{x}} = \mathbf{A}\mathbf{x}$ is dealt with in Section 4 of Unit 6, while finding eigenvalues and eigenvectors is covered in Unit 5.

Example 2

Suppose that in the mechanical system in Example 1, each particle has mass 0.1 kg, and each spring has stiffness $0.2\,\mathrm{N\,m^{-1}}$. Determine the general solution of the equation of motion.

Solution

From the given data, we have $k/m_1 = k/m_2 = 0.2/0.1 = 2$, so the dynamic matrix of the system is

$$\mathbf{A} = \begin{pmatrix} -2k/m_1 & k/m_1 \\ k/m_2 & -2k/m_2 \end{pmatrix} = \begin{pmatrix} -4 & 2 \\ 2 & -4 \end{pmatrix}.$$

To find each normal mode, and hence the general solution of the equation of motion for the system, the first step is to determine the eigenvalues and eigenvectors of the dynamic matrix. The eigenvalues can be found by forming the characteristic equation, and solving the resulting quadratic equation:

$$\begin{vmatrix} -4-\lambda & 2 \\ 2 & -4-\lambda \end{vmatrix} = \lambda^2 + 8\lambda + 12 = (\lambda+2)(\lambda+6) = 0.$$

So the eigenvalues are -2 and -6.

- For $\lambda = -2$, the eigenvector equations are

  $$\begin{pmatrix} -4-(-2) & 2 \\ 2 & -4-(-2) \end{pmatrix} \begin{pmatrix} v_1 \\ v_2 \end{pmatrix} = \begin{pmatrix} 0 \\ 0 \end{pmatrix},$$

  or equivalently

  $$-2v_1 + 2v_2 = 0,$$
  $$2v_1 - 2v_2 = 0.$$

  These equations both reduce to $v_2 = v_1$, so $(1 \quad 1)^T$ is an eigenvector.

- For $\lambda = -6$, the eigenvector equations are

  $$\begin{pmatrix} -4-(-6) & 2 \\ 2 & -4-(-6) \end{pmatrix} \begin{pmatrix} v_1 \\ v_2 \end{pmatrix} = \begin{pmatrix} 0 \\ 0 \end{pmatrix},$$

  or equivalently

  $$2v_1 + 2v_2 = 0,$$
  $$2v_1 + 2v_2 = 0.$$

  These equations both reduce to $v_2 = -v_1$, so $(1 \quad -1)^T$ is an eigenvector.

As you saw above, each eigenvalue/eigenvector pair gives a normal mode of the system of the form $\mathbf{v}\cos(\omega t + \phi)$, where $\mathbf{v}$ is the eigenvector and $\omega = \sqrt{-\lambda}$, with λ the eigenvalue. For $\lambda = -2$ we have the normal mode angular frequency $\omega_1 = \sqrt{2}$, and for $\lambda = -6$ we have $\omega_2 = \sqrt{6}$. Therefore two linearly independent normal modes of the system are

$$\mathbf{x}(t) = \begin{pmatrix} 1 \\ 1 \end{pmatrix} \cos(\sqrt{2}t + \phi_1) \quad \text{and} \quad \mathbf{x}(t) = \begin{pmatrix} 1 \\ -1 \end{pmatrix} \cos(\sqrt{6}t + \phi_2).$$

Hence by Theorem 1, the general solution of the equation of motion is

$$\mathbf{x}(t) = \begin{pmatrix} x_1(t) \\ x_2(t) \end{pmatrix} = C_1 \begin{pmatrix} 1 \\ 1 \end{pmatrix} \cos(\sqrt{2}t + \phi_1) + C_2 \begin{pmatrix} 1 \\ -1 \end{pmatrix} \cos(\sqrt{6}t + \phi_2),$$

where C_1, C_2, ϕ_1 and ϕ_2 are arbitrary constants.

Exercise 3

Determine the general solution of the equation of motion for the mechanical system considered in Example 1 if the particles have masses 0.1 kg and 0.2 kg, respectively, and each spring has stiffness $0.2\,\mathrm{N\,m^{-1}}$. Give your calculations to three decimal places.

The method employed in Examples 1 and 2 can be generalised to any oscillating mechanical system (without damping or forcing), and takes the form of the following general procedure.

Procedure 1 Analysing oscillating mechanical systems

To analyse the motion of a given oscillating mechanical system (without damping or forcing) with n degrees of freedom, proceed as follows.

◀ Draw picture ▶

1. Model the system using particles in conjunction with model springs, model rods, etc. Draw a sketch of the physical situation and annotate it with any relevant information.

◀ Choose coordinates ▶

2. Choose n coordinates (denoted by, say, $x_1, \ldots, x_n$) and corresponding origins.

◀ State assumption(s) ▶

3. State any assumptions that you make about the objects and the forces acting on them.

◀ Draw force diagram(s) ▶

4. Draw a force diagram for each particle separately.

◀ Apply Newton's 2nd law ▶

5. Apply Newton's second law to each particle separately, and resolve each force in the directions of appropriate axes to obtain n linear second-order differential equations of motion.

◀ Solve equation(s) ▶

6. Write the equations in matrix form as $\ddot{\mathbf{x}} = \mathbf{A}\mathbf{x}$, where $\mathbf{A}$ is the $n \times n$ dynamic matrix of the system.

7. Find the eigenvalues $\lambda_1, \ldots, \lambda_n$ and a corresponding set of linearly independent eigenvectors $\mathbf{v}_1, \ldots, \mathbf{v}_n$ of $\mathbf{A}$. (The eigenvectors $\mathbf{v}_1, \ldots, \mathbf{v}_n$ are often referred to as the **normal mode eigenvectors** of the system.)
8. Determine the normal mode angular frequencies $\omega_1, \ldots, \omega_n$ from the formula $\omega_i = \sqrt{-\lambda_i}$.
9. Write down the general solution of the equations of motion in the form

$$\mathbf{x}(t) = C_1\mathbf{v}_1\cos(\omega_1 t + \phi_1) + \cdots + C_n\mathbf{v}_n\cos(\omega_n t + \phi_n), \quad (10)$$

where $C_1, \ldots, C_n$ and $\phi_1, \ldots, \phi_n$ are arbitrary constants.
10. Interpret the solution in terms of the original problem.

If the eigenvalues are not distinct, then the procedure may break down. If any of them are zero, then the procedure needs to be adapted as described in Subsection 2.2. If any of them are positive or complex, then you have made a mistake!

◀Interpret solution▶

We make the following notes about the application of this procedure.

- As mentioned earlier, the eigenvalues of the dynamic matrix of an oscillating mechanical system (without damping and forcing) are always real and non-positive. However, if the eigenvalues are not distinct or if any of them are zero, Procedure 1 will break down as Theorem 1 does not apply in such circumstances.
- If we know $x_i(0)$ and $\dot{x}_i(0)$ for each i, then a particular solution of the equations of motion may be obtained. But obtaining such a solution can involve solving a system of $2n$ equations – even for a mechanical system with just two degrees of freedom, this could mean solving a system of four equations.
- When we do not have particular values for the C_i and ϕ_i, it is still possible to obtain some information about the behaviour of the system. For example, knowledge of the normal mode angular frequencies ω_i enables us to determine the frequencies f_i and periods τ_i of the normal modes from the formulas $f_i = \omega_i/(2\pi)$ and $\tau_i = 2\pi/\omega_i$.

As you will see later, if the mechanical system starts from rest, then each ϕ_i can be taken to be zero, and a system of only n equations then needs to be solved.

See Unit 9.

We now apply Procedure 1 to the double pendulum.

Example 3

Consider the motion of a double pendulum where the lower pendulum is attached to the bob of the upper pendulum and where both pendulums are constrained to move in the same vertical plane (see Figure 12). The stems of the pendulums have lengths l_1 and l_2, and may be modelled as light model rods. The pendulum bobs have masses m_1 and m_2, and may be modelled as particles. The angles made by the stems of the pendulums to the vertical are denoted by θ_1 and θ_2 (measured anticlockwise from the vertical). The oscillations of both pendulums are sufficiently small for the approximations $\sin\theta_i \simeq \theta_i$ and $\cos\theta_i \simeq 1$ $(i = 1, 2)$ to be applicable at all times during the motion of the system.

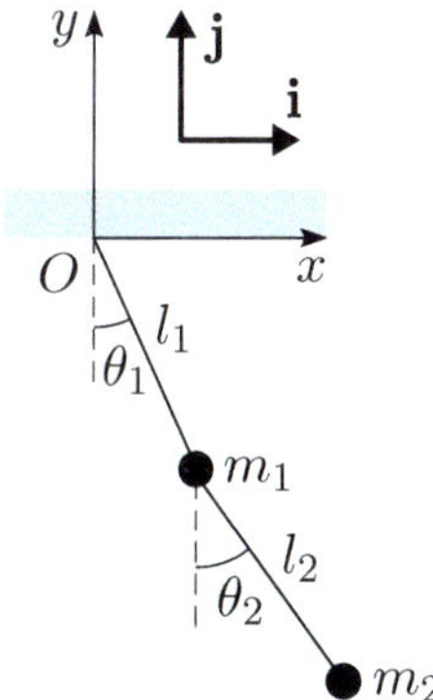

Figure 12 A double pendulum

(a) Derive linear equations of motion for the double pendulum.

(b) Suppose that $l_1 = l_2$ and $m_1 = m_2$. What is the general solution of the equations of motion?

(c) If each pendulum has length 50 cm, estimate the periods of the two normal modes.

Solution

◀ Choose coordinates ▶

(a) The system has two degrees of freedom – one associated with each pendulum. A suitable choice of coordinates would be the angles θ_1 and θ_2, which are marked on Figure 12, as these completely specify the configuration of the system at any given time. However, in order to use Newton's second law to generate equations of motion, we also need a set of Cartesian coordinate axes, which we will take to have origin O at the point of attachment of the upper pendulum, with the x-axis pointing horizontally to the right and the y-axis pointing vertically upwards, as in Figure 12.

◀ State assumption(s) ▶

We assume that the rods are model rods (with no mass) and the bobs are considered as particles. We assume that the only forces acting on the bobs are the gravitational forces and the tension forces in the rods. We further assume that the displacements are small.

◀ Draw force diagram(s) ▶

The force diagrams for the system are shown in Figure 13, with $\mathbf{W}_1 = -m_1 g\mathbf{j}$ and $\mathbf{W}_2 = -m_2 g\mathbf{j}$ denoting the weights of the particles, and

We model the forces on the particles due to the model rods as tension forces in the same way that we model forces on a particle due to model strings.

$$\mathbf{T}_1 = -|\mathbf{T}_1|\sin\theta_1\,\mathbf{i} + |\mathbf{T}_1|\cos\theta_1\,\mathbf{j},$$
$$\mathbf{T}_2 = |\mathbf{T}_2|\sin\theta_2\,\mathbf{i} - |\mathbf{T}_2|\cos\theta_2\,\mathbf{j},$$
$$\mathbf{T}_3 = -|\mathbf{T}_3|\sin\theta_3\,\mathbf{i} + |\mathbf{T}_3|\cos\theta_3\,\mathbf{j}$$

denoting the tension forces exerted on the particles by the model rods.

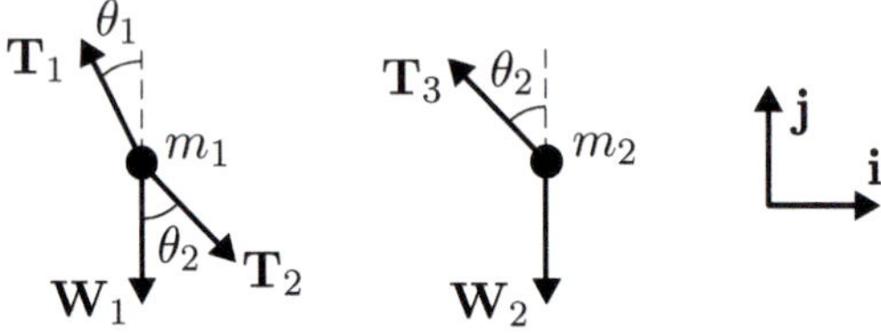

Figure 13 Force diagrams for the two particles

◀ Apply Newton's 2nd law ▶

Applying Newton's second law to the two particles separately, we obtain

$$m_1\ddot{\mathbf{r}}_1 = \mathbf{W}_1 + \mathbf{T}_1 + \mathbf{T}_2, \quad m_2\ddot{\mathbf{r}}_2 = \mathbf{W}_2 + \mathbf{T}_3,$$

where $\mathbf{r}_1 = x_1\mathbf{i} + y_1\mathbf{j}$ and $\mathbf{r}_2 = x_2\mathbf{i} + y_2\mathbf{j}$ are the position vectors of the two pendulum bobs relative to O.

Resolving in the **i**- and **j**-directions gives

$$\begin{aligned}
m_1\ddot{x}_1 &= -|\mathbf{T}_1|\sin\theta_1 + |\mathbf{T}_2|\sin\theta_2,\\
m_1\ddot{y}_1 &= -m_1g + |\mathbf{T}_1|\cos\theta_1 - |\mathbf{T}_2|\cos\theta_2,\\
m_2\ddot{x}_2 &= -|\mathbf{T}_3|\sin\theta_2,\\
m_2\ddot{y}_2 &= -m_2g + |\mathbf{T}_3|\cos\theta_2.
\end{aligned}$$

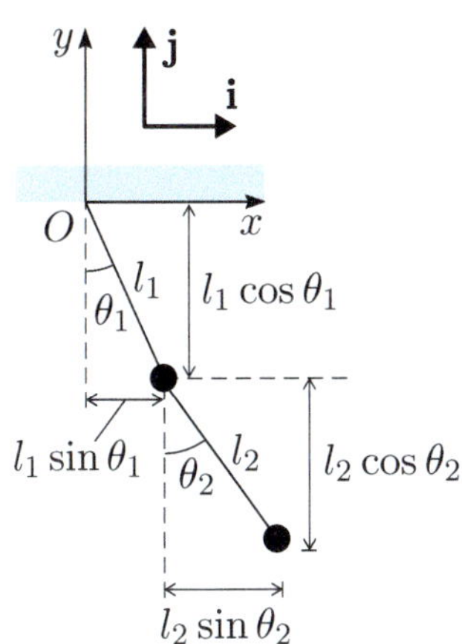

Figure 14 Relationships between linear and angular coordinates

From Figure 14 we can deduce the following relationships between the linear coordinates and the angular coordinates:

$$\begin{aligned}
x_1 &= l_1\sin\theta_1, \quad x_2 = l_1\sin\theta_1 + l_2\sin\theta_2,\\
y_1 &= -l_1\cos\theta_1, \quad y_2 = -l_1\cos\theta_1 - l_2\cos\theta_2.
\end{aligned}$$

Now since the oscillations of the pendulums are small, we can use the small-angle approximations specified in the question, that is, $\sin\theta_i \simeq \theta_i$ and $\cos\theta_i \simeq 1$ $(i = 1, 2)$, in the previous equations, to obtain

$$m_1\ddot{x}_1 \simeq -|\mathbf{T}_1|\theta_1 + |\mathbf{T}_2|\theta_2, \tag{11}$$

$$m_1\ddot{y}_1 \simeq -m_1g + |\mathbf{T}_1| - |\mathbf{T}_2|, \tag{12}$$

$$m_2\ddot{x}_2 \simeq -|\mathbf{T}_3|\theta_2, \tag{13}$$

$$m_2\ddot{y}_2 \simeq -m_2g + |\mathbf{T}_3|, \tag{14}$$

and

$$\begin{aligned}
x_1 &\simeq l_1\theta_1, \quad x_2 \simeq l_1\theta_1 + l_2\theta_2,\\
y_1 &\simeq -l_1, \quad y_2 \simeq -l_1 - l_2.
\end{aligned}$$

(These small-angle approximations will ensure that the resulting differential equations are linear.)

From $x_1 \simeq l_1\theta_1$, an approximation for the horizontal acceleration can be derived by differentiation as $\ddot{x}_1 \simeq l_1\ddot{\theta}_1$. Similarly, differentiating $x_2 \simeq l_1\theta_1 + l_2\theta_2$ gives $\ddot{x}_2 \simeq l_1\ddot{\theta}_1 + l_2\ddot{\theta}_2$. As the expressions above for y_1 and y_2 are constants, it follows that $\ddot{y}_1 \simeq 0$ and $\ddot{y}_2 \simeq 0$. (The physical interpretation of this is that for small oscillations the pendulum bobs do not move vertically.) These approximations for the horizontal and vertical accelerations can then be substituted into equations (11)–(14) to give

Differentiation of these approximations is valid provided that θ_1 and θ_2 not only are small, but also do not vary too rapidly.

$$m_1l_1\ddot{\theta}_1 \simeq -|\mathbf{T}_1|\theta_1 + |\mathbf{T}_2|\theta_2, \tag{15}$$

$$0 \simeq -m_1g + |\mathbf{T}_1| - |\mathbf{T}_2|, \tag{16}$$

$$m_2(l_1\ddot{\theta}_1 + l_2\ddot{\theta}_2) \simeq -|\mathbf{T}_3|\theta_2, \tag{17}$$

$$0 \simeq -m_2g + |\mathbf{T}_3|. \tag{18}$$

From equation (18) we obtain $|\mathbf{T}_3| \simeq m_2g$. By Newton's third law we have $|\mathbf{T}_2| = |\mathbf{T}_3|$, therefore $|\mathbf{T}_2| \simeq m_2g$. This can be substituted into equation (16) to yield $|\mathbf{T}_1| \simeq (m_1 + m_2)g$. If we then substitute for $|\mathbf{T}_1|$, $|\mathbf{T}_2|$ and $|\mathbf{T}_3|$ in equations (15) and (17), we have

Newton's third law tells us that the forces exerted at either end of a model rod must be equal in magnitude but opposite in direction. Recall the case of a model string in Unit 2.

$$\begin{aligned}
m_1l_1\ddot{\theta}_1 &\simeq -(m_1 + m_2)g\theta_1 + m_2g\theta_2,\\
m_2(l_1\ddot{\theta}_1 + l_2\ddot{\theta}_2) &\simeq -m_2g\theta_2.
\end{aligned}$$

To put these equations into a standard form, that is, with only one second derivative on the left-hand side of each, first divide by the masses to get

$$l_1\ddot{\theta}_1 \simeq -\left(1+\frac{m_2}{m_1}\right)g\theta_1 + \frac{m_2}{m_1}g\theta_2, \tag{19}$$

$$l_1\ddot{\theta}_1 + l_2\ddot{\theta}_2 \simeq -g\theta_2. \tag{20}$$

Next subtract equation (19) from equation (20) to eliminate the $l_1\ddot{\theta}_1$ term from the left-hand side, giving

$$\begin{aligned} l_2\ddot{\theta}_2 &\simeq -g\theta_2 - \left(-\left(1+\frac{m_2}{m_1}\right)g\theta_1 + \frac{m_2}{m_1}g\theta_2\right) \\ &= \left(1+\frac{m_2}{m_1}\right)g\theta_1 - \left(1+\frac{m_2}{m_1}\right)g\theta_2. \end{aligned} \tag{21}$$

Now divide equation (19) by l_1, and equation (21) by l_2, to obtain equations with one second derivative on the left-hand side of each. Thus we arrive at the following linear second-order differential equations of motion for the double pendulum when it is undergoing small oscillations:

Since this is a *model* of the motion, there is no need to retain the $\simeq$ signs.

$$\ddot{\theta}_1 = -\left(1+\frac{m_2}{m_1}\right)\frac{g}{l_1}\theta_1 + \frac{m_2}{m_1}\frac{g}{l_1}\theta_2,$$

$$\ddot{\theta}_2 = \left(1+\frac{m_2}{m_1}\right)\frac{g}{l_2}\theta_1 - \left(1+\frac{m_2}{m_1}\right)\frac{g}{l_2}\theta_2.$$

◀ Solve equation(s) ▶

(b) In the case where $l_1 = l_2\ (= l)$ and $m_1 = m_2$, the equations of motion become

$$\ddot{\theta}_1 = -\frac{2g}{l}\theta_1 + \frac{g}{l}\theta_2,$$

$$\ddot{\theta}_2 = \frac{2g}{l}\theta_1 - \frac{2g}{l}\theta_2.$$

To solve these equations, we write them in matrix form as

$$\begin{pmatrix}\ddot{\theta}_1\\ \ddot{\theta}_2\end{pmatrix} = \frac{g}{l}\begin{pmatrix}-2 & 1\\ 2 & -2\end{pmatrix}\begin{pmatrix}\theta_1\\ \theta_2\end{pmatrix}.$$

Following Procedure 6 in Unit 6, we now need to find the eigenvalues and eigenvectors of the dynamic matrix

$$\mathbf{A} = \frac{g}{l}\begin{pmatrix}-2 & 1\\ 2 & -2\end{pmatrix}.$$

However, this is equivalent to finding the eigenvalues and eigenvectors of

Recall from Unit 5 that if λ is an eigenvalue of $\mathbf{M}$ with eigenvector $\mathbf{v}$, then $p\lambda$ is an eigenvalue of $p\mathbf{M}$ with the same eigenvector $\mathbf{v}$.

$$\mathbf{B} = \begin{pmatrix}-2 & 1\\ 2 & -2\end{pmatrix},$$

provided that we remember to scale these eigenvalues by the factor g/l later. The characteristic equation of $\mathbf{B}$ is

$$\begin{vmatrix}-2-\lambda & 1\\ 2 & -2-\lambda\end{vmatrix} = \lambda^2 + 4\lambda + 2 = 0.$$

Therefore the eigenvalues of $\mathbf{B}$ are $\lambda = -2 \pm \sqrt{2}$.

- For $\lambda = -2 + \sqrt{2}$, the eigenvector equations are

$$\begin{pmatrix} -2 - (-2 + \sqrt{2}) & 1 \\ 2 & -2 - (-2 + \sqrt{2}) \end{pmatrix} \begin{pmatrix} v_1 \\ v_2 \end{pmatrix} = \begin{pmatrix} 0 \\ 0 \end{pmatrix},$$

or equivalently

$$\begin{aligned} -\sqrt{2}v_1 + \quad v_2 &= 0, \\ 2v_1 - \sqrt{2}v_2 &= 0. \end{aligned}$$

Both of these equations simplify to $v_2 = \sqrt{2}v_1$, so $(1 \quad \sqrt{2})^T$ is an eigenvector.

- For $\lambda = -2 - \sqrt{2}$, the eigenvector equations are

$$\begin{pmatrix} -2 - (-2 - \sqrt{2}) & 1 \\ 2 & -2 - (-2 - \sqrt{2}) \end{pmatrix} \begin{pmatrix} v_1 \\ v_2 \end{pmatrix} = \begin{pmatrix} 0 \\ 0 \end{pmatrix},$$

or equivalently

$$\begin{aligned} \sqrt{2}v_1 + \quad v_2 &= 0, \\ 2v_1 + \sqrt{2}v_2 &= 0. \end{aligned}$$

Both of these equations simplify to $v_2 = -\sqrt{2}v_1$, so $(1 \quad -\sqrt{2})^T$ is an eigenvector.

After we have scaled by g/l, we obtain the eigenvalues of the dynamic matrix $\mathbf{A}$ as $\lambda_1 = -g(2 - \sqrt{2})/l$ and $\lambda_2 = -g(2 + \sqrt{2})/l$. Since $\omega_i = \sqrt{-\lambda_i}$, the normal mode angular frequencies are

Note that $2 > \sqrt{2}$, so both eigenvalues are negative.

$$\omega_1 = \sqrt{\frac{g(2 - \sqrt{2})}{l}}, \quad \omega_2 = \sqrt{\frac{g(2 + \sqrt{2})}{l}}.$$

From equation (10), the general solution of the equations of motion can then be written as

$$\begin{pmatrix} \theta_1(t) \\ \theta_2(t) \end{pmatrix} = C_1 \begin{pmatrix} 1 \\ \sqrt{2} \end{pmatrix} \cos(\omega_1 t + \phi_1) + C_2 \begin{pmatrix} 1 \\ -\sqrt{2} \end{pmatrix} \cos(\omega_2 t + \phi_2),$$

where C_1, C_2, ϕ_1 and ϕ_2 are constants that depend on the initial conditions.

(c) The information that each pendulum has length 50 cm, that is, $l_1 = l_2 = 0.5$, enables us to calculate the normal mode angular frequencies as

◀ Interpret solution ▶

$$\omega_1 = \sqrt{\frac{g(2 - \sqrt{2})}{0.5}} \simeq 3.390, \quad \omega_2 = \sqrt{\frac{g(2 + \sqrt{2})}{0.5}} \simeq 8.185.$$

The periods of oscillation of the two normal modes of the double pendulum are

$$\tau_1 = \frac{2\pi}{\omega_1} \simeq 1.853, \quad \tau_2 = \frac{2\pi}{\omega_2} \simeq 0.768.$$

So this model predicts that the periods of oscillation of the normal modes are approximately 0.8 s and 1.9 s.

Notice how, in this example, we used small-angle approximations to produce a system of *linear* differential equations; otherwise the equations would have been non-linear.

Exercise 4

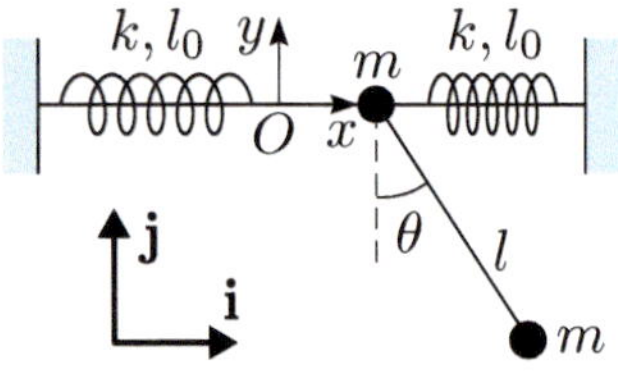

Figure 15 shows a mechanical system in which a particle of mass m is free to move along a smooth straight horizontal bar. The particle is subject to the forces exerted by two identical model springs of stiffness k and natural length l_0, and to the force exerted by a pendulum of length l suspended from the particle. The bar is twice the natural length of each spring, so the system is in equilibrium when the springs are at their natural lengths, that is, not extended or compressed.

Figure 15 A spring pendulum

The pendulum stem may be modelled as a light model rod and its bob as a particle of mass m. Assume that the oscillations of the pendulum are small, and hence that small-angle approximations (as in Example 3) are valid.

The displacement x_1 of the sliding particle is measured from its equilibrium position O, and the displacement of the pendulum is given by the angle θ (measured anticlockwise from the vertical), as shown in Figure 15.

(a) Derive linear equations of motion for the system in terms of x_1 and θ.

(b) If $k = 1$, $m = 1$ and $l = 1$, find the normal mode angular frequencies to two decimal places.

1.3 Interpretation of normal mode eigenvectors

For oscillating systems with more than one degree of freedom, how do we know the initial conditions required to make a system oscillate in a normal mode? This question is answered in this subsection, by using the normal mode eigenvectors. Before we do this, we derive a result that is useful in describing systems where all the particles start from rest (which are the easiest starting conditions in practice). We illustrate the argument in relation to a system with two degrees of freedom, though the argument can be generalised to a system with any number of degrees of freedom.

For a system with two degrees of freedom, the general solution of the equation of motion for the system (equation (10)) becomes

$$\mathbf{x}(t) = C_1\mathbf{v}_1\cos(\omega_1 t + \phi_1) + C_2\mathbf{v}_2\cos(\omega_2 t + \phi_2).$$

Differentiating this gives

$$\dot{\mathbf{x}}(t) = -\omega_1 C_1\mathbf{v}_1\sin(\omega_1 t + \phi_1) - \omega_2 C_2\mathbf{v}_2\sin(\omega_2 t + \phi_2).$$

Initially, at $t = 0$,

$$\dot{\mathbf{x}}(0) = -\omega_1 C_1\mathbf{v}_1\sin\phi_1 - \omega_2 C_2\mathbf{v}_2\sin\phi_2.$$

Applying the condition that all the particles start from rest, that is, $\dot{\mathbf{x}}(0) = \mathbf{0}$, gives

$$\mathbf{0} = -\omega_1 C_1 \mathbf{v}_1 \sin\phi_1 - \omega_2 C_2 \mathbf{v}_2 \sin\phi_2.$$

Notice that this initial condition is satisfied if $\phi_1 = \phi_2 = 0$. We will now show that we may assume that $\phi_1 = \phi_2 = 0$.

Since the two normal mode eigenvectors are linearly independent, the only way that a linear combination of them can be zero is if all the coefficients are zero, that is, if

$$\omega_1 C_1 \sin\phi_1 = 0 \quad \text{and} \quad \omega_2 C_2 \sin\phi_2 = 0.$$

Since the normal mode angular frequencies are non-zero, we have

You will see mechanical systems with zero normal mode angular frequencies later.

$$C_1 \sin\phi_1 = 0 \quad \text{and} \quad C_2 \sin\phi_2 = 0.$$

From the first equation, either $C_1 = 0$ or $\sin\phi_1 = 0$. If $C_1 = 0$, then the normal mode does not contribute to the motion $\mathbf{x}(t)$ and the value of ϕ_1 is irrelevant, so we can choose $\phi_1 = 0$. If $\sin\phi_1 = 0$, then either $\phi_1 = 0$ or $\phi_1 = \pi$. If $\phi_1 = \pi$, then the term $C_1\mathbf{v}_1\cos(\omega t + \pi)$ becomes $-C_1\mathbf{v}_1\cos(\omega t)$, and we can include the minus sign in the normal mode eigenvector $\mathbf{v}_1$ and choose $\phi_1 = 0$. So the first equation implies that either $\phi_1 = 0$ or we have a free choice and may take $\phi_1 = 0$. Similarly, we may take $\phi_2 = 0$. This is an important result that is worth remembering.

Recall that for normal modes we take $-\pi < \phi \leq \pi$.

Systems starting from rest

If one of the initial conditions of a mechanical system is that all the particles start from rest, then it follows that all the phase angles are zero.

Now we use this result to find a suitable initial condition for normal mode motion when all the particles start from rest. As before, consider a system with two degrees of freedom to illustrate the argument. If the system starts from rest, then by the above result all the phase angles are zero. So the general motion of the two-degrees-of-freedom system can be written as

$$\mathbf{x}(t) = C_1\mathbf{v}_1\cos(\omega_1 t) + C_2\mathbf{v}_2\cos(\omega_2 t). \tag{22}$$

Initially, at $t = 0$, this becomes

$$\mathbf{x}(0) = C_1\mathbf{v}_1 + C_2\mathbf{v}_2. \tag{23}$$

Suppose that the system is initially at a position corresponding to the first normal mode eigenvector, that is, at a position $\mathbf{x}(0) = k\mathbf{v}_1$ for some constant $k \neq 0$. Then substituting for $\mathbf{x}(0)$ in equation (23), we obtain

Recall, from Unit 5, that if $\mathbf{v}$ is an eigenvector, then so is $k\mathbf{v}$ for any non-zero constant k.

$$k\mathbf{v}_1 = C_1\mathbf{v}_1 + C_2\mathbf{v}_2. \tag{24}$$

Since the eigenvectors $\mathbf{v}_1$ and $\mathbf{v}_2$ are linearly independent, we can equate their respective coefficients on either side of equation (24), thereby finding that $C_1 = k \neq 0$ and $C_2 = 0$. Substituting into equation (22) gives

$$\mathbf{x}(t) = k\mathbf{v}_1\cos(\omega_1 t). \tag{25}$$

Therefore if we start the system from rest at a position $\mathbf{x}(0) = k\mathbf{v}_1$, it will move with the normal mode motion given by equation (25). Similarly, starting the system from rest at $\mathbf{x}(0) = k\mathbf{v}_2$ results in the normal mode motion given by $\mathbf{x}(t) = k\mathbf{v}_2 \cos(\omega_2 t)$. Generalising this argument to any number of degrees of freedom gives the following result.

Normal modes of systems starting from rest

If all the particles within a mechanical system start from rest at positions given by the elements of a normal mode eigenvector, then the system will oscillate in the corresponding normal mode.

The following example illustrates how this result can be used.

Example 4

In Example 2 the following general solution was derived for the positions of two particles, each of mass 0.1 kg, joined to one another and to two fixed walls by three identical model springs of stiffness $0.2\,\mathrm{N\,m^{-1}}$ (see Figure 10):

$$\mathbf{x}(t) = C_1 \begin{pmatrix} 1 \\ 1 \end{pmatrix} \cos(\sqrt{2}t + \phi_1) + C_2 \begin{pmatrix} 1 \\ -1 \end{pmatrix} \cos(\sqrt{6}t + \phi_2).$$

What initial conditions give rise to normal mode motion in this system when both particles start from rest?

Solution

When the two particles start from rest, so that $\dot{\mathbf{x}}(0) = \mathbf{0}$, we can deduce that $\phi_1 = \phi_2 = 0$, whatever the subsequent motion. For the normal mode with angular frequency $\sqrt{2}$, a normal mode eigenvector is $(1 \quad 1)^T$. The positions of the particles are measured from their respective equilibrium positions. So one initial condition for normal mode motion is that each particle starts from rest at a distance 1 cm, say, to the right of its equilibrium position. However, any constant multiple of $(1 \quad 1)^T$ is also an eigenvector, so we can say rather more: a normal mode results if both particles start from rest after being given the *same* displacement. Therefore suitable initial conditions include not only starting both particles from rest 1 cm to the right of their equilibrium positions, but also starting both particles 1 cm to the left, or 10 cm to the right (provided that the natural length of the spring is significantly greater than 10 cm), and so on.

For the second normal mode, a normal mode eigenvector is $(1 \quad -1)^T$. Hence another initial condition for normal mode motion is that the first particle starts from rest at a distance 1 cm to the right of its equilibrium position, and the second particle starts from rest 1 cm to the left of its equilibrium position. Again, however, we can say more: a normal mode motion results whenever the particles start from rest after being displaced from equilibrium by equal distances in opposite directions.

Exercise 5

For the double pendulum analysed in Example 3, what initial conditions give rise to normal mode motion when both bobs start from rest?

Another feature of the motion can be read from the normal mode eigenvectors. The signs of the components of the eigenvectors indicate whether the normal mode motion is in-phase or phase-opposed.

For example, consider the double pendulum and Example 3. The normal mode eigenvector corresponding to the smallest normal mode frequency is $(1 \quad \sqrt{2})^T$. The components of this normal mode eigenvector have the same sign, so the normal mode motion is in-phase.

Note that $(-1 \quad -\sqrt{2})^T$ is also a normal mode eigenvector for this motion, and its components are not positive, but they have the same sign.

The other normal mode eigenvector of the double pendulum was calculated to be $(1 \quad -\sqrt{2})^T$ in Example 3. Here the two components have opposite signs and the normal mode motion is phase-opposed. Again, this is generally true.

We summarise this as follows.

Theorem 2

For a normal mode motion, the relative motion of a pair of particles with non-zero coordinates is determined by the components of the corresponding normal mode eigenvector. The motion is:

- in-phase if the components have the same sign
- phase-opposed if the components have opposite signs.

Example 5

Consider the oscillating mechanical system with three degrees of freedom shown in Figure 16. Three particles are constrained to move in a straight line between two fixed walls. The particles are attached to one another and to the walls by four identical model springs that are aligned with the direction of movement of the particles. The positions of the particles are measured from their equilibrium positions, as shown in the figure.

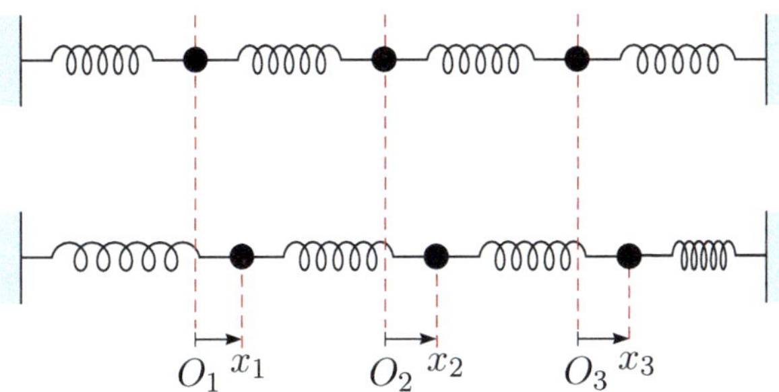

Figure 16 An oscillating system with three degrees of freedom

A normal mode eigenvector of the lowest angular frequency normal mode is given as $(1 \quad \sqrt{2} \quad 1)^T$. Draw sketches of the mechanical system at the following times while it is oscillating according to this normal mode:

(a) when all the particles are instantaneously at rest at the same time

(b) one-sixth of a cycle after the time in part (a)

(c) one-quarter of a cycle after the time in part (a)

(d) half a cycle after the time in part (a).

Is the motion in-phase or phase-opposed?

Solution

As a starting point for the system, consider a time instant when all three particles are instantaneously at rest at the same time. An equation of motion for the given normal mode is $\mathbf{x}(t) = k\mathbf{v}\cos\omega t$, where k is a constant and $\mathbf{v} = (1 \quad \sqrt{2} \quad 1)^T$. The required sketch is shown in Figure 17.

For a system starting from rest, we can take $\phi = 0$.

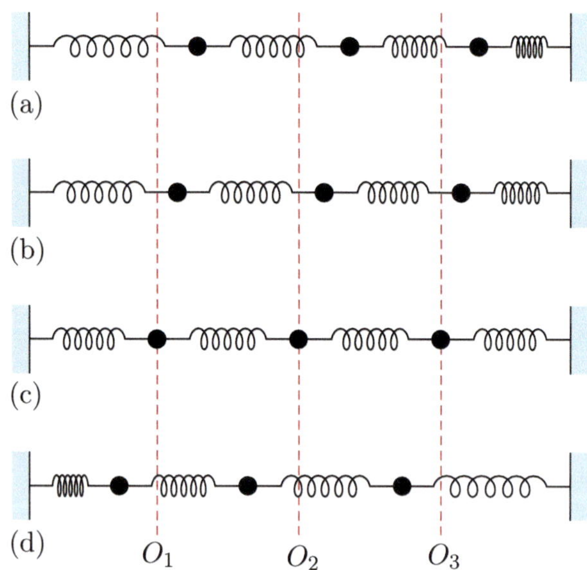

Figure 17 The oscillating system at various times during the motion

(a) In this instance the positions of the particles relative to their equilibrium positions are given by $k\mathbf{v}$, and are sketched in Figure 17(a). Thus if particle 1 is at, say, $x_1 = 2$, then particle 2 will be at $x_2 = 2\sqrt{2}$, and particle 3 will be at $x_3 = 2$, as determined by $k\mathbf{v} = 2\mathbf{v} = (2 \quad 2\sqrt{2} \quad 2)^T$.

(b) After a sixth of a cycle (i.e. when $\omega t = \frac{2\pi}{6} = \frac{\pi}{3}$), the equation of motion yields $\mathbf{x}(t) = k\mathbf{v}\cos\omega t = k\mathbf{v}\cos\frac{\pi}{3} = \frac{1}{2}k\mathbf{v}$, so each particle is half as far from its equilibrium position as it was in part (a), as sketched in Figure 17(b).

(c) After a quarter of a cycle (i.e. when $\omega t = \frac{2\pi}{4} = \frac{\pi}{2}$), $\mathbf{x}(t) = k\mathbf{v}\cos\frac{\pi}{2} = \mathbf{0}$, so each particle is at its equilibrium position, as sketched in Figure 17(c).

(d) After half a cycle (i.e. when $\omega t = \frac{2\pi}{2} = \pi$), $\mathbf{x}(t) = k\mathbf{v}\cos\pi = -k\mathbf{v}$, so each particle is the same distance from its equilibrium position as it was in part (a), but in the opposite direction, as sketched in Figure 17(d).

Note that although the particles are at their equilibrium positions, the *system* is not in equilibrium since the velocities are non-zero. (In fact, the speeds are at a maximum.)

Since all the coordinates of the given normal mode eigenvector have the same sign, all three particles move in-phase with one another, as the sketches in Figure 17 illustrate.

Exercise 6

Repeat Example 5 for the normal modes with normal mode eigenvectors $(1 \quad 0 \quad -1)^T$ and $(1 \quad -\sqrt{2} \quad 1)^T$.

This section concludes by asking you to apply the methods described in this section to another mechanical system.

Exercise 7

Consider the mechanical system with two degrees of freedom, shown in Figure 18, in which two identical particles of mass m are attached to one another and to a fixed wall by two model springs of stiffnesses $3k$ and $2k$, each with natural length l_0. The particles are constrained to move across a frictionless horizontal surface in a straight line along the line of the springs. Let x_1 and x_2 be the displacements of the two particles along this line, measured from the equilibrium positions of the particles as shown in the figure.

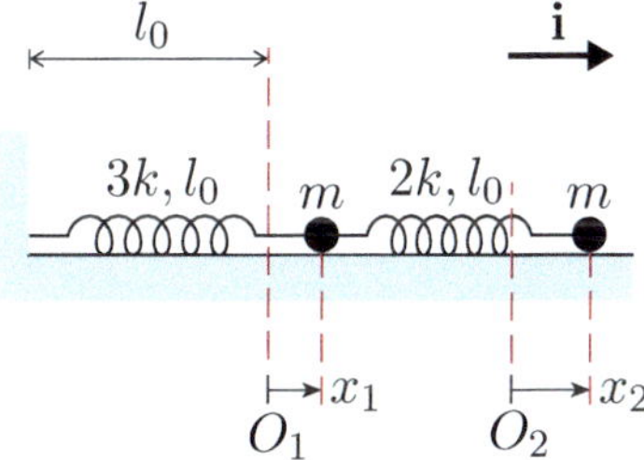

Figure 18 A system with two degrees of freedom

(a) Derive the equations of motion for this system.

(b) If $k/m = 1$, determine the normal mode eigenvectors and the angular frequencies. Hence write down the general solution of the matrix equation of motion.

(c) Write down equations from which the constants could be determined in the general solution obtained in part (b), given the initial condition that both particles start from rest, with displacements d_1 and d_2, respectively.

2 One-dimensional systems

In this section we continue to look at mechanical systems where particles are restricted to move along a line (such systems have the number of degrees of freedom equal to the number of particles).

A quicker method for deriving the equations of motion is developed in this section that has the advantage that the equations of motion can be derived without first needing to calculate the equilibrium positions of particles.

Subsection 2.1 considers systems that have one or two springs fixed at one end. The equations of motion derived are similar to those derived in Section 1 and are not investigated further. Subsection 2.2 considers free systems that have no components fixed. Here the solution of the equations of motion is different, so we investigate it.

2.1 Displacement from equilibrium

In this subsection you will see how to obtain the homogeneous equations of motion for a mechanical system (i.e. the equation obtained by taking the origin at the equilibrium position), without first having to find an equilibrium position of the system.

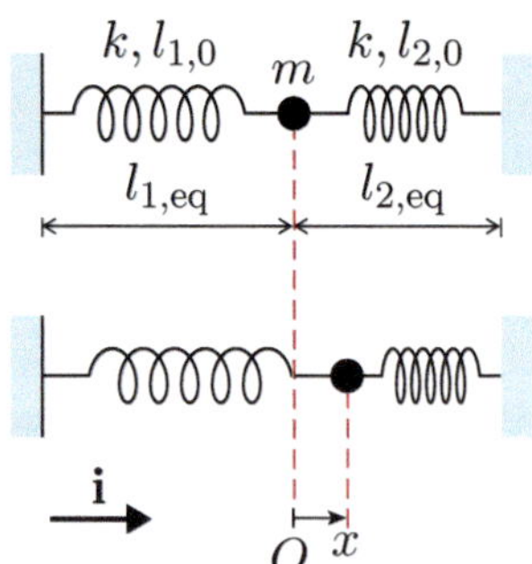

Figure 19 A particle and two springs

Consider the system made up of a particle of mass m attached to two walls by two model springs of stiffnesses k_1 and k_2, as shown in Figure 19, with the particle constrained to move in one dimension along the line of the springs (friction may be neglected). Because the springs have different stiffnesses, the equilibrium position of the system is not obvious.

We denote the forces exerted by the springs when the system is in equilibrium by $\mathbf{H}_{1,\text{eq}}$ and $\mathbf{H}_{2,\text{eq}}$, and the corresponding forces when the system is displaced a distance x from the equilibrium position by $\mathbf{H}_1$ and $\mathbf{H}_2$. At equilibrium,

$$\mathbf{H}_{1,\text{eq}} + \mathbf{H}_{2,\text{eq}} = \mathbf{0}. \tag{26}$$

Applying Hooke's law, we obtain

$$\mathbf{H}_1 = k_1((l_{1,\text{eq}} + x) - l_{1,0})(-\mathbf{i}), \quad \mathbf{H}_{1,\text{eq}} = k_1(l_{1,\text{eq}} - l_{1,0})(-\mathbf{i}),$$
$$\mathbf{H}_2 = k_2((l_{2,\text{eq}} - x) - l_{2,0})\mathbf{i}, \qquad \mathbf{H}_{2,\text{eq}} = k_2(l_{2,\text{eq}} - l_{2,0})\mathbf{i},$$

where $l_{1,\text{eq}}$ and $l_{2,\text{eq}}$ are the equilibrium lengths of the springs, $l_{1,0}$ and $l_{2,0}$ are the natural lengths of the springs, and $\mathbf{i}$ is a unit vector in the positive x-direction marked on Figure 19. From these equations, the changes in the forces exerted by the springs when the particle is displaced a distance x from the equilibrium position are given by

The symbol Δ is an upper-case delta. It is quite commonly used to denote a change in a quantity (just as a lower-case delta, δ, denotes a small change in a quantity).

$$\Delta\mathbf{H}_1 = \mathbf{H}_1 - \mathbf{H}_{1,\text{eq}} = -k_1x\mathbf{i}, \tag{27}$$
$$\Delta\mathbf{H}_2 = \mathbf{H}_2 - \mathbf{H}_{2,\text{eq}} = -k_2x\mathbf{i}. \tag{28}$$

Then

$$\Delta\mathbf{H}_1 + \Delta\mathbf{H}_2 = (\mathbf{H}_1 + \mathbf{H}_2) - (\mathbf{H}_{1,\text{eq}} + \mathbf{H}_{2,\text{eq}}) = -(k_1 + k_2)x\mathbf{i},$$

and on substituting from equation (26), we obtain

$$\Delta\mathbf{H}_1 + \Delta\mathbf{H}_2 = \mathbf{H}_1 + \mathbf{H}_2 = -(k_1 + k_2)x\mathbf{i}.$$

So the total force $\mathbf{H}_1 + \mathbf{H}_2$ on the particle when it is displaced a distance x from the equilibrium position is the same as the sum of the changes in the forces, $\Delta\mathbf{H}_1 + \Delta\mathbf{H}_2$. Furthermore, if we apply Newton's second law to the particle when it is displaced a distance x from equilibrium, we find that

$$m\ddot{\mathbf{r}} = \mathbf{H}_1 + \mathbf{H}_2 = \Delta\mathbf{H}_1 + \Delta\mathbf{H}_2 = -(k_1 + k_2)x\mathbf{i},$$

where $\mathbf{r} = x\mathbf{i}$ is the position vector, and the origin is at the equilibrium position. Resolving in the $\mathbf{i}$-direction and rearranging gives the equation of motion as

$$m\ddot{x} + (k_1 + k_2)x = 0. \tag{29}$$

Thus we have found the equation of motion without knowing the equilibrium position of the particle, knowing only the changes in the forces acting as given by equations (27) and (28).

This method of obtaining an equation of motion can be generalised to any number of forces acting on any particle that has been disturbed from an equilibrium position. If forces $\mathbf{F}_i$ $(i = 1, \ldots, n)$ act on a particle disturbed from an equilibrium position, and $\mathbf{F}_{i,\text{eq}}$ $(i = 1, \ldots, n)$ denote the corresponding forces when the particle is in equilibrium, while $\Delta\mathbf{F}_i$ $(i = 1, \ldots, n)$ denote the changes in the forces from equilibrium, then

$$\sum_{i=1}^{n} \Delta\mathbf{F}_i = \sum_{i=1}^{n} (\mathbf{F}_i - \mathbf{F}_{i,\text{eq}}) = \sum_{i=1}^{n} \mathbf{F}_i - \sum_{i=1}^{n} \mathbf{F}_{i,\text{eq}} = \sum_{i=1}^{n} \mathbf{F}_i,$$

since $\sum_{i=1}^{n} \mathbf{F}_{i,\text{eq}} = \mathbf{0}$. So the total force $\sum_{i=1}^{n} \mathbf{F}_i$ acting on the particle is the same as the total change in the forces $\sum_{i=1}^{n} \Delta\mathbf{F}_i$, which leads to the following useful result based on Newton's second law.

$\sum_{i=1}^{n} \mathbf{F}_{i,\text{eq}} = \mathbf{0}$ is the equilibrium condition for particles, from Unit 2.

Particle displaced from equilibrium

If a particle of mass m, displaced from an equilibrium position, is acted on by forces $\mathbf{F}_i$ $(i = 1, \ldots, n)$, then its acceleration $\ddot{\mathbf{r}}$ is given by

$$m\ddot{\mathbf{r}} = \sum_{i=1}^{n} \Delta\mathbf{F}_i, \tag{30}$$

where $\mathbf{r}$ is the displacement of the particle from its equilibrium position, and $\Delta\mathbf{F}_i = \mathbf{F}_i - \mathbf{F}_{i,\text{eq}}$ $(i = 1, \ldots, n)$, where $\mathbf{F}_{i,\text{eq}}$ is the force corresponding to $\mathbf{F}_i$ when the particle is at its equilibrium position.

Applying equation (30) is equivalent to applying Newton's second law.

Equation (30) can provide a useful shortcut in the derivation of the equation of motion for a mechanical system involving a displacement from equilibrium, if the changes in the forces acting can be calculated easily. For a spring displaced along its length, you have seen that the changes in the forces can be calculated easily using equations (27) and (28).

The general result that applies to any coordinate system for a change of length Δl is

See Section 3 for formulas that apply in other cases.

$$\Delta\mathbf{H} = k\,\Delta l\,\widehat{\mathbf{s}}, \tag{31}$$

where k is the stiffness of the spring, $\Delta l = l - l_{\text{eq}}$ is the increase in the spring's length from its equilibrium value, and $\widehat{\mathbf{s}}$ is a constant unit vector directed from the particle to the centre of the spring.

You saw in Unit 9 that the period of oscillations of a mechanical system is the same whether the system is oriented horizontally or vertically.

Another force for which this shortcut is useful is the force $\mathbf{W}$ due to gravity, that is, the weight of an object, since this force is constant for a given object near the Earth's surface, so $\Delta\mathbf{W} = \mathbf{0}$. Thus the weights of the components of a mechanical system have no effect on the system's equation of motion when the components are displaced from their equilibrium positions. If a normal reaction force $\mathbf{N}$ balances the weight of a particle (such as when objects are placed on a smooth horizontal table), then $\mathbf{N}$ is constant thus $\Delta\mathbf{N} = \mathbf{0}$.

This result also applies to systems involving model rods, as you will see in Unit 19.

These results have useful consequences for one- or two-dimensional mechanical systems that can be modelled using only model springs and particles. It means that in deriving the equations of motion for such systems, provided that displacements are measured from the equilibrium positions, we can ignore all weights and normal reactions, irrespective of the orientation of the system – in effect, whatever its orientation, the system behaves as if it is in a horizontal plane. We will adopt the policy of ignoring weights and normal reactions for such systems for the remainder of this unit.

Let us now see how these ideas can be applied, by looking at an example similar to the one considered in Example 1.

Example 6

Consider the mechanical system shown in Figure 20. This system consists of two particles of masses m_1 and m_2, which are constrained to move in a straight line across a frictionless horizontal surface while attached by model springs of stiffnesses k_1, k_2 and k_3 to each other and to two fixed walls. Derive the equation of motion for each of the particles.

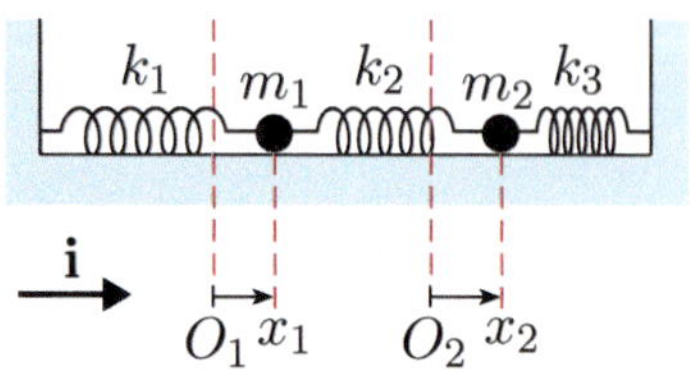

Figure 20 A spring–mass system. (Where natural lengths are not needed for calculations, we may not mark them on diagrams from here onwards.)

Solution

Unlike the situation in Example 1, the springs have different stiffnesses and we do not know the equilibrium positions of the components of the system. Hence we need to adopt a different approach from that used in Example 1.

The first step is to draw a force diagram for each particle (Figure 21). We use the usual notation for the spring forces and, as discussed above, we ignore the weights and normal reactions, which balance.

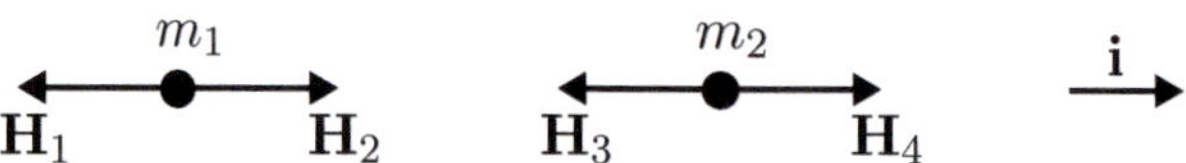

Figure 21 Force diagrams showing only the additional forces

Applying equation (30) to each particle gives

$$m_1\ddot{\mathbf{r}}_1 = \Delta\mathbf{H}_1 + \Delta\mathbf{H}_2, \tag{32}$$

$$m_2\ddot{\mathbf{r}}_2 = \Delta\mathbf{H}_3 + \Delta\mathbf{H}_4, \tag{33}$$

where $\mathbf{r}_1 = x_1\mathbf{i}$ and $\mathbf{r}_2 = x_2\mathbf{i}$ are the position vectors of the two particles with respect to their (unknown) equilibrium positions, and $\mathbf{i}$ is a unit vector in the direction of positive x_1 and x_2.

Noting that the additional extension of the middle spring is $x_2 - x_1$, we use equation (31) to obtain

$$\begin{aligned}
\Delta\mathbf{H}_1 &= k_1x_1(-\mathbf{i}) = -k_1x_1\mathbf{i},\\
\Delta\mathbf{H}_2 &= k_2(x_2 - x_1)\mathbf{i},\\
\Delta\mathbf{H}_3 &= k_2(x_2 - x_1)(-\mathbf{i}) = -k_2(x_2 - x_1)\mathbf{i},\\
\Delta\mathbf{H}_4 &= k_3(-x_2)\mathbf{i} = -k_3x_2\mathbf{i},
\end{aligned}$$

where the minus sign in the last expression indicates that the third spring has been compressed by the amount x_2 (or extended by an amount $-x_2$) from its equilibrium point, as in equations (27) and (28).

Substituting these expressions into equations (32) and (33) gives

$$\begin{aligned}
m_1\ddot{\mathbf{r}}_1 &= -k_1x_1\mathbf{i} + k_2(x_2 - x_1)\mathbf{i},\\
m_2\ddot{\mathbf{r}}_2 &= -k_2(x_2 - x_1)\mathbf{i} - k_3x_2\mathbf{i}.
\end{aligned}$$

Resolving in the $\mathbf{i}$-direction then gives

$$m_1\ddot{x}_1 = -(k_1 + k_2)x_1 + k_2x_2, \tag{34}$$

$$m_2\ddot{x}_2 = k_2x_1 - (k_2 + k_3)x_2. \tag{35}$$

These are the equations of motion for the system illustrated in Figure 20.

Exercise 8

Use equation (30) to derive the equations of motion for the system in Exercise 7 (see Figure 18).

2.2 Free motion

The one-dimensional mechanical systems studied so far in this unit have all been constrained by being attached to fixed walls. This subsection looks at one-dimensional mechanical systems without fixed points, where free motion (in one dimension) is possible. You will see that the absence of fixed points manifests itself in the dynamic matrix having a zero eigenvalue, and you will see how this affects the normal modes of the system. We begin with an example.

Example 7

This example is concerned with the motions of a hydrogen molecule, in particular the vibrations within it. This molecule consists of two atoms of equal mass joined by a bond. We model this as two particles of equal mass m, joined by a model spring of stiffness k. The system is considered to be one-dimensional, so only those vibrations that are along the straight line joining the two atoms are considered.

(a) Derive the equation of motion for this system.

(b) Determine the general solution of this equation of motion for the system in terms of m and k.

(c) Interpret this general solution.

(d) Experimentally, the frequency of vibration for a hydrogen molecule is 1.3×10^{14} Hz. If the mass of a hydrogen atom is 1.67×10^{-27} kg, estimate the stiffness of the bond between the two hydrogen atoms in the molecule.

Solution

◀ Draw picture ▶

(a) A sketch of the model of the hydrogen molecule is shown in Figure 22.

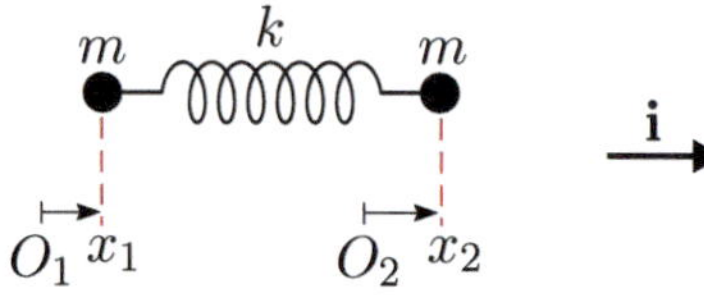

Figure 22 Model of a hydrogen molecule

◀ Choose coordinates ▶

Choose axes aligned with the spring that joins the two particles, with two separate origins at the equilibrium positions of the two particles (i.e. separated by a distance equal to the natural length of the model spring), as in Figure 22. Using the usual notation, we can draw the force diagrams (one for each particle) in Figure 23.

◀ Draw force diagram(s) ▶

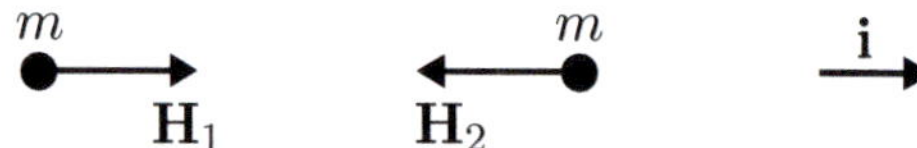

Figure 23 Force diagrams for the two atoms

◀ Apply Newton's 2nd law ▶

Applying Newton's second law (in the form of equation (30)) to each particle, we obtain

$$m\ddot{\mathbf{r}}_1 = \Delta\mathbf{H}_1, \quad m\ddot{\mathbf{r}}_2 = \Delta\mathbf{H}_2, \tag{36}$$

where $\mathbf{r}_1 = x_1\mathbf{i}$ and $\mathbf{r}_2 = x_2\mathbf{i}$, and $\mathbf{i}$ is a unit vector in the direction of positive x_1 and x_2.

Then from equation (31),

$$\Delta\mathbf{H}_1 = k(x_2 - x_1)\mathbf{i},$$
$$\Delta\mathbf{H}_2 = k(x_2 - x_1)(-\mathbf{i}) = -k(x_2 - x_1)\mathbf{i}.$$

Substituting these expressions into equations (36) gives

$$m\ddot{\mathbf{r}}_1 = k(x_2 - x_1)\mathbf{i},$$
$$m\ddot{\mathbf{r}}_2 = -k(x_2 - x_1)\mathbf{i}.$$

Resolving in the **i**-direction and using matrix notation, we obtain the equation of motion that represents the vibrations of a hydrogen molecule:

$$\begin{pmatrix}\ddot{x}_1\\ \ddot{x}_2\end{pmatrix} = \begin{pmatrix}-k/m & k/m\\ k/m & -k/m\end{pmatrix}\begin{pmatrix}x_1\\ x_2\end{pmatrix}.$$

(b) To find the eigenvalues and eigenvectors of the dynamic matrix, it is convenient to take out the common factor k/m and work with the matrix $\begin{pmatrix}-1 & 1\\ 1 & -1\end{pmatrix}$. The eigenvalues are found by solving

◀Solve equation(s)▶

$$\begin{vmatrix}-1-\lambda & 1\\ 1 & -1-\lambda\end{vmatrix} = \lambda^2 + 2\lambda = 0.$$

So the eigenvalues are $\lambda = 0$ and $\lambda = -2$. We now proceed to find the eigenvectors.

- For $\lambda = 0$, the eigenvector equations are

  $$\begin{pmatrix}-1-0 & 1\\ 1 & -1-0\end{pmatrix}\begin{pmatrix}v_1\\ v_2\end{pmatrix} = \begin{pmatrix}0\\ 0\end{pmatrix},$$

  or equivalently

  $$-v_1 + v_2 = 0,$$
  $$v_1 - v_2 = 0.$$

  Both of these equations reduce to $v_2 = v_1$, so $(1 \quad 1)^T$ is an eigenvector.

- For $\lambda = -2$, the eigenvector equations are

  $$\begin{pmatrix}-1-(-2) & 1\\ 1 & -1-(-2)\end{pmatrix}\begin{pmatrix}v_1\\ v_2\end{pmatrix} = \begin{pmatrix}0\\ 0\end{pmatrix},$$

  or equivalently

  $$v_1 + v_2 = 0,$$
  $$v_1 + v_2 = 0.$$

  Both of these equations reduce to $v_2 = -v_1$, so $(1 \quad -1)^T$ is an eigenvector.

To obtain the eigenvalues of the dynamic matrix, we must multiply the above eigenvalues by the factor k/m. Hence the dynamic matrix has eigenvalues 0 and $-2k/m$.

In previous examples all the eigenvalues were negative, so all the terms in the general solution of the equation of motion were sinusoidal. But this time we have one zero eigenvalue and one negative eigenvalue.

See Procedure 6 of Unit 6.

You may recall that a zero eigenvalue gives rise to a linear term in the general solution of the equation of motion, so in this case the general solution has the form

$$\begin{pmatrix} x_1(t) \\ x_2(t) \end{pmatrix} = \begin{pmatrix} 1 \\ 1 \end{pmatrix} (A + Bt) + C \begin{pmatrix} 1 \\ -1 \end{pmatrix} \cos(\omega t + \phi),$$

where A, B, C and ϕ are constants, and $\omega = \sqrt{2k/m}$. Therefore there is just one normal mode, with normal mode angular frequency ω, which is given by the sinusoidal term, and one other 'mode', which is given by the linear term.

◀ Interpret solution ▶

(c) As just noted, the sinusoidal term in the general solution represents a normal mode. Since a corresponding normal mode eigenvector is $(1 \ \ -1)^T$, the system will vibrate in this normal mode if it starts from rest with the first particle a distance d from its origin and the second particle a distance $-d$ from its origin, for any non-zero value of d. The angular frequency of the vibrations will be $\omega = \sqrt{2k/m}$.

The linear term in the general solution represents a solution of the form $\mathbf{x}(t) = \mathbf{v}(A + Bt)$, where $\mathbf{v} = (1 \ \ 1)^T$; this corresponds to the linear motion of the system. Now $\dot{\mathbf{x}}(t) = \mathbf{v}B$, so if the system starts from rest, that is, $\dot{\mathbf{x}}(0) = \mathbf{v}B = 0$, it follows that $B = 0$, giving $\mathbf{x}(t) = \mathbf{v}A = (A \ \ A)^T$. Thus if the particles start from rest, equidistant from their equilibrium positions and in the same direction, then they will remain static for all t. Alternatively, putting $\mathbf{x}(0) = \mathbf{0}$, we obtain $\mathbf{x}(0) = \mathbf{v}A = \mathbf{0}$, so $A = 0$, hence $\mathbf{x}(t) = \mathbf{v}Bt$ and $\dot{\mathbf{x}}(t) = \mathbf{v}B = (B \ \ B)^T$. The system will move in this mode if the particles start at their equilibrium positions with the same velocity, in which case they will continue to move indefinitely at the same velocity. In both of these 'linear modes', the particles remain separated by a distance equal to the natural length of the spring (consequently there are no forces exerted by the spring, and therefore no vibrations).

More specifically in terms of the hydrogen molecule, the normal mode given by the sinusoidal term relates to the vibrations of the atoms within the molecule, while the 'mode' given by the linear term relates to the translational motion of the molecule.

(d) From the given frequency of vibration, we can calculate the normal mode angular frequency as

$$\omega = 2\pi f = 2\pi \times 1.3 \times 10^{14} \simeq 8.2 \times 10^{14}.$$

But $\omega = \sqrt{2k/m}$, so

$$k = \tfrac{1}{2} m\omega^2 = \tfrac{1}{2} \times 1.67 \times 10^{-27} \times (8.2 \times 10^{14})^2 \simeq 557.$$

Hence the stiffness of the bond in the hydrogen molecule is about $560\,\mathrm{N\,m^{-1}}$.

Hanging a 1 kg object on a spring of this stiffness would extend it by only 2 cm. This suggests that the atoms in a hydrogen molecule are very tightly bonded.

In part (c) of Example 7 you saw how a zero eigenvalue leads to a linear term in the general solution of the equation of motion, and how this term can be interpreted as representing linear motion, with the constituent parts of the system staying a fixed distance apart, and with each constituent part either moving with the same velocity or remaining at rest.

This type of motion, where all the constituent parts move as one, is given a special name.

Rigid body motion is the motion that occurs when all the particles of a system move with the distances between them remaining invariant; that is, the particles move as if part of a rigid body.

Rigid bodies were introduced in Unit 2. You may recall that the rigid body model assumes that whatever forces act on the body, it does not change shape or vibrate.

If a mechanical system is not constrained by being attached to a fixed object (e.g. the ground or a wall) but is free to move, then it is capable of executing rigid body motion. For such a system, one or more of the eigenvalues of the dynamic matrix will be zero. For one-dimensional systems, there is at most one type of rigid body motion (translation along the axis), so there is at most one zero eigenvalue for any dynamic matrix. For two-dimensional systems, there are three possible distinct (i.e. linearly independent) rigid body motions (translation along either axis, plus a rotation), so the situation is more complicated, with up to three zero eigenvalues.

The free motion of two-dimensional systems is discussed in Unit 19.

Although a zero eigenvalue does not lead to a normal mode as defined in Section 1, the resulting mode has similar properties and is often loosely referred to as a normal mode, as in the next two exercises. In this unit, if you are asked to determine or analyse normal modes, you should always include these 'linear modes' in your answer.

Sometimes a 'linear mode' is referred to as a 'trivial normal mode'.

Exercise 9

Consider a railway engine connected to two trucks, as shown in Figure 24. The trucks and the engine are modelled as particles of equal mass m, and the couplings are modelled as model springs of equal stiffness k. Assume that the train travels along a horizontal straight frictionless track and that the engine provides no motive force. The positions of the components of the system are measured from the equilibrium positions shown in the figure.

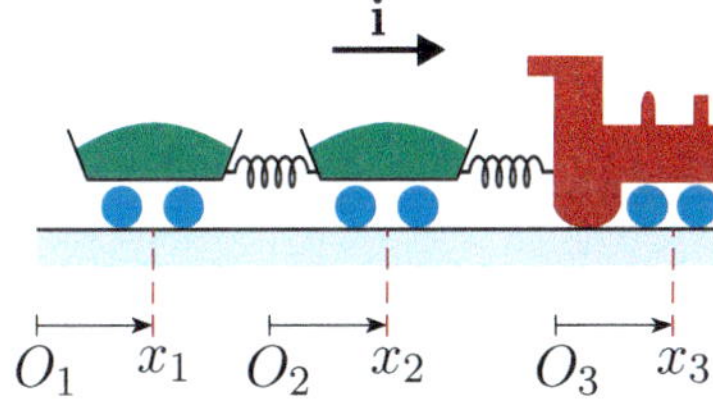

Figure 24 A railway engine and two trucks

(a) Derive the equation of motion for this system.

(b) Use the following information to obtain the general solution of the equation of motion:

$$\begin{pmatrix} -1 & 1 & 0 \\ 1 & -2 & 1 \\ 0 & 1 & -1 \end{pmatrix} \begin{pmatrix} 1 \\ 1 \\ 1 \end{pmatrix} = \begin{pmatrix} 0 \\ 0 \\ 0 \end{pmatrix},$$

$$\begin{pmatrix} -1 & 1 & 0 \\ 1 & -2 & 1 \\ 0 & 1 & -1 \end{pmatrix} \begin{pmatrix} 1 \\ 0 \\ -1 \end{pmatrix} = \begin{pmatrix} -1 \\ 0 \\ 1 \end{pmatrix},$$

$$\begin{pmatrix} -1 & 1 & 0 \\ 1 & -2 & 1 \\ 0 & 1 & -1 \end{pmatrix} \begin{pmatrix} 1 \\ -2 \\ 1 \end{pmatrix} = \begin{pmatrix} -3 \\ 6 \\ -3 \end{pmatrix}.$$

(c) For each normal mode, state the initial conditions that give rise to it.

Exercise 10

This exercise considers a mechanical system that is not one-dimensional, but it belongs here because it is a simple system that exhibits free motion. Since the system is not one-dimensional, the method introduced in this section does not apply and you will need to use the more general approach of Section 1 (see Exercise 4).

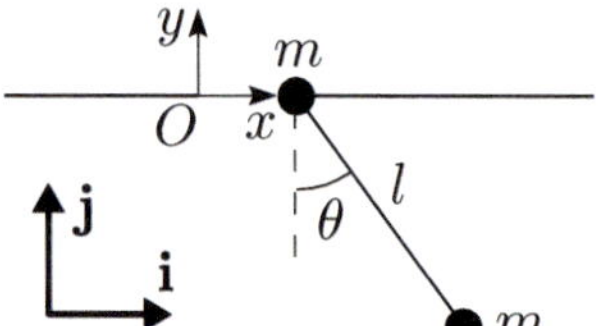

Figure 25 A slider pendulum

Figure 25 shows a particle of mass m that slides freely along a smooth straight horizontal bar and has a pendulum suspended from it. Model the pendulum stem as a light model rod of length l, and the bob as a particle of mass m. Assume that the oscillations of the pendulum are small. Measure the displacement x_1 of the sliding particle from a fixed point O on the bar, and the displacement of the pendulum by the angle θ (measured anticlockwise from the vertical), as shown in the figure.

(a) Derive linear equations of motion for the system, in terms of x_1 and θ.

(b) Find the normal mode angular frequencies and eigenvectors, and hence write down the general solution of the matrix equation of motion.

(c) How does the period of this pendulum compare with the period of a simple pendulum of the same length (with equation of motion $\ddot{\theta} = -(g/l)\theta$)?

(d) For each of the normal modes, draw a sketch of the motion.

3 Modelling a guitar string

In this section a series of models that approximate the behaviour of a guitar string are analysed.

Vibrations of a stretched string, such as a guitar string, that are aligned with the string are **longitudinal**, while vibrations at right angles to the string are **transverse**. In this context, the **fundamental** is the normal mode of a system that has the lowest non-zero normal mode angular frequency.

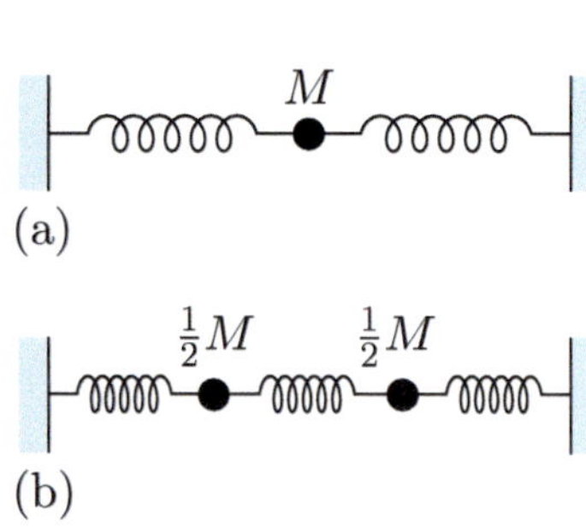

Figure 26 Modelling a guitar string

You might think that we could model a guitar string as a model spring, but this model would not be able to predict the vibrations of the string because, by definition, a model spring has no mass thus, in effect, there would be nothing to vibrate! To model the vibrations of a guitar string, the string has to be represented as having both mass and elasticity. The simplest such model is shown in Figure 26(a). In this diagram, all the mass of the string is 'lumped' together as a single particle in the middle of the string, and the elasticity is divided into two identical model springs. Figure 26(b) shows a revised model, where the string is represented as two particles of equal mass joined by three identical model springs. Since the revised model in Figure 26(b) is, in some sense, closer to the real situation where the mass is continuously distributed along the string, we would expect this to be a better model. Similarly, by increasing the number of particles and springs, we would expect to produce better and better models.

In the models of a guitar string described above, the properties of a real object are lumped together into discrete components (e.g. particles and model springs). This type of model is given a special name.

A model of a real-world system in which properties of the system are lumped together into discrete components (e.g. particles and model springs) is called a **lumped parameter model**.

To make the discussion less abstract, we will consider a particular string on the guitar, namely the E string, which has fundamental frequency 323 Hz. We will take an example of an E string that has stiffness $4000\,\mathrm{N\,m^{-1}}$, mass 0.25 g, and natural length 0.633 m. When this string is tuned to the note E, its length is 0.65 m. We will model the E string by means of lumped parameter models: the purpose of the modelling is to construct a model of the E string that successfully predicts the fundamental frequency of the string as 323 Hz.

A different sort of model of the E string, as an elastic heavy continuous string, is discussed in Unit 14.

Before we begin the modelling, we introduce a quantity that will prove useful.

The **equilibrium tension in a model spring**, T_{eq}, is defined by the formula

$$T_{\mathrm{eq}} = k(l_{\mathrm{eq}} - l_0), \tag{37}$$

where k is the stiffness of the spring, l_0 is its natural length, and l_{eq} is its equilibrium length.

The equilibrium tension in a model spring is a scalar quantity that is equal to plus or minus the magnitude of the force exerted by the spring, with the sign depending on whether the spring is extended or compressed when in equilibrium. For the E string modelled as a single model spring, we have

$$T_{\mathrm{eq}} = 4000(0.65 - 0.633) = 68,$$

so the equilibrium tension in the E string is 68 N.

It was noted above that a single-spring model is inadequate when considering the *vibrations* of a guitar string, but we can use a single-spring model for a guitar string that remains *in equilibrium*.

In modelling the E string by means of lumped parameter models, we will need to ensure that the equilibrium tension in each model spring is the same as the equilibrium tension in the E string modelled as a single spring. This means that although l_{eq}, l_0 and k will change as we introduce more springs, T_{eq} will remain the same, at 68 N. Hence as we use more springs, the stiffness of these springs increases by the same factor by which their natural length and equilibrium length decrease.

To keep the modelling simple, in any given model we will use identical model springs.

A basic model of the E string is derived in Subsection 3.1, and this is revised in Subsection 3.2. Subsection 3.3 investigates briefly how further revisions might improve the model.

3.1 A first model

We start by developing the basic model of the guitar string that lumps all the mass at the centre of the string.

Example 8

Suppose that the E string is modelled with all its mass (0.25 g) lumped together as a particle at the centre of the string, with the particle connected by model springs to endpoints 0.65 m apart, as shown in Figure 27. The equilibrium length of each of the two springs is half the equilibrium length of the E string, so $l_{eq} = 0.65/2 = 0.325$, but the equilibrium tension in each spring is the same as the equilibrium tension in the E string (68 N, as calculated above).

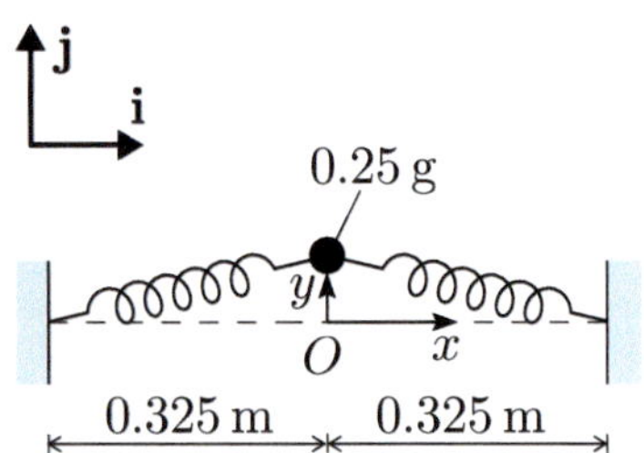

Figure 27 Guitar string modelled as a single particle and two springs

In order to simplify the analysis, the following assumptions are made.

- The oscillations are in a horizontal plane. This means that we have a *two-dimensional* mechanical system modelled solely by springs and particles. Consequently (by the argument in Subsection 2.1), if we measure displacements from the equilibrium position of the particle, the force of gravity can be ignored in deriving the equation of motion.
- The particle is displaced in a direction perpendicular to the equilibrium alignment of the springs (i.e. along the y-axis marked on Figure 27); therefore the vibrations are transverse. This assumption means that the system behaves as if it has one degree of freedom.
- The oscillations are small, that is, the displacement y is small compared with the length of the string.

Calculate the fundamental frequency of vibration.

Solution

◀ Choose coordinates ▶

Choose the axes that are shown in Figure 27, with the origin at the equilibrium position of the particle (i.e. the point halfway between the endpoints).

◀ State assumption(s) ▶

◀ Draw force diagram(s) ▶

The only two forces affecting the particle's motion are those due to the two springs (as gravity can be ignored), so the force diagram (using the usual notation) is as follows.

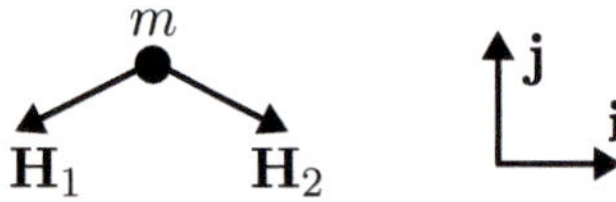

Figure 28 Force diagram for the single-particle model

◀ Apply Newton's 2nd law ▶

Newton's second law applied to the particle gives

$$m\ddot{\mathbf{r}} = \mathbf{H}_1 + \mathbf{H}_2, \tag{38}$$

where $\mathbf{r} = y\mathbf{j}$, and $\mathbf{j}$ is a unit vector in the positive y-direction.

Now the forces acting on the particle must be modelled. The displacement of the particle is not aligned with the springs, therefore the changes in the forces acting cannot be modelled using equation (31). Instead, we use the full form of Hooke's law, $\mathbf{H} = k(l - l_0)\,\widehat{\mathbf{s}}$.

You will see below that an alternative strategy is to make use of equation (30).

First, consider $\mathbf{H}_1$. From Figure 29 you can see that a unit vector from the particle to the centre of the spring is $\widehat{\mathbf{s}} = -(l_{\text{eq}}\mathbf{i} + y\mathbf{j})/l$, where $l = \sqrt{l_{\text{eq}}^2 + y^2}$. Thus, by Hooke's law,

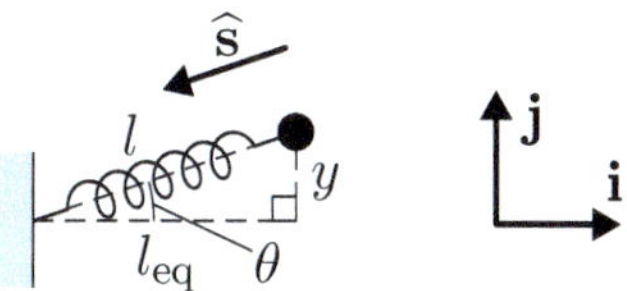

Figure 29 Defining the unit vector $\widehat{\mathbf{s}}$

$$\mathbf{H}_1 = k(l - l_0)\left(\frac{-(l_{\text{eq}}\mathbf{i} + y\mathbf{j})}{l}\right).$$

Under the assumption that the oscillations are small compared with the length of the string, we can use the approximation $l \simeq l_{\text{eq}}$ to simplify this equation to

$$\mathbf{H}_1 \simeq -k(l_{\text{eq}} - l_0)\left(\mathbf{i} + \frac{y}{l_{\text{eq}}}\mathbf{j}\right).$$

We can then simplify further by using the equation for the equilibrium tension in a model spring, $T_{\text{eq}} = k(l_{\text{eq}} - l_0)$, to obtain

$$\mathbf{H}_1 \simeq -T_{\text{eq}}\left(\mathbf{i} + \frac{y}{l_{\text{eq}}}\mathbf{j}\right). \qquad (39)$$

Similarly, for $\mathbf{H}_2$ we have $\widehat{\mathbf{s}} = (l_{\text{eq}}\mathbf{i} - y\mathbf{j})/l$, hence

$$\mathbf{H}_2 \simeq T_{\text{eq}}\left(\mathbf{i} - \frac{y}{l_{\text{eq}}}\mathbf{j}\right).$$

Substituting into equation (38), we obtain the model

$$m\ddot{\mathbf{r}} = -T_{\text{eq}}\left(\mathbf{i} + \frac{y}{l_{\text{eq}}}\mathbf{j}\right) + T_{\text{eq}}\left(\mathbf{i} - \frac{y}{l_{\text{eq}}}\mathbf{j}\right) = -T_{\text{eq}}\frac{2y}{l_{\text{eq}}}\mathbf{j}.$$

As this is a *model* of the motion, we revert to = rather than ≃ signs.

Resolving in the $\mathbf{j}$-direction and rearranging gives

$$\ddot{y} = -\frac{2T_{\text{eq}}}{ml_{\text{eq}}}\,y.$$

◀ Solve equation(s) ▶

Writing this in the form $\ddot{y} + \omega^2 y = 0$, where $\omega^2 = 2T_{\text{eq}}/(ml_{\text{eq}})$, we can see that this is the equation for simple harmonic motion (as in Unit 9). We can therefore write its general solution as

$$y(t) = A\cos(\omega t + \phi),$$

where the angular frequency ω is given by

$$\omega = \sqrt{\frac{2T_{\text{eq}}}{ml_{\text{eq}}}}, \qquad (40)$$

and A and ϕ are constants that can be determined from the initial conditions.

◀ Interpret solution ▶

Substituting in the data in the question, we find that the angular frequency for this vibration is

$$\omega = \sqrt{\frac{2 \times 68}{(0.25 \times 10^{-3}) \times 0.325}} \simeq 1294.$$

The fundamental frequency of the guitar string is then obtained by converting the above angular frequency to a frequency:

$$f = \frac{\omega}{2\pi} = \frac{1294}{2\pi} \simeq 206.$$

So the fundamental frequency of vibration predicted by this first model of the E string is approximately 206 Hz.

The frequency predicted in Example 8 is not close to 323 Hz, the experimentally determined fundamental frequency of the E string. This suggests that the model needs modifying. Accordingly, modifications are made in the next exercise and, in a different way, in the next subsection.

Exercise 11

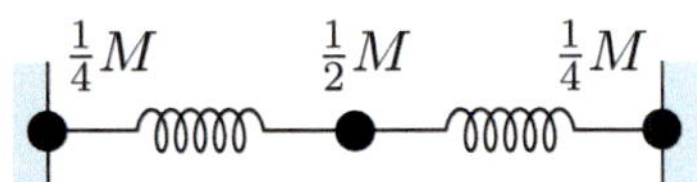

Figure 30 An alternative strategy for distributing the mass of the guitar string

The model of a guitar string used in Example 8 is not the best one-degree-of-freedom model. A better model, based on the argument that the ends of the string are not moving, involves distributing the total mass, M say, between a particle of mass $m = \frac{1}{2}M$ in the middle of the string and particles of mass $\frac{1}{4}M$ at either end, with two model springs in between, as shown in Figure 30. Only the particle of mass $\frac{1}{2}M$ in the middle vibrates.

Use equation (40) to find the fundamental frequency of vibration predicted by this model. Comment on your answer.

In Example 8 the particle was displaced by a small amount, $y\mathbf{j}$, and the force exerted by the left-hand spring on the particle was given by equation (39) as

$$\mathbf{H}_1 \simeq -T_{\text{eq}}\left(\mathbf{i} + \frac{y}{l_{\text{eq}}}\mathbf{j}\right) = -T_{\text{eq}}\mathbf{i} - \frac{T_{\text{eq}}}{l_{\text{eq}}}y\,\mathbf{j}.$$

Now $-T_{\text{eq}}\mathbf{i}$ is simply the force exerted by the spring when the system is in equilibrium, so the approximate change in the force from its equilibrium value is

$$\Delta\mathbf{H}_1 \simeq -\frac{T_{\text{eq}}}{l_{\text{eq}}}y\,\mathbf{j}.$$

We could have used this approximation and the corresponding one for $\Delta\mathbf{H}_2$, in conjunction with equation (30), to obtain the equation of motion in Example 8.

This expression for the approximate change in the force due to a spring holds for any model spring when a small displacement $y\mathbf{j}$ is made at one end and at right angles to the equilibrium alignment of the spring. Hence we have the useful approximation

$$\Delta\mathbf{H} \simeq -\frac{T_{\text{eq}}}{l_{\text{eq}}}y\,\mathbf{j}. \tag{41}$$

We will make frequent use of this approximation in the next subsection, where we try to further increase the accuracy of the fundamental frequency by modelling the guitar string as n particles and $n+1$ springs, as n increases.

3.2 Revised models

In the previous subsection we looked at two lumped parameter models of the E string of the guitar, both of which behaved as if they had one degree of freedom. In this subsection we will try to improve the accuracy of our models by considering lumped parameter models of the E string that behave as if they have two or more degrees of freedom.

Example 9

The next model of the E string has two identical particles and three identical model springs, as shown in Figure 31.

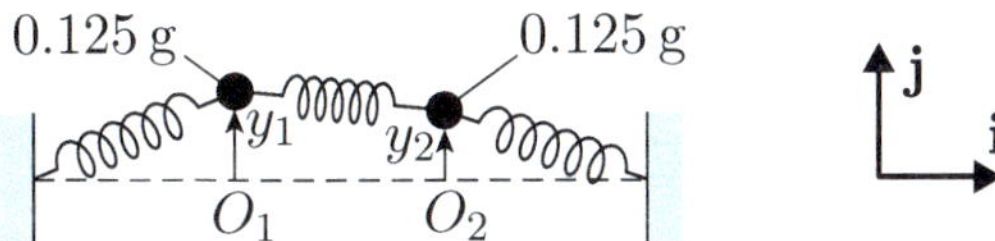

Figure 31 Two-particle model of the guitar string

The mass of the string is evenly distributed between the two particles (i.e. 0.125 g each), and the equilibrium length of each spring is $l_{eq} = 0.65/3 \simeq 0.217$. The equilibrium tension in the springs, T_{eq}, remains the same (68 N) as in previous models. The three assumptions made in Example 8 are used again here.

Calculate the fundamental frequency of vibration.

Solution

Choose the axes that are shown in Figure 31, with the origins at the equilibrium positions of the particles.

◀ Choose coordinates ▶

The only two forces affecting each particle's motion are those due to the two springs attached to it (as gravity can be ignored). The force diagrams for the two particles (using the usual notation) are shown in Figure 32.

◀ State assumption(s) ▶

◀ Draw force diagram(s) ▶

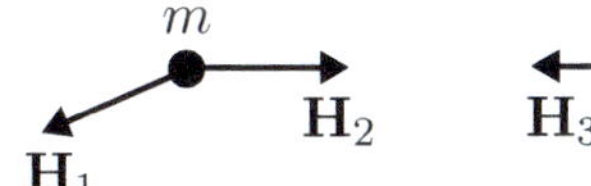

Figure 32 Force diagrams for the two particles

Applying equation (30) to the two particles in turn, we obtain

◀ Apply Newton's 2nd law ▶

$$m\ddot{\mathbf{r}}_1 = \Delta\mathbf{H}_1 + \Delta\mathbf{H}_2, \tag{42}$$

$$m\ddot{\mathbf{r}}_2 = \Delta\mathbf{H}_3 + \Delta\mathbf{H}_4, \tag{43}$$

where $\mathbf{r}_1 = y_1\mathbf{j}$ and $\mathbf{r}_2 = y_2\mathbf{j}$ are the displacements of the first and second particles, respectively, and $\mathbf{j}$ is a unit vector in the positive y-direction.

Now the forces must be modelled. From approximation (41) we have

Since we are modelling here, we write = rather than $\simeq$ when using (41).

$$\Delta\mathbf{H}_1 = -\frac{T_{eq}}{l_{eq}}\,y_1\,\mathbf{j}, \quad \Delta\mathbf{H}_4 = -\frac{T_{eq}}{l_{eq}}\,y_2\,\mathbf{j}.$$

The computations of the other forces are complicated by the fact that both ends of the central spring are displaced. For the force $\mathbf{H}_2$, the spring is displaced at the left-hand end by $y_1 - y_2$ relative to the right-hand end, so from approximation (41) we have

If $y_1 > y_2$, the relative displacement is $(y_1 - y_2)\mathbf{j}$. If $y_2 > y_1$, the relative displacement is $(y_2 - y_1)(-\mathbf{j}) = (y_1 - y_2)\mathbf{j}$.

$$\Delta\mathbf{H}_2 = -\frac{T_{\text{eq}}}{l_{\text{eq}}}(y_1 - y_2)\,\mathbf{j}.$$

Similarly, we obtain

$$\Delta\mathbf{H}_3 = -\frac{T_{\text{eq}}}{l_{\text{eq}}}(y_2 - y_1)\,\mathbf{j} = \frac{T_{\text{eq}}}{l_{\text{eq}}}(y_1 - y_2)\,\mathbf{j}.$$

This expression for $\Delta\mathbf{H}_3$ can also be obtained by using the fact that $\mathbf{H}_3 = -\mathbf{H}_2$ (since they are forces exerted by the same model spring), hence $\Delta\mathbf{H}_3 = -\Delta\mathbf{H}_2$.

Substitution into equations (42) and (43) gives

$$m\ddot{\mathbf{r}}_1 = -\frac{T_{\text{eq}}}{l_{\text{eq}}}\,y_1\,\mathbf{j} - \frac{T_{\text{eq}}}{l_{\text{eq}}}(y_1 - y_2)\,\mathbf{j},$$

$$m\ddot{\mathbf{r}}_2 = \frac{T_{\text{eq}}}{l_{\text{eq}}}(y_1 - y_2)\,\mathbf{j} - \frac{T_{\text{eq}}}{l_{\text{eq}}}\,y_2\,\mathbf{j}.$$

Resolving in the $\mathbf{j}$-direction and rearranging, we obtain

$$\ddot{y}_1 = -2\frac{T_{\text{eq}}}{ml_{\text{eq}}}\,y_1 + \frac{T_{\text{eq}}}{ml_{\text{eq}}}\,y_2,$$

$$\ddot{y}_2 = \frac{T_{\text{eq}}}{ml_{\text{eq}}}\,y_1 - 2\frac{T_{\text{eq}}}{ml_{\text{eq}}}\,y_2,$$

which can be written in matrix form as

$$\begin{pmatrix}\ddot{y}_1\\ \ddot{y}_2\end{pmatrix} = \frac{T_{\text{eq}}}{ml_{\text{eq}}}\begin{pmatrix}-2 & 1\\ 1 & -2\end{pmatrix}\begin{pmatrix}y_1\\ y_2\end{pmatrix}.$$

◀ Solve equation(s) ▶

To solve this, the eigenvalues of the matrix $\begin{pmatrix}-2 & 1\\ 1 & -2\end{pmatrix}$ must be found by solving

$$\begin{vmatrix}-2-\lambda & 1\\ 1 & -2-\lambda\end{vmatrix} = \lambda^2 + 4\lambda + 3 = (\lambda+1)(\lambda+3) = 0,$$

to give $\lambda = -1$ and $\lambda = -3$. So the eigenvalues of the dynamic matrix are $-T_{\text{eq}}/(ml_{\text{eq}})$ and $-3T_{\text{eq}}/(ml_{\text{eq}})$.

Hence the normal mode angular frequencies are

$$\omega_1 = \sqrt{\frac{T_{\text{eq}}}{ml_{\text{eq}}}} \quad\text{and}\quad \omega_2 = \sqrt{\frac{3T_{\text{eq}}}{ml_{\text{eq}}}}.$$

◀ Interpret solution ▶

The objective of this analysis is to calculate the fundamental frequency of vibration. This corresponds to the smaller normal mode angular frequency, thus is given by

$$f = \frac{\omega_1}{2\pi} = \frac{1}{2\pi}\sqrt{\frac{68}{(0.125\times 10^{-3})\times 0.217}} \simeq 252.$$

Therefore the fundamental frequency is approximately 252 Hz.

The frequency 252 Hz obtained in Example 9 is still a long way below the 323 Hz experimental value, but it is much better than the 206 Hz predicted by the model with one degree of freedom in Example 8. Further revisions of the model are considered in Exercises 12 and 13.

Exercise 12

The model in Example 9 may be improved by taking the two moving particles each to have mass $\frac{1}{3}M$, while the remaining $\frac{1}{3}M$ is modelled by particles placed at either end of the string ($\frac{1}{6}M$ at each end); see Figure 33. Only the two particles of mass $\frac{1}{3}M$ vibrate. Find the fundamental frequency of vibration predicted by this model. Comment on your answer.

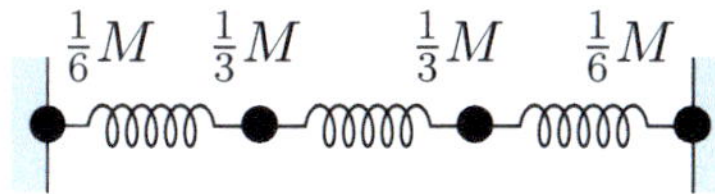

Figure 33 An alternative two-particle model of the guitar string

Exercise 13

Now consider the next refinement in developing lumped parameter models of the E string of the guitar, that is, the model shown in Figure 34, which behaves as if it has three degrees of freedom. In this model, the mass is distributed evenly between three identical particles, with four identical model springs linking the particles to one another and to the endpoints. All the particles vibrate. This model is a refinement of those in Examples 8 and 9, and uses the same simplifying assumptions and data.

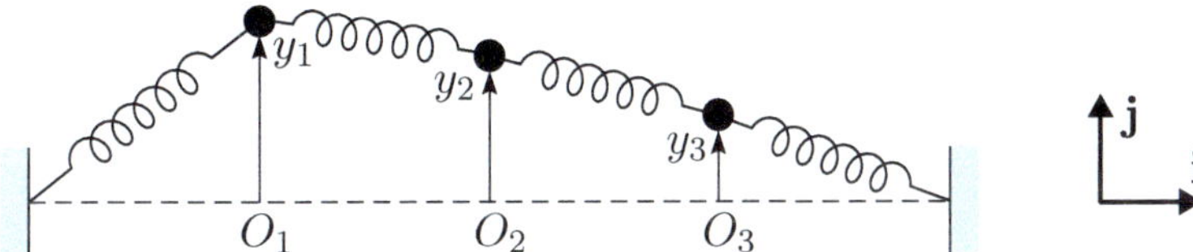

Figure 34 Three-particle model of the guitar string

(a) Use the axes shown in Figure 34 and the methods of Example 9 to derive the matrix form of the equation of motion for the system.

(b) The eigenvalues of the matrix

$$\begin{pmatrix} -2 & 1 & 0 \\ 1 & -2 & 1 \\ 0 & 1 & -2 \end{pmatrix}$$

are -0.586, -2 and -3.414. Using this information, calculate the fundamental frequency of vibration predicted by this model. Comment on your answer.

3.3 Approaching the limit

So far in this section we have considered various lumped parameter models of a guitar string. This subsection links together the results of the previous two subsections and investigates the behaviour of the models as the number of degrees of freedom increases. However, we will look only at models where *all* the mass of the string is assumed to vibrate; we will not consider the models of Exercises 11 and 12 where some of the mass does not vibrate.

In Example 8, a one-particle model was studied and the fundamental frequency was calculated to be approximately 206 Hz. In Example 9, a two-particle model was studied and the fundamental frequency was calculated to be approximately 252 Hz. Exercise 13 calculated the fundamental frequency for a three-particle model as approximately 273 Hz. What happens as the number of particles increases beyond three? Does the predicted fundamental frequency become closer and closer to the experimental value of 323 Hz as the number of particles is increased?

To answer these questions, we look at the equations of motion that were derived for the various models, and see if a pattern emerges.

In Example 8, the equation of motion was

$$\ddot{y} = -\frac{2T_{\text{eq}}}{ml_{\text{eq}}}\, y.$$

In order to make the comparisons easier, we will change the notation slightly. The y here is the y-coordinate of the first particle, so we will call it y_1. The m is the total mass of the guitar string, and we will call this M. The l_{eq} is the equilibrium length of each spring, and is the distance between the endpoints, L, divided by 2. Therefore the equation of motion can be written as

$$\ddot{y}_1 = \frac{T_{\text{eq}}}{M(L/2)} \times (-2) \times y_1. \qquad (44)$$

In Example 9, the equation of motion was

$$\begin{pmatrix} \ddot{y}_1 \\ \ddot{y}_2 \end{pmatrix} = \frac{T_{\text{eq}}}{ml_{\text{eq}}} \begin{pmatrix} -2 & 1 \\ 1 & -2 \end{pmatrix} \begin{pmatrix} y_1 \\ y_2 \end{pmatrix}.$$

Writing this in terms of the same parameters as equation (44) gives

$$\begin{pmatrix} \ddot{y}_1 \\ \ddot{y}_2 \end{pmatrix} = \frac{T_{\text{eq}}}{(M/2)(L/3)} \begin{pmatrix} -2 & 1 \\ 1 & -2 \end{pmatrix} \begin{pmatrix} y_1 \\ y_2 \end{pmatrix}. \qquad (45)$$

The equation of motion from Exercise 13 can be written in terms of M and L as

$$\begin{pmatrix} \ddot{y}_1 \\ \ddot{y}_2 \\ \ddot{y}_3 \end{pmatrix} = \frac{T_{\text{eq}}}{(M/3)(L/4)} \begin{pmatrix} -2 & 1 & 0 \\ 1 & -2 & 1 \\ 0 & 1 & -2 \end{pmatrix} \begin{pmatrix} y_1 \\ y_2 \\ y_3 \end{pmatrix}. \qquad (46)$$

By examining the pattern emerging from equations (44), (45) and (46), you should be able to predict that the equation of motion for the corresponding four-particle model is

$$\begin{pmatrix} \ddot{y}_1 \\ \ddot{y}_2 \\ \ddot{y}_3 \\ \ddot{y}_4 \end{pmatrix} = \frac{T_{\text{eq}}}{(M/4)(L/5)} \begin{pmatrix} -2 & 1 & 0 & 0 \\ 1 & -2 & 1 & 0 \\ 0 & 1 & -2 & 1 \\ 0 & 0 & 1 & -2 \end{pmatrix} \begin{pmatrix} y_1 \\ y_2 \\ y_3 \\ y_4 \end{pmatrix}.$$

This pattern enables a computer program to be written to compute the fundamental frequency f of the corresponding n-particle model, for any positive integer value of n, and then to plot the results on a graph. Such a graph is shown in Figure 35. It can be seen that the predicted fundamental frequency approaches the experimental value of 323 Hz. The convergence to this value is rapid at first, but slows down as n increases: a model with ten degrees of freedom predicts 307 Hz, a model with twenty degrees of freedom predicts 318 Hz, and a model with fifty degrees of freedom predicts 320 Hz.

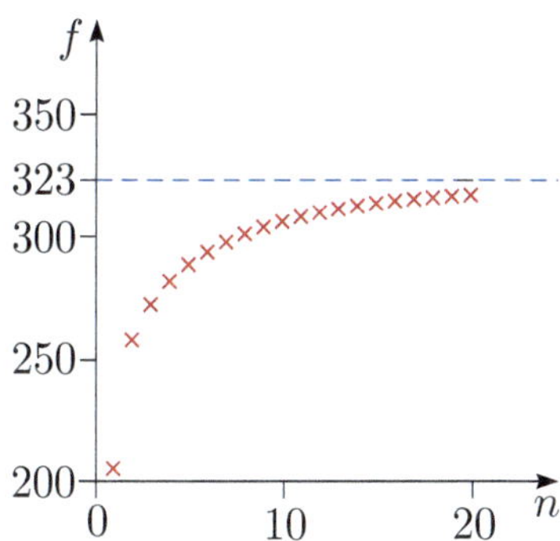

Figure 35 Predicted frequency as the number of particles increases

Up to now in this section we have looked at the vibrations of the E string of a guitar under the assumption that the string is displaced only in a direction perpendicular to its equilibrium alignment, that is, that there is only transverse vibration. But in practice, there is some *longitudinal vibration* too, that is, some vibration along the length of the string. The following exercises examine this longitudinal vibration.

Exercise 14

If you look back at the start of Subsection 2.1, you should recognise that the system shown in Figure 19 is similar to the one-particle lumped parameter model of the E string considered in Example 8. However, in the former case it was assumed that the particle was constrained to move in one dimension along the line of the springs, that is, we modelled longitudinal vibrations, whereas in Example 8 we modelled transverse vibrations. Now, by using the ideas of Subsection 2.1, we can model the longitudinal vibrations in the lumped parameter models of the E string.

Example 6 did this for what was, in effect, a two-particle lumped parameter model. In this exercise you are asked to extend the ideas of Subsection 2.1 to model the longitudinal vibrations of the three-particle lumped parameter model of the E string, as illustrated in Figure 36. (You considered the transverse vibrations in Exercise 13.)

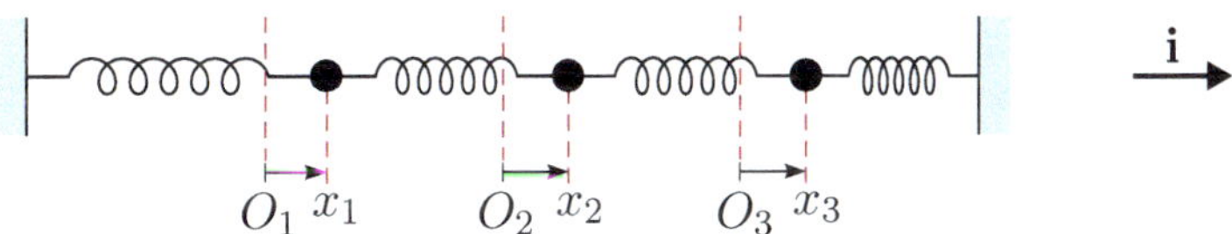

Figure 36 Three-particle system with longitudinal vibrations

(a) Derive the matrix form of the equation of motion for the longitudinal vibrations of this three-particle model.

(b) Verify that $(1 \quad \sqrt{2} \quad 1)^T$, $(1 \quad 0 \quad -1)^T$ and $(1 \quad -\sqrt{2} \quad 1)^T$ are eigenvectors of the dynamic matrix for this model, and hence deduce the eigenvalues of the dynamic matrix.

You met these eigenvectors in Example 5 and Exercise 6.

(c) Determine the normal mode angular frequencies of the longitudinal vibrations of this model, given that the mass of each particle is $\frac{1}{3}(0.25 \times 10^{-3})$ kg, and that each model spring has natural length and equilibrium length equal to a quarter of the corresponding length of the E string (so for a fixed equilibrium tension of 68 N, each model spring has stiffness $4 \times 4000 = 16\,000\,\mathrm{N\,m^{-1}}$).

From equation (37), $T_{\text{eq}} = k(l_{\text{eq}} - l_0)$, so for a *constant* T_{eq}, dividing l_{eq} and l_0 by 4 means multiplying k by 4.

Exercise 15

The aim of this question is to calculate the fundamental frequency of longitudinal vibration of the E string of a guitar by using an argument similar to the one in this subsection. Successively more sophisticated models of the longitudinal vibrations of the E string are given by equation (29) (with $k_1 = k_2 = k$ and $x_1 = x$), equations (34) and (35) (with $k_1 = k_2 = k_3 = k$ and $m_1 = m_2 = m$), and the equations of motion from the Solution to Exercise 14, as follows:

$$m\ddot{x}_1 = -2kx_1, \tag{47}$$

$$\begin{cases} m\ddot{x}_1 = -2kx_1 + \ kx_2, \\ m\ddot{x}_2 = \qquad kx_1 - 2kx_2, \end{cases} \tag{48}$$

$$\begin{cases} m\ddot{x}_1 = -2kx_1 + \ kx_2, \\ m\ddot{x}_2 = \qquad kx_1 - 2kx_2 + \ kx_3, \\ m\ddot{x}_3 = \qquad\qquad\quad kx_2 - 2kx_3. \end{cases} \tag{49}$$

Suppose that the n-degrees-of-freedom model of the longitudinal vibrations represents the guitar string (mass M, equilibrium length L, natural length L_0 and stiffness K) by $n+1$ identical model springs connecting n particles of equal mass.

(a) Write equation (47) in terms of properties of the guitar string, that is, T_{eq}, M, L and L_0.

(*Hint*: Use equation (37) to eliminate the stiffness k from the equation, bearing in mind that the equilibrium tension in each spring is the same as the equilibrium tension in the guitar string.)

(b) Repeat part (a) for the sets of equations (48) and (49), giving your answers in matrix form.

(c) Use your answers to parts (a) and (b) to deduce the equation of motion for the corresponding four-degrees-of-freedom model of the longitudinal vibrations of the guitar string.

(d) By comparing your answers with the results of this subsection, show that the ratio of the fundamental frequency for the longitudinal vibrations to the fundamental frequency for the transverse vibrations of the guitar string is $\sqrt{L/(L-L_0)}$.

If the fundamental frequency of the transverse vibrations is 323 Hz, calculate the fundamental frequency of the longitudinal vibrations of the guitar string.

Learning outcomes

After studying this unit, you should be able to:

- understand the terms degrees of freedom, normal mode, normal mode angular frequency, in-phase motion, phase-opposed motion and rigid body motion, all in the context of an oscillating mechanical system
- derive the equation of motion for simple one- and two-dimensional oscillating mechanical systems (without damping or forcing)
- solve the equation of motion for simple one- and two-dimensional oscillating mechanical systems by finding the eigenvalues and corresponding normal mode eigenvectors of the dynamic matrix
- interpret the normal mode eigenvectors of a simple oscillating mechanical system, in terms of the initial conditions required for the system to oscillate in a normal mode
- model a simple oscillating mechanical system by taking the equilibrium positions of the particles of the system as the origins of coordinates, and using formulas involving the changes in forces
- understand how to model a simple oscillating mechanical system as a lumped parameter model.

Solutions to exercises

Solution to Exercise 1

(a) This system has two degrees of freedom.

(b) This system has two degrees of freedom.

(c) This system has two degrees of freedom. (The additional spring does not alter the number of coordinates needed.)

(d) This system has one degree of freedom.

(e) This system has three degrees of freedom.

(f) This system has three degrees of freedom. (The additional spring does not alter the number of coordinates needed.)

Solution to Exercise 2

(a) These graphs represent normal mode motion, since both x_1 and x_2 vary sinusoidally with the same frequency (just over three cycles completed in 10 seconds) and phase angle. The two particles are in-phase.

(b) These three graphs represent normal mode motion, since the three coordinates vary sinusoidally with the same frequency (again, just over three cycles completed in 10 seconds) and phase angle. For the pairs of particles:

x_1 and x_2 are phase-opposed,

x_1 and x_3 are in-phase,

x_2 and x_3 are phase-opposed.

(c) These graphs do not represent normal mode motion. (The period of the second graph is more than twice that of the first.)

Solution to Exercise 3

From the given data, $k/m_1 = 2$ and $k/m_2 = 1$. So the dynamic matrix of the system is

$$\begin{pmatrix} -4 & 2 \\ 1 & -2 \end{pmatrix}.$$

The eigenvalues are found by solving the quadratic equation

$$\begin{vmatrix} -4-\lambda & 2 \\ 1 & -2-\lambda \end{vmatrix} = \lambda^2 + 6\lambda + 6 = 0,$$

which gives

$$\lambda_1 = -3 + \sqrt{3} \simeq -1.268 \quad \text{and} \quad \lambda_2 = -3 - \sqrt{3} \simeq -4.732.$$

- For $\lambda_1 = -3 + \sqrt{3}$, the eigenvector equations are

$$\begin{aligned}(-1-\sqrt{3})v_1 + \qquad\quad 2v_2 &= 0,\\ v_1 + (1-\sqrt{3})v_2 &= 0,\end{aligned}$$

which both reduce to the same equation. Putting $v_1 = 1$ gives $v_2 \simeq 1.366$, so $(1 \quad 1.366)^T$ is an eigenvector.

- For $\lambda_2 = -3 - \sqrt{3}$, the eigenvector equations are

$$\begin{aligned}(-1+\sqrt{3})v_1 + \qquad\quad 2v_2 &= 0,\\ v_1 + (1+\sqrt{3})v_2 &= 0,\end{aligned}$$

which both reduce to the same equation. Putting $v_1 = 1$ gives $v_2 \simeq -0.366$, so $(1 \quad -0.366)^T$ is an eigenvector.

The normal mode angular frequencies are

$$\omega_1 = \sqrt{-\lambda_1} \simeq 1.126 \quad \text{and} \quad \omega_2 = \sqrt{-\lambda_2} \simeq 2.175.$$

So the general solution of the equation of motion for the specified mechanical system can be written as

$$\mathbf{x} = C_1 \begin{pmatrix} 1 \\ 1.366 \end{pmatrix} \cos(1.126t + \phi_1) + C_2 \begin{pmatrix} 1 \\ -0.366 \end{pmatrix} \cos(2.175t + \phi_2),$$

where C_1, C_2, ϕ_1 and ϕ_2 are constants that can be determined from the initial conditions.

Solution to Exercise 4

(a) Using the usual notation, the force diagrams for the particles can be drawn as follows.

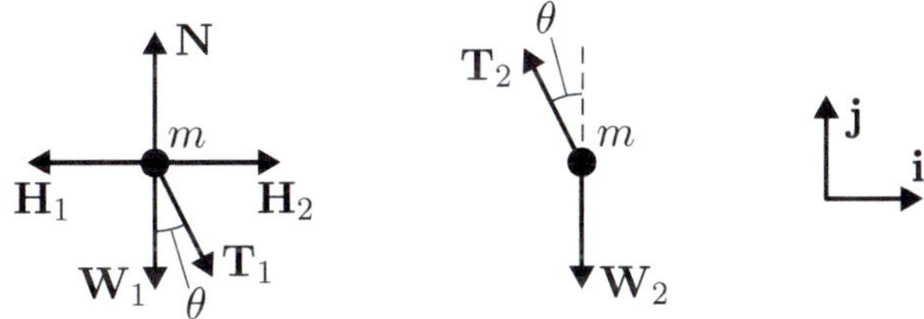

Applying Newton's second law to each particle gives

$$\begin{aligned} m\ddot{\mathbf{r}}_1 &= \mathbf{H}_1 + \mathbf{H}_2 + \mathbf{T}_1 + \mathbf{W}_1 + \mathbf{N},\\ m\ddot{\mathbf{r}}_2 &= \mathbf{T}_2 + \mathbf{W}_2,\end{aligned}$$

where $\mathbf{r}_1 = x_1\mathbf{i} + y_1\mathbf{j}$ and $\mathbf{r}_2 = x_2\mathbf{i} + y_2\mathbf{j}$ are the position vectors, relative to O, of the sliding particle and the pendulum bob, with $\mathbf{i}$ and $\mathbf{j}$ being Cartesian unit vectors in the positive x- and y-directions, respectively.

Now use Hooke's law to model the spring forces. For the force $\mathbf{H}_1$, the length of the spring is $l_0 + x_1$ (since the equilibrium position is at the natural length) and $\widehat{\mathbf{s}} = -\mathbf{i}$, so by Hooke's law,

$$\mathbf{H}_1 = k((l_0 + x_1) - l_0)(-\mathbf{i}) = -kx_1\mathbf{i}.$$

Similarly, for the force $\mathbf{H}_2$, the length of the spring is $l_0 - x_1$ and $\widehat{\mathbf{s}} = \mathbf{i}$, so by Hooke's law,

$$\mathbf{H}_2 = k((l_0 - x_1) - l_0)\mathbf{i} = -kx_1\mathbf{i}.$$

These can be substituted into the above equations, along with $\mathbf{W}_1 = \mathbf{W}_2 = -mg\mathbf{j}$, to yield

$$\begin{aligned} m\ddot{\mathbf{r}}_1 &= -kx_1\mathbf{i} - kx_1\mathbf{i} + \mathbf{T}_1 - mg\mathbf{j} + \mathbf{N}, \\ m\ddot{\mathbf{r}}_2 &= \mathbf{T}_2 - mg\mathbf{j}. \end{aligned}$$

Resolving in the **i**- and **j**-directions gives

$$\begin{aligned} m\ddot{x}_1 &= -2kx_1 + |\mathbf{T}_1|\sin\theta, \\ m\ddot{y}_1 &= -|\mathbf{T}_1|\cos\theta - mg + |\mathbf{N}|, \\ m\ddot{x}_2 &= -|\mathbf{T}_2|\sin\theta, \\ m\ddot{y}_2 &= |\mathbf{T}_2|\cos\theta - mg. \end{aligned}$$

From Figure 15, we can deduce the following relationships between the linear and angular coordinates:

$$x_2 = x_1 + l\sin\theta, \quad y_1 = 0, \quad y_2 = -l\cos\theta.$$

Now we use the small-angle approximations $\sin\theta \simeq \theta$, $\cos\theta \simeq 1$, so $x_2 \simeq x_1 + l\theta$, $y_2 \simeq -l$ (and hence $\ddot{y}_2 \simeq 0$), which on substitution in the equations of motion give

$$\begin{aligned} &m\ddot{x}_1 \simeq -2kx_1 + |\mathbf{T}_1|\theta, \\ &0 \simeq -|\mathbf{T}_1| - mg + |\mathbf{N}|, \\ &m(\ddot{x}_1 + l\ddot{\theta}) \simeq -|\mathbf{T}_2|\theta, \\ &0 \simeq |\mathbf{T}_2| - mg. \end{aligned}$$

From the last of these approximations we have $|\mathbf{T}_2| \simeq mg$, and since the forces exerted at either end of a model rod are equal in magnitude, $|\mathbf{T}_1| = |\mathbf{T}_2| \simeq mg$. So the first and third of the above equations become

$$\begin{aligned} &m\ddot{x}_1 \simeq -2kx_1 + mg\theta, \\ &m(\ddot{x}_1 + l\ddot{\theta}) \simeq -mg\theta. \end{aligned}$$

(As a check, if $\ddot{x}_1 = 0$ (e.g. the upper particle is fixed), then the last equation gives $\ddot{\theta} \simeq -(g/l)\,\theta$, which is the simple harmonic motion equation that is expected for a simple pendulum.)

Subtract the first displayed approximation from the second to obtain another approximation with only one second derivative on the left-hand side:

$$ml\ddot{\theta} \simeq -mg\theta - (-2kx_1 + mg\theta) = 2kx_1 - 2mg\theta.$$

Hence the linear equations of motion for the system are

$$\ddot{x}_1 = -\frac{2k}{m}x_1 + g\theta, \quad \ddot{\theta} = \frac{2k}{lm}x_1 - \frac{2g}{l}\theta.$$

(b) Writing the equations in matrix form gives

$$\begin{pmatrix} \ddot{x}_1 \\ \ddot{\theta} \end{pmatrix} = \begin{pmatrix} -2k/m & g \\ 2k/(lm) & -2g/l \end{pmatrix} \begin{pmatrix} x_1 \\ \theta \end{pmatrix}.$$

From the given data, $k/m = 1/1 = 1$ and $l = 1$, so the dynamic matrix becomes

$$\begin{pmatrix} -2 & g \\ 2 & -2g \end{pmatrix}.$$

The eigenvalues of this matrix are found by solving

$$\begin{vmatrix} -2-\lambda & g \\ 2 & -2g-\lambda \end{vmatrix} = \lambda^2 + 2(1+g)\lambda + 2g = 0,$$

to give

$$\lambda = \frac{-2(1+g) \pm \sqrt{4(1+g)^2 - 8g}}{2} = -1 - g \pm \sqrt{1+g^2}.$$

Substituting $g = 9.81$ into this gives the two eigenvalues as -0.95 and -20.67 (to two decimal places). As $\omega = \sqrt{-\lambda}$, these eigenvalues correspond to the two normal mode angular frequencies $0.97\,\text{rad s}^{-1}$ and $4.55\,\text{rad s}^{-1}$ (to two decimal places). The corresponding periods of normal mode oscillations, $T = 2\pi/\omega$, are $6.48\,\text{s}$ and $1.38\,\text{s}$, respectively.

Solution to Exercise 5

The normal mode eigenvectors derived in Example 3 were $(1 \quad \sqrt{2})^T$ and $(1 \quad -\sqrt{2})^T$. The coordinates chosen to describe the system are the angles θ_1 and θ_2 (see Figure 12), so the eigenvectors refer to these angles. The model is based on *small* angular displacements, so initial conditions where $\theta_1 = 1$ and $\theta_2 = \sqrt{2} \simeq 1.4$ (in radians) are too large. Since both of the pendulum bobs start from rest, we obtain the following possible initial conditions for normal mode motion by scaling the eigenvectors:

normal mode 1: $\theta_1 = 0.1$, $\theta_2 = 0.14$ (to 2 d.p.);

normal mode 2: $\theta_1 = 0.1$, $\theta_2 = -0.14$ (to 2 d.p.).

Angles of 10° and 14° would also be acceptable.

For mode 1, any initial condition in which the ratio of the initial θ_2 displacement to the initial θ_1 displacement is $\sqrt{2}$ is acceptable, provided that both θ_1 and θ_2 are small. Similarly, for mode 2, any initial condition where the ratio is $-\sqrt{2}$ is acceptable, again provided that θ_1 and θ_2 are small.

Solution to Exercise 6

For $(1 \quad 0 \quad -1)^T$, the central particle is stationary, while the other two particles are phase-opposed.

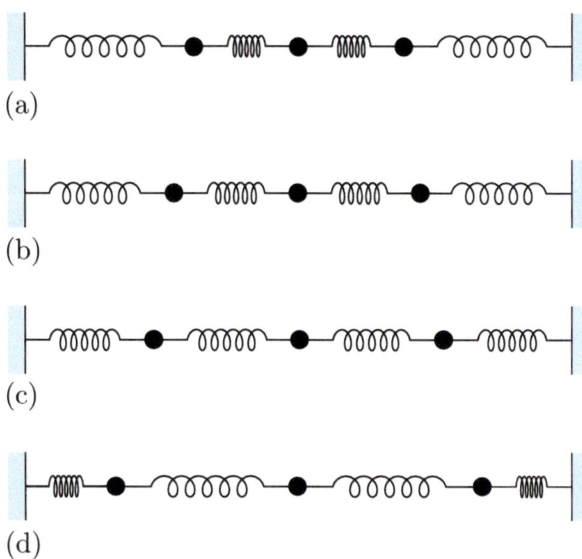

For $(1 \quad -\sqrt{2} \quad 1)^T$, the particle on the left and the central particle are phase-opposed, as are the particle on the right and the central particle. The particles on the left and right are in-phase.

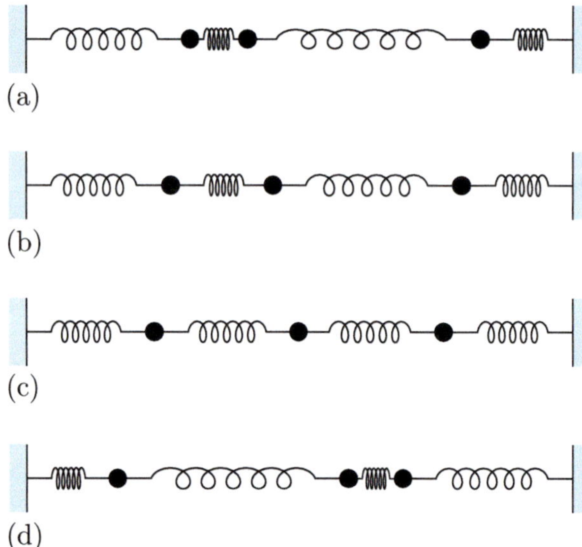

Solution to Exercise 7

(a) The force diagrams for the particles are as follows, where the symbols have their usual meanings.

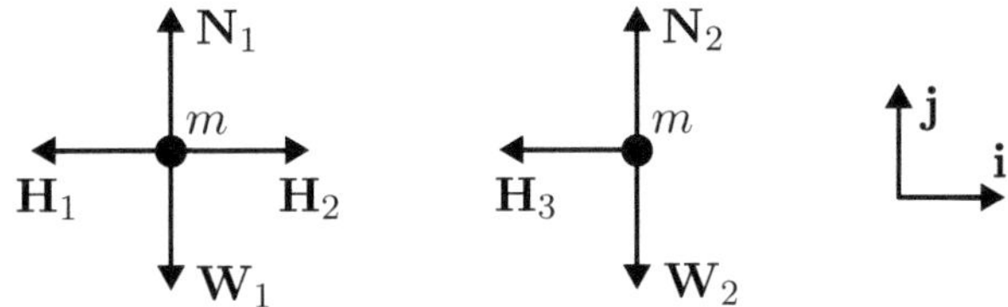

Applying Newton's second law to the two particles separately gives

$$m\ddot{\mathbf{r}}_1 = \mathbf{H}_1 + \mathbf{H}_2 + \mathbf{W}_1 + \mathbf{N}_1,$$
$$m\ddot{\mathbf{r}}_2 = \mathbf{H}_3 + \mathbf{W}_2 + \mathbf{N}_2,$$

where $\mathbf{r}_1 = x_1\mathbf{i}$ and $\mathbf{r}_2 = x_2\mathbf{i}$, and $\mathbf{i}$ is a unit vector in the direction of positive x_1 and x_2.

Now use Hooke's law to model the spring forces. For the force $\mathbf{H}_1$, the length of the spring is $l = l_0 + x_1$ (since the equilibrium position is at the natural length), the stiffness is $3k$ and $\widehat{\mathbf{s}} = -\mathbf{i}$, so by Hooke's law we have

$$\mathbf{H}_1 = 3k((l_0 + x_1) - l_0)\widehat{\mathbf{s}} = -3kx_1\mathbf{i}.$$

For the force $\mathbf{H}_2$, the length of the spring is $l = l_0 + x_2 - x_1$ (since the equilibrium position is at the natural length), the stiffness is $2k$ and $\widehat{\mathbf{s}} = \mathbf{i}$, so by Hooke's law we have

$$\mathbf{H}_2 = 2k((l_0 + x_2 - x_1) - l_0)\widehat{\mathbf{s}} = 2k(x_2 - x_1)\mathbf{i}.$$

For the force $\mathbf{H}_3$, the length of the spring is $l = l_0 + x_2 - x_1$, the stiffness is $2k$ and $\widehat{\mathbf{s}} = -\mathbf{i}$, so by Hooke's law we have

$$\mathbf{H}_3 = 2k((l_0 + x_2 - x_1) - l_0)\widehat{\mathbf{s}} = -2k(x_2 - x_1)\mathbf{i}.$$

(Of course, $\mathbf{H}_3$ can most easily be obtained by using the fact that a model spring exerts forces of equal magnitude and opposite direction at either end, thus $\mathbf{H}_3 = -\mathbf{H}_2$.)

Substituting these expressions into the Newton's second law equations gives

$$m\ddot{\mathbf{r}}_1 = -3kx_1\mathbf{i} + 2k(x_2 - x_1)\mathbf{i} - mg\mathbf{j} + |\mathbf{N}_1|\,\mathbf{j},$$
$$m\ddot{\mathbf{r}}_2 = -2k(x_2 - x_1)\mathbf{i} - mg\mathbf{j} + |\mathbf{N}_2|\,\mathbf{j}.$$

Resolving in the $\mathbf{i}$-direction gives the equations of motion:

$$m\ddot{x}_1 = -5kx_1 + 2kx_2,$$
$$m\ddot{x}_2 = 2kx_1 - 2kx_2.$$

(b) If $k/m = 1$, then the equations of motion become (in matrix form)

$$\begin{pmatrix} \ddot{x}_1 \\ \ddot{x}_2 \end{pmatrix} = \begin{pmatrix} -5 & 2 \\ 2 & -2 \end{pmatrix} \begin{pmatrix} x_1 \\ x_2 \end{pmatrix}.$$

To find the eigenvalues, we solve

$$\begin{vmatrix} -5-\lambda & 2 \\ 2 & -2-\lambda \end{vmatrix} = \lambda^2 + 7\lambda + 6 = 0,$$

and obtain $\lambda = -1$ or $\lambda = -6$.

- For $\lambda = -1$, the eigenvector equations are

$$\begin{pmatrix} -4 & 2 \\ 2 & -1 \end{pmatrix} \begin{pmatrix} v_1 \\ v_2 \end{pmatrix} = \begin{pmatrix} 0 \\ 0 \end{pmatrix},$$

or equivalently

$$\begin{aligned} -4v_1 + 2v_2 &= 0, \\ 2v_1 - v_2 &= 0. \end{aligned}$$

These equations both reduce to $v_2 = 2v_1$, so $(1 \quad 2)^T$ is an eigenvector.

- For $\lambda = -6$, the eigenvector equations are

$$\begin{pmatrix} 1 & 2 \\ 2 & 4 \end{pmatrix} \begin{pmatrix} v_1 \\ v_2 \end{pmatrix} = \begin{pmatrix} 0 \\ 0 \end{pmatrix},$$

or equivalently

$$\begin{aligned} v_1 + 2v_2 &= 0, \\ 2v_1 + 4v_2 &= 0. \end{aligned}$$

These equations both reduce to $v_2 = -\frac{1}{2}v_1$, so $(2 \quad -1)^T$ is an eigenvector.

The normal mode angular frequencies are calculated from the eigenvalues by using $\omega = \sqrt{-\lambda}$, thus $\omega_1 = 1$ and $\omega_2 = \sqrt{6}$. From equation (10), the general solution of the matrix equation of motion has the form

$$\mathbf{x}(t) = C_1\mathbf{v}_1 \cos(\omega_1 t + \phi_1) + C_2\mathbf{v}_2(\cos\omega_2 t + \phi_2).$$

Substituting for $\mathbf{v}_1$, $\mathbf{v}_2$, ω_1 and ω_2 gives the general solution as

$$\begin{pmatrix} x_1(t) \\ x_2(t) \end{pmatrix} = C_1 \begin{pmatrix} 1 \\ 2 \end{pmatrix} \cos(t + \phi_1) + C_2 \begin{pmatrix} 2 \\ -1 \end{pmatrix} \cos(\sqrt{6}t + \phi_2).$$

(c) The given initial condition has both particles starting from rest, so we can take $\phi_1 = \phi_2 = 0$. The constants C_1 and C_2 can be determined by substituting $x_1 = d_1$ and $x_2 = d_2$ when $t = 0$ into the general solution of the equation of motion:

$$\begin{pmatrix} d_1 \\ d_2 \end{pmatrix} = C_1 \begin{pmatrix} 1 \\ 2 \end{pmatrix} + C_2 \begin{pmatrix} 2 \\ -1 \end{pmatrix}.$$

Note that

$$\begin{pmatrix} 1 & 2 \\ 2 & -1 \end{pmatrix}^{-1} = \frac{1}{5}\begin{pmatrix} 1 & 2 \\ 2 & -1 \end{pmatrix}.$$

These can be solved to give C_1 and C_2 in terms of d_1 and d_2 as

$$\begin{pmatrix} C_1 \\ C_2 \end{pmatrix} = d_1 \begin{pmatrix} 0.2 \\ 0.4 \end{pmatrix} + d_2 \begin{pmatrix} 0.4 \\ -0.2 \end{pmatrix}.$$

These values of C_1 and C_2 can be substituted into the general solution to determine the particular solution satisfying the initial conditions.

Solution to Exercise 8

Draw a force diagram for each particle (omitting the weights and normal reactions), using the usual notation for the spring forces.

m m $\mathbf{i}$

$\mathbf{H}_1$ $\mathbf{H}_2$ $\mathbf{H}_3$

Apply equation (30) to each particle, to obtain

$$m\ddot{\mathbf{r}}_1 = \Delta\mathbf{H}_1 + \Delta\mathbf{H}_2,$$
$$m\ddot{\mathbf{r}}_2 = \Delta\mathbf{H}_3,$$

where $\mathbf{r}_1 = x_1\mathbf{i}$ and $\mathbf{r}_2 = x_2\mathbf{i}$, as in Exercise 7.

Using equation (31) gives

$$\Delta\mathbf{H}_1 = 3kx_1(-\mathbf{i}) = -3kx_1\mathbf{i},$$
$$\Delta\mathbf{H}_2 = 2k(x_2 - x_1)\mathbf{i},$$
$$\Delta\mathbf{H}_3 = 2k(x_2 - x_1)(-\mathbf{i}) = -2k(x_2 - x_1)\mathbf{i}.$$

Substituting into the original equations gives

$$m\ddot{\mathbf{r}}_1 = -3kx_1\mathbf{i} + 2k(x_2 - x_1)\mathbf{i},$$
$$m\ddot{\mathbf{r}}_2 = -2k(x_2 - x_1)\mathbf{i}.$$

Resolving in the $\mathbf{i}$-direction gives the equations of motion:

$$m\ddot{x}_1 = -5kx_1 + 2kx_2,$$
$$m\ddot{x}_2 = 2kx_1 - 2kx_2.$$

(In this example, at equilibrium the springs have their natural lengths, so $\mathbf{H}_{1,\text{eq}} = \mathbf{H}_{2,\text{eq}} = \mathbf{H}_{3,\text{eq}} = \mathbf{0}$, hence $\Delta\mathbf{H}_1 = \mathbf{H}_1$, $\Delta\mathbf{H}_2 = \mathbf{H}_2$ and $\Delta\mathbf{H}_3 = \mathbf{H}_3$.)

Solution to Exercise 9

(a) With the usual notation, omitting the weights and normal reactions, which balance, the three force diagrams (one for each particle) are as follows.

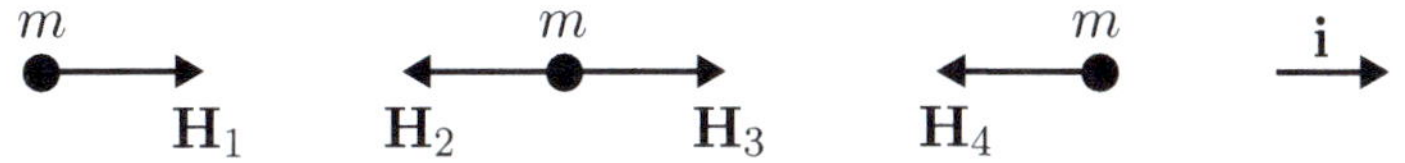

Applying equation (30) to each particle gives

$$m\ddot{\mathbf{r}}_1 = \Delta\mathbf{H}_1,$$
$$m\ddot{\mathbf{r}}_2 = \Delta\mathbf{H}_2 + \Delta\mathbf{H}_3,$$
$$m\ddot{\mathbf{r}}_3 = \Delta\mathbf{H}_4,$$

where $\mathbf{r}_1 = x_1\mathbf{i}$, $\mathbf{r}_2 = x_2\mathbf{i}$ and $\mathbf{r}_3 = x_3\mathbf{i}$, and $\mathbf{i}$ is a unit vector in the direction of positive x_1, x_2 and x_3.

Then from equation (31) we obtain

$$\begin{aligned}
\Delta\mathbf{H}_1 &= k(x_2 - x_1)\mathbf{i},\\
\Delta\mathbf{H}_2 &= k(x_2 - x_1)(-\mathbf{i}) = -k(x_2 - x_1)\mathbf{i},\\
\Delta\mathbf{H}_3 &= k(x_3 - x_2)\mathbf{i},\\
\Delta\mathbf{H}_4 &= k(x_3 - x_2)(-\mathbf{i}) = -k(x_3 - x_2)\mathbf{i}.
\end{aligned}$$

Substituting these expressions into the original equations gives

$$\begin{aligned}
m\ddot{\mathbf{r}}_1 &= k(x_2 - x_1)\mathbf{i},\\
m\ddot{\mathbf{r}}_2 &= -k(x_2 - x_1)\mathbf{i} + k(x_3 - x_2)\mathbf{i},\\
m\ddot{\mathbf{r}}_3 &= -k(x_3 - x_2)\mathbf{i}.
\end{aligned}$$

On resolving in the **i**-direction, we obtain the equations of motion

$$\begin{aligned}
m\ddot{x}_1 &= -kx_1 + kx_2,\\
m\ddot{x}_2 &= kx_1 - 2kx_2 + kx_3,\\
m\ddot{x}_3 &= kx_2 - kx_3,
\end{aligned}$$

or in matrix form,

$$\begin{pmatrix}\ddot{x}_1\\ \ddot{x}_2\\ \ddot{x}_3\end{pmatrix} = \frac{k}{m}\begin{pmatrix}-1 & 1 & 0\\ 1 & -2 & 1\\ 0 & 1 & -1\end{pmatrix}\begin{pmatrix}x_1\\ x_2\\ x_3\end{pmatrix}.$$

(b) The given data tell us that the matrix has eigenvalues 0, -1 and -3, with corresponding eigenvectors $(1\ \ 1\ \ 1)^T$, $(1\ \ 0\ \ -1)^T$ and $(1\ \ -2\ \ 1)^T$. So the normal mode angular frequencies are 0, $\sqrt{k/m}$ and $\sqrt{3k/m}$. (Do not forget the scaling factor k/m for the dynamic matrix.)

Therefore the general solution of the equation of motion is

$$\begin{pmatrix}x_1(t)\\ x_2(t)\\ x_3(t)\end{pmatrix} = \begin{pmatrix}1\\ 1\\ 1\end{pmatrix}(A + Bt) + C_1\begin{pmatrix}1\\ 0\\ -1\end{pmatrix}\cos\left(\sqrt{\frac{k}{m}}\,t + \phi_1\right) + C_2\begin{pmatrix}1\\ -2\\ 1\end{pmatrix}\cos\left(\sqrt{\frac{3k}{m}}\,t + \phi_2\right),$$

where A, B, C_1, C_2, ϕ_1 and ϕ_2 are constants that can be determined using the initial conditions.

(c) For the first normal mode, that is, the linear mode with eigenvector $(1\ \ 1\ \ 1)^T$, a suitable initial condition is that the constituent parts all start from the given equilibrium positions with the same velocity (or alternatively that they remain at rest, each at the same displacement from its given equilibrium position).

For the second normal mode, with eigenvector $(1\ \ 0\ \ -1)^T$, a suitable initial condition is that the constituent parts all start from rest with the leftmost truck and the engine displaced equal distances to the right and left of their given equilibrium positions, respectively, while the central truck starts (and stays) at its given equilibrium position.

For the third normal mode, with normal mode eigenvector $(1 \quad -2 \quad 1)^T$, a suitable initial condition is that the constituent parts all start from rest with the leftmost truck and the engine both displaced the same distance to the right of their given equilibrium positions, while the central truck is displaced twice that distance to the left of its given equilibrium position.

Solution to Exercise 10

(a) With the usual notation, the force diagrams for the particles are as follows.

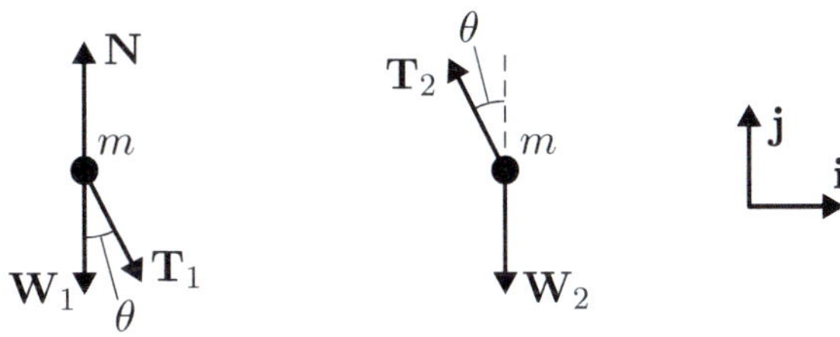

Applying Newton's second law to each particle gives

$$m\ddot{\mathbf{r}}_1 = \mathbf{T}_1 + \mathbf{W}_1 + \mathbf{N},$$
$$m\ddot{\mathbf{r}}_2 = \mathbf{T}_2 + \mathbf{W}_2,$$

where $\mathbf{r}_1 = x_1\mathbf{i} + y_1\mathbf{j}$ and $\mathbf{r}_2 = x_2\mathbf{i} + y_2\mathbf{j}$ are the position vectors, relative to the fixed point O, of the sliding particle and the pendulum bob, respectively, and $\mathbf{i}$ and $\mathbf{j}$ are Cartesian unit vectors in the positive x- and y-directions, respectively.

We have $\mathbf{W}_1 = \mathbf{W}_2 = -mg\mathbf{j}$, and applying Newton's second law gives

$$m\ddot{\mathbf{r}}_1 = \mathbf{T}_1 - mg\mathbf{j} + \mathbf{N},$$
$$m\ddot{\mathbf{r}}_2 = \mathbf{T}_2 - mg\mathbf{j}.$$

Resolving in the $\mathbf{i}$- and $\mathbf{j}$-directions gives

$$m\ddot{x}_1 = |\mathbf{T}_1| \sin\theta,$$
$$m\ddot{y}_1 = -|\mathbf{T}_1| \cos\theta - mg + |\mathbf{N}|,$$
$$m\ddot{x}_2 = -|\mathbf{T}_2| \sin\theta,$$
$$m\ddot{y}_2 = |\mathbf{T}_2| \cos\theta - mg.$$

Now, from the figure in the margin, the relationships between the linear and angular coordinates are

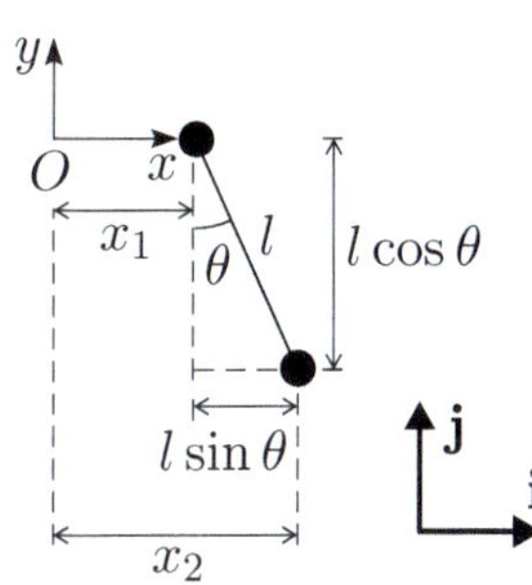

$$x_2 = x_1 + l\sin\theta, \quad y_1 = 0, \quad y_2 = -l\cos\theta.$$

Using the small-angle approximations $\sin\theta \simeq \theta$ and $\cos\theta \simeq 1$, we have $x_2 \simeq x_1 + l\theta$ and $y_2 \simeq -l$ (so $\ddot{y}_2 \simeq 0$), which on substitution in the resolved equations give

$$m\ddot{x}_1 \simeq |\mathbf{T}_1|\theta,$$
$$0 \simeq -|\mathbf{T}_1| - mg + |\mathbf{N}|,$$
$$m(\ddot{x}_1 + l\ddot{\theta}) \simeq -|\mathbf{T}_2|\theta,$$
$$0 \simeq |\mathbf{T}_2| - mg.$$

From the last of these approximations, $|\mathbf{T}_2| \simeq mg$, and since the forces exerted at either end of a model rod are equal in magnitude, $|\mathbf{T}_1| = |\mathbf{T}_2| \simeq mg$. So the first and third equations above become

$$m\ddot{x}_1 \simeq mg\theta,$$
$$m(\ddot{x}_1 + l\ddot{\theta}) \simeq -mg\theta.$$

On subtracting the first of these equations from the second, we obtain

$$ml\ddot{\theta} \simeq -mg\theta - mg\theta = -2mg\theta.$$

Hence the linear equations of motion for the system are

$$\ddot{x}_1 = g\theta, \quad \ddot{\theta} = -\frac{2g}{l}\theta.$$

(These equations could also have been obtained by putting $k = 0$ – because there is no spring here – into the result of Exercise 4(a).)

(b) Writing the equations of motion in matrix form gives

$$\begin{pmatrix} \ddot{x}_1 \\ \ddot{\theta} \end{pmatrix} = \begin{pmatrix} 0 & g \\ 0 & -2g/l \end{pmatrix} \begin{pmatrix} x_1 \\ \theta \end{pmatrix}.$$

The eigenvalues are found by solving

$$\begin{vmatrix} 0 - \lambda & g \\ 0 & -2g/l - \lambda \end{vmatrix} = \lambda(\lambda + 2g/l) = 0.$$

So the eigenvalues are $\lambda = 0$ and $\lambda = -2g/l$.

- For $\lambda = 0$, the eigenvector equations are

$$\begin{pmatrix} 0 - 0 & g \\ 0 & -2g/l - 0 \end{pmatrix} \begin{pmatrix} v_1 \\ v_2 \end{pmatrix} = \begin{pmatrix} 0 \\ 0 \end{pmatrix},$$

or equivalently $gv_2 = 0$ and $2gv_2/l = 0$. Therefore $v_2 = 0$ (there is no restriction on v_1), so $(1 \quad 0)^T$ is an eigenvector.

- For $\lambda = -2g/l$, the eigenvector equations are

$$\begin{pmatrix} 0 - (-2g/l) & g \\ 0 & -2g/l - (-2g/l) \end{pmatrix} \begin{pmatrix} v_1 \\ v_2 \end{pmatrix} = \begin{pmatrix} 0 \\ 0 \end{pmatrix},$$

or equivalently

$$(2g/l)v_1 + gv_2 = 0,$$
$$0 = 0.$$

The first equation gives the condition $v_2 = -2v_1/l$, so $(1 \quad -2/l)^T$ is an eigenvector. The normal mode angular frequencies are calculated from the eigenvalues as $\omega_1 = 0$ and $\omega_2 = \sqrt{2g/l}$.

With these eigenvectors and normal mode angular frequencies, the general solution of the matrix equation of motion is

$$\begin{pmatrix} x_1(t) \\ \theta(t) \end{pmatrix} = \begin{pmatrix} 1 \\ 0 \end{pmatrix} (A + Bt) + C \begin{pmatrix} 1 \\ -2/l \end{pmatrix} \cos\left(\sqrt{\frac{2g}{l}}\, t + \phi\right),$$

where A, B, C and ϕ are constants that can be determined from the initial conditions.

(c) The angular frequency of a simple pendulum is $\sqrt{g/l}$, so the angular frequency of this pendulum, which is $\sqrt{2g/l}$, is larger by a factor of $\sqrt{2}$. Therefore the period of this pendulum is smaller by a factor of $\sqrt{2}$ than that of the simple pendulum.

(d) For the rigid body motion of the system, the whole system is translated along the horizontal bar (with $\theta = 0$), as shown below.

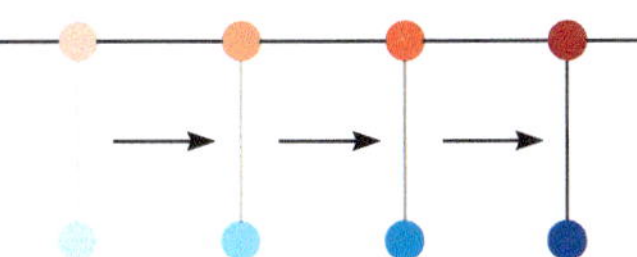

For the non-zero angular frequency, the motion of both particles is oscillatory with the same frequency. The displacement ratio is negative, so the motion is phase-opposed, as shown below.

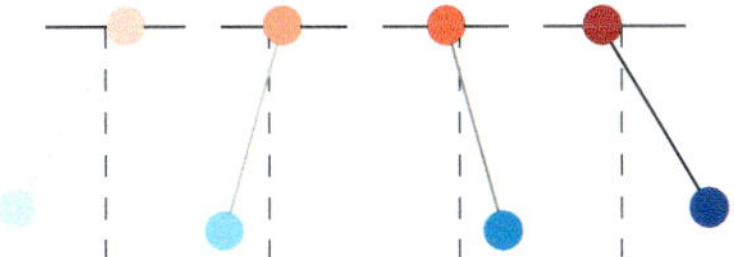

Solution to Exercise 11

The question suggests using equation (40), which is applicable because the model of the guitar string is the same as that in Example 8; only the parameter m is different. Instead of the mass of the whole spring, 0.25 g, we substitute half the mass of the spring, 0.125 g, because only the central particle (of mass $\frac{1}{2}M$) vibrates; the other parameters have the same values as before. Then

$$\omega = \sqrt{\frac{2 \times 68}{(0.125 \times 10^{-3}) \times 0.325}} \simeq 1830.$$

This gives a fundamental frequency of

$$f = \frac{\omega}{2\pi} = \frac{1830}{2\pi} \simeq 291.$$

So this model predicts a fundamental frequency of approximately 291 Hz, which is much closer to the experimental value of 323 Hz than the 206 Hz predicted in Example 8.

Solution to Exercise 12

The basic two-particle model applies here because only the two central particles vibrate. This means that the model used in Example 9 is relevant; only a parameter (the mass of the particles) has changed. Therefore the analysis is the same as in that example, leading to

$$\omega_1 = \sqrt{\frac{68}{((0.25/3) \times 10^{-3}) \times 0.217}} = 1939.$$

Hence the fundamental frequency is predicted to be

$$f = \frac{\omega_1}{2\pi} = \frac{1939}{2\pi} \simeq 309.$$

The predicted fundamental frequency of approximately 309 Hz is closer to the experimental value of 323 Hz than the 252 Hz predicted by Example 9, and is closer than the 291 Hz predicted in Exercise 11, so this model is an improvement on the previous ones.

Solution to Exercise 13

(a) The force diagrams for the three particles (using the usual notation) are as follows.

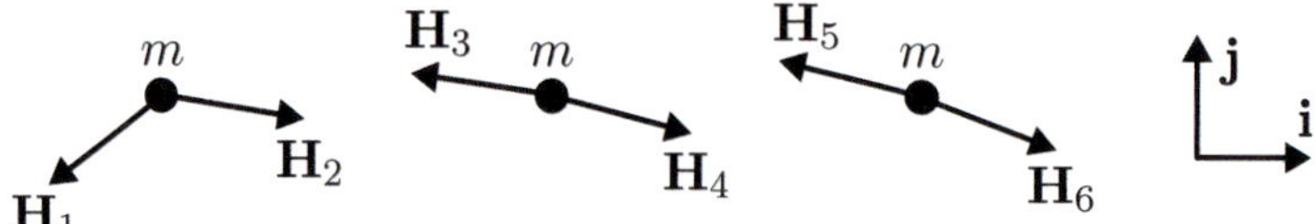

Applying equation (30) to each particle gives

$$m\ddot{\mathbf{r}}_1 = \Delta\mathbf{H}_1 + \Delta\mathbf{H}_2,$$
$$m\ddot{\mathbf{r}}_2 = \Delta\mathbf{H}_3 + \Delta\mathbf{H}_4,$$
$$m\ddot{\mathbf{r}}_3 = \Delta\mathbf{H}_5 + \Delta\mathbf{H}_6,$$

where $\mathbf{r}_1 = y_1\mathbf{j}$, $\mathbf{r}_2 = y_2\mathbf{j}$ and $\mathbf{r}_3 = y_3\mathbf{j}$ are the displacements of the particles, and $\mathbf{j}$ is a unit vector in the positive y-direction.

Using expression (41) to model the forces exerted by the outermost springs, we find

$$\Delta\mathbf{H}_1 = -\frac{T_{\text{eq}}}{l_{\text{eq}}}\, y_1\,\mathbf{j}, \quad \Delta\mathbf{H}_6 = -\frac{T_{\text{eq}}}{l_{\text{eq}}}\, y_3\,\mathbf{j}.$$

The computations of the other forces are complicated by the fact that both ends of the springs are displaced. For the force $\mathbf{H}_2$, the spring is displaced by $y_1 - y_2$ at the left-hand end relative to the right-hand end, so from equation (41),

$$\Delta\mathbf{H}_2 = -\frac{T_{\text{eq}}}{l_{\text{eq}}}\,(y_1 - y_2)\,\mathbf{j}.$$

Similarly,

$$\Delta\mathbf{H}_3 = -\Delta\mathbf{H}_2 = \frac{T_{\text{eq}}}{l_{\text{eq}}}\,(y_1 - y_2)\,\mathbf{j},$$

$$\Delta\mathbf{H}_4 = -\frac{T_{\text{eq}}}{l_{\text{eq}}}\,(y_2 - y_3)\,\mathbf{j},$$

$$\Delta\mathbf{H}_5 = -\Delta\mathbf{H}_4 = \frac{T_{\text{eq}}}{l_{\text{eq}}}\,(y_2 - y_3)\,\mathbf{j}.$$

These can be substituted into the original equations to obtain

$$m\ddot{\mathbf{r}}_1 = -\frac{T_{\text{eq}}}{l_{\text{eq}}}\,y_1\,\mathbf{j} - \frac{T_{\text{eq}}}{l_{\text{eq}}}\,(y_1 - y_2)\,\mathbf{j},$$

$$m\ddot{\mathbf{r}}_2 = \frac{T_{\text{eq}}}{l_{\text{eq}}}\,(y_1 - y_2)\,\mathbf{j} - \frac{T_{\text{eq}}}{l_{\text{eq}}}\,(y_2 - y_3)\,\mathbf{j},$$

$$m\ddot{\mathbf{r}}_3 = \frac{T_{\text{eq}}}{l_{\text{eq}}}\,(y_2 - y_3)\,\mathbf{j} - \frac{T_{\text{eq}}}{l_{\text{eq}}}\,y_3\,\mathbf{j}.$$

Resolving in the $\mathbf{j}$-direction and rearranging gives

$$\ddot{y}_1 = -2\frac{T_{\text{eq}}}{ml_{\text{eq}}}\,y_1 + \frac{T_{\text{eq}}}{ml_{\text{eq}}}\,y_2,$$

$$\ddot{y}_2 = \frac{T_{\text{eq}}}{ml_{\text{eq}}}\,y_1 - 2\frac{T_{\text{eq}}}{ml_{\text{eq}}}\,y_2 + \frac{T_{\text{eq}}}{ml_{\text{eq}}}\,y_3,$$

$$\ddot{y}_3 = \frac{T_{\text{eq}}}{ml_{\text{eq}}}\,y_2 - 2\frac{T_{\text{eq}}}{ml_{\text{eq}}}\,y_3,$$

which can be written in matrix form as

$$\begin{pmatrix} \ddot{y}_1 \\ \ddot{y}_2 \\ \ddot{y}_3 \end{pmatrix} = \frac{T_{\text{eq}}}{ml_{\text{eq}}} \begin{pmatrix} -2 & 1 & 0 \\ 1 & -2 & 1 \\ 0 & 1 & -2 \end{pmatrix} \begin{pmatrix} y_1 \\ y_2 \\ y_3 \end{pmatrix}.$$

(b) The given eigenvalue of smallest magnitude is -0.586, and this corresponds to the fundamental frequency. Therefore the eigenvalue of the dynamic matrix that has the smallest magnitude is

$$\lambda_1 = -0.586 \times \frac{T_{\text{eq}}}{ml_{\text{eq}}}.$$

For this system, $T_{\text{eq}} = 68$, $m = (0.25 \times 10^{-3})/3$ and $l_{\text{eq}} = 0.65/4$, so the corresponding angular frequency is

$$\omega_1 = \sqrt{\frac{0.586 \times 68}{(0.25 \times 10^{-3}/3) \times (0.65/4)}} \simeq 1715,$$

hence the fundamental frequency is

$$f_1 = \frac{\omega_1}{2\pi} = \frac{1715}{2\pi} \simeq 273.$$

So this model has a fundamental frequency of approximately 273 Hz. This is still much smaller than the experimental value of 323 Hz, but it is better than the predictions from the models with one and two degrees of freedom in Examples 8 and 9; however, it is still not as close to the experimental value as the value found in Exercise 12, or even that in Exercise 11.

Solution to Exercise 14

(a) The first step is to draw a force diagram for each particle, using the usual notation.

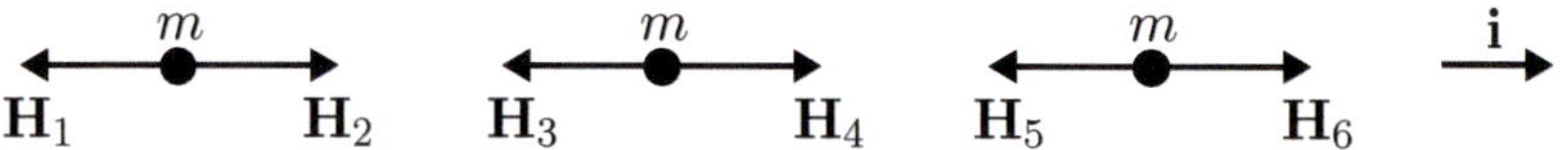

Applying equation (30) to each particle gives

$$m\ddot{\mathbf{r}}_1 = \Delta\mathbf{H}_1 + \Delta\mathbf{H}_2,$$
$$m\ddot{\mathbf{r}}_2 = \Delta\mathbf{H}_3 + \Delta\mathbf{H}_4,$$
$$m\ddot{\mathbf{r}}_3 = \Delta\mathbf{H}_5 + \Delta\mathbf{H}_6,$$

where $\mathbf{r}_1 = x_1\mathbf{i}$, $\mathbf{r}_2 = x_2\mathbf{i}$ and $\mathbf{r}_3 = x_3\mathbf{i}$ are the position vectors of the particles, and $\mathbf{i}$ is a unit vector in the positive x-direction. Modelling the forces using equation (31), we obtain

$$\Delta\mathbf{H}_1 = kx_1(-\mathbf{i}),$$
$$\Delta\mathbf{H}_2 = k(x_2 - x_1)\mathbf{i},$$
$$\Delta\mathbf{H}_3 = k(x_2 - x_1)(-\mathbf{i}),$$
$$\Delta\mathbf{H}_4 = k(x_3 - x_2)\mathbf{i},$$
$$\Delta\mathbf{H}_5 = k(x_3 - x_2)(-\mathbf{i}),$$
$$\Delta\mathbf{H}_6 = -kx_3\mathbf{i}.$$

Substituting into the original equations gives

$$m\ddot{\mathbf{r}}_1 = -kx_1\mathbf{i} + k(x_2 - x_1)\mathbf{i},$$
$$m\ddot{\mathbf{r}}_2 = -k(x_2 - x_1)\mathbf{i} + k(x_3 - x_2)\mathbf{i},$$
$$m\ddot{\mathbf{r}}_3 = -k(x_3 - x_2)\mathbf{i} - kx_3\mathbf{i}.$$

Resolving in the $\mathbf{i}$-direction gives the equations of motion for the longitudinal vibrations of the three-particle model as

$$m\ddot{x}_1 = -2kx_1 + kx_2,$$
$$m\ddot{x}_2 = kx_1 - 2kx_2 + kx_3,$$
$$m\ddot{x}_3 = kx_2 - 2kx_3,$$

which can be written in matrix form as

$$\begin{pmatrix}\ddot{x}_1\\ \ddot{x}_2\\ \ddot{x}_3\end{pmatrix} = \frac{k}{m}\begin{pmatrix}-2 & 1 & 0\\ 1 & -2 & 1\\ 0 & 1 & -2\end{pmatrix}\begin{pmatrix}x_1\\ x_2\\ x_3\end{pmatrix}.$$

(b) To verify that the given vectors are eigenvectors, evaluate the products:

$$\begin{pmatrix}-2 & 1 & 0\\ 1 & -2 & 1\\ 0 & 1 & -2\end{pmatrix}\begin{pmatrix}1\\ \sqrt{2}\\ 1\end{pmatrix} = \begin{pmatrix}-2+\sqrt{2}\\ 2-2\sqrt{2}\\ \sqrt{2}-2\end{pmatrix} = (-2+\sqrt{2})\begin{pmatrix}1\\ \sqrt{2}\\ 1\end{pmatrix},$$

$$\begin{pmatrix} -2 & 1 & 0 \\ 1 & -2 & 1 \\ 0 & 1 & -2 \end{pmatrix} \begin{pmatrix} 1 \\ 0 \\ -1 \end{pmatrix} = \begin{pmatrix} -2 \\ 0 \\ 2 \end{pmatrix} = -2 \begin{pmatrix} 1 \\ 0 \\ -1 \end{pmatrix},$$

$$\begin{pmatrix} -2 & 1 & 0 \\ 1 & -2 & 1 \\ 0 & 1 & -2 \end{pmatrix} \begin{pmatrix} 1 \\ -\sqrt{2} \\ 1 \end{pmatrix} = \begin{pmatrix} -2-\sqrt{2} \\ 2+2\sqrt{2} \\ -\sqrt{2}-2 \end{pmatrix} = (-2-\sqrt{2}) \begin{pmatrix} 1 \\ -\sqrt{2} \\ 1 \end{pmatrix}.$$

It follows that the eigenvalues of the matrix above are $-2+\sqrt{2}$, -2 and $-2-\sqrt{2}$, and the eigenvalues of the dynamic matrix are k/m times these.

(c) Using the given data,

$$\frac{k}{m} = \frac{16\,000}{(0.25 \times 10^{-3})/3} = 192 \times 10^6.$$

Hence, from $\omega = \sqrt{-\lambda}$, the normal mode angular frequencies are

$$\omega_1 = \sqrt{192 \times 10^6 \times (2-\sqrt{2})} \simeq 10\,605,$$

$$\omega_2 = \sqrt{192 \times 10^6 \times 2} \simeq 19\,596,$$

$$\omega_3 = \sqrt{192 \times 10^6 \times (2+\sqrt{2})} \simeq 25\,603.$$

(These angular frequencies are much larger than those for the transverse vibrations, and so the periods of vibration are much shorter.)

Solution to Exercise 15

(a) Using the hint, substitute for k from $T_{\text{eq}} = k(l_{\text{eq}} - l_0)$ in equation (47) to obtain

$$\ddot{x}_1 = \frac{-2T_{\text{eq}}}{m(l_{\text{eq}} - l_0)}\, x_1.$$

Now, in the present case, the equilibrium length of the spring is half the equilibrium length of the guitar string (i.e. $l_{\text{eq}} = L/2$), and the natural length is half the natural length of the guitar string (i.e. $l_0 = L_0/2$), while the mass of the particle is the mass of the guitar string (i.e. $m = M$). This gives the following equation of motion for the longitudinal vibration of the guitar string:

$$\ddot{x}_1 = \frac{2T_{\text{eq}}}{M(L - L_0)}\, (-2)x_1.$$

(b) As in part (a), use the formula $T_{\text{eq}} = k(l_{\text{eq}} - l_0)$ to eliminate k from the equations.

For the two-degrees-of-freedom model represented by equation (48), there are three identical springs, so their equilibrium and natural lengths are a third of the equilibrium and natural length of the guitar string, that is, $l_{\text{eq}} = L/3$ and $l_0 = L_0/3$. Therefore the stiffness of each spring is $k = 3T_{\text{eq}}/(L - L_0)$. The mass of each particle is half the mass of the guitar string, so $m = M/2$.

Thus the equations of motion become, in matrix form,

$$\begin{pmatrix}\ddot{x}_1\\ \ddot{x}_2\end{pmatrix} = \frac{2\times 3\times T_{\text{eq}}}{M(L-L_0)}\begin{pmatrix}-2 & 1\\ 1 & -2\end{pmatrix}\begin{pmatrix}x_1\\ x_2\end{pmatrix}.$$

For the three-degrees-of-freedom model represented by equation (49), there are four springs (so $l_{\text{eq}} = L/4$ and $l_0 = L_0/4$) and three particles (so $m = M/3$). Consequently, we have

$$\begin{pmatrix}\ddot{x}_1\\ \ddot{x}_2\\ \ddot{x}_3\end{pmatrix} = \frac{3\times 4\times T_{\text{eq}}}{M(L-L_0)}\begin{pmatrix}-2 & 1 & 0\\ 1 & -2 & 1\\ 0 & 1 & -2\end{pmatrix}\begin{pmatrix}x_1\\ x_2\\ x_3\end{pmatrix}.$$

(c) For the four-degrees-of-freedom model, there are four particles and five springs, which have the effect that the 3 in the above formula becomes a 4, and the 4 becomes a 5. The vectors have four components, and the matrix is a 4×4 matrix. So the equation of motion is

$$\begin{pmatrix}\ddot{x}_1\\ \ddot{x}_2\\ \ddot{x}_3\\ \ddot{x}_4\end{pmatrix} = \frac{4\times 5\times T_{\text{eq}}}{M(L-L_0)}\begin{pmatrix}-2 & 1 & 0 & 0\\ 1 & -2 & 1 & 0\\ 0 & 1 & -2 & 1\\ 0 & 0 & 1 & -2\end{pmatrix}\begin{pmatrix}x_1\\ x_2\\ x_3\\ x_4\end{pmatrix}.$$

(d) By comparing the above equations of motion with the results derived earlier, it can be seen that the only difference between the transverse and longitudinal vibrational models lies in the constant that multiplies the matrix on the right-hand side of the equation of motion. In fact, the only difference is that the constant in the transverse case includes a factor $1/L$, whereas in the longitudinal case the corresponding factor is $1/(L-L_0)$. Hence the dynamic matrix for the longitudinal model is $(1/(L-L_0))/(1/L) = L/(L-L_0)$ times the dynamic matrix for the transverse model, and the eigenvalues are similarly related.

Writing f_{T}, ω_{T} and λ_{T} (respectively, f_{L}, ω_{L} and λ_{L}) for the fundamental frequency, the corresponding normal mode angular frequency and the corresponding eigenvalue of the dynamic matrix for the transverse model (respectively, longitudinal model), we have $\lambda_{\text{L}} = \lambda_{\text{T}}L/(L-L_0)$ from the above argument. Hence

$$\omega_{\text{L}} = \sqrt{-\lambda_{\text{L}}} = \sqrt{\frac{-\lambda_{\text{T}}L}{L-L_0}} = \sqrt{\frac{L}{L-L_0}}\sqrt{-\lambda_{\text{T}}} = \sqrt{\frac{L}{L-L_0}}\,\omega_{\text{T}},$$

which yields

$$f_{\text{L}} = \frac{\omega_{\text{L}}}{2\pi} = \sqrt{\frac{L}{L-L_0}}\frac{\omega_{\text{T}}}{2\pi} = \sqrt{\frac{L}{L-L_0}}\,f_{\text{T}},$$

that is, the ratio of f_{L} to f_{T} is $\sqrt{L/(L-L_0)}$.

In the case of the E string of the guitar, $L = 0.65$ and $L_0 = 0.633$, so the ratio is

$$f_{\text{L}}/f_{\text{T}} = \sqrt{0.65/(0.65-0.633)} \simeq 6.2.$$

Therefore the fundamental frequency of the longitudinal vibrations is approximately $6.2\times 323 \simeq 2000\,\text{Hz}$.

Index